Migrant Marketplaces

Migrant Marketplaces

Food and Italians in North and South America

ELIZABETH ZANONI

UNIVERSITY OF ILLINOIS PRESS
Urbana, Chicago, and Springfield

Library of Congress Cataloging-in-Publication Data
Names: Zanoni, Elizabeth, author.
Title: Migrant marketplaces : food and Italians in North and South
    America / Elizabeth Zanoni.
Description: [Urbana] : University of Illinois Press, [2018] | Includes
    bibliographical references and index.
Identifiers: LCCN 2017031761| ISBN 9780252041655 (hardcover : alk.
    paper) | ISBN 9780252083297 (pbk. : alk. paper)
Subjects: LCSH: Italians—Food—United States—History. |
    Italians—Food—Argentina—History. | Italy—Emigration and
    immigration. | Consumers' preferences—United States—History.
    | Consumers' preferences—Argentina—History. | Emigration and
    immigration—Economic aspects—History.
Classification: LCC GT2850 .Z36 2018 | DDC 641.59/251073—dc23
LC record available at https://lccn.loc.gov/2017031761

To Leo, Cathy, and Dominic Zanoni

Contents

Acknowledgments

While writing this book, I have learned from and relied on numerous individuals and institutions. First, my deepest gratitude goes to Donna Gabaccia for her endless intellectual and personal generosity during my doctoral work at the University of Minnesota and beyond. Her knowledge of transnational, migration, and gender history and her moral support were indispensable tools for seeing this project through from inception to completion. Erika Lee also offered her guidance, providing me critical feedback and encouragement all along the way. In Donna Gabaccia and Erika Lee I found unparalleled mentors, inspiring me through their scholarship and professional and personal lives. I also want to thank Mary Jo Maynes, Sarah Chambers, and Jennifer Pierce for their careful readings and keen insights.

Several fellowships and institutions at the University of Minnesota (UMN) furnished essential support for this book. The James W. Nelson Graduate Fellowship in Immigration Studies and the UNICO National Graduate Fellowship in Italian American Studies allowed me to conduct research at the Immigration History Research Center (IHRC). As it has been for generations of scholars, the IHRC was for me a vibrant intellectual home, a productive space for interdisciplinary and international exchanges about human mobility. I am grateful for these exchanges, for they indelibly shaped my research agenda. I want to thank in particular IHRC staff and friends, especially Daniel Necas, Halyna Myroniuk, and Haven Hawley. Also, the Hella Mears Graduate Fellowship for German and European Studies at the Center for German and European Studies sent me to Italy for research, and a Graduate School Research Grant from the UMN Graduate School supported four months of archival work in Argentina. Finally, the UMN Graduate School's Doctoral Dissertation Fellowship gave me invaluable, uninterrupted time to finish my work.

In Italy, helpful archivists and personnel at the Archivio Centrale dello Stato, the Biblioteca Storica Nazionale dell'Agricoltura, the Istituto Nazionale di Statistica, and the Biblioteca Nazionale Centrale di Roma lent assistance in navigating their rich collections. In Rome, Stefano Luconi and Matteo Sanfilippo granted me helpful encouragement and research advice. While I was not in the archive, I enjoyed the wonderful company of my roommate, Giuliana Candia, whose own work on immigration inspired great conversation. A very special thank you goes to my family in Rome, especially Nadia, Valeria, and Aris Marches, who wholeheartedly welcomed me into their home, families, and lives. Due to their kindness, Via Pietro Paolo Rubens served as my second home, a place of relaxation, company, and delicious meals, and a jumping-off point for our numerous adventures across the Italian peninsula. Trips to Cloz, Trentino-Alto Adige, the hometown of my grandparents Oreste and Annetta Zanoni, permitted me time to explore my own family's migration story; there, Maria Pia Zuech served as a gracious host, lovely companion, and expert cook. I thank her and all the Zanonis and Zuechs in Cloz for embracing their distant *cugina americana* from Michigan.

In Buenos Aires, Argentina, I benefited from the guidance of staff and archivists at the Centro de Estudios Migratorios Latinoamericanos (CEMLA), the Biblioteca Nacional de la República Argentina, and the Centro de Documentación e Información at the Ministerio de Hacienda y Finanzas Públicas. Alicia Bernasconi and Mónica López at CEMLA were particularly friendly and helpful, as was the staff of Biblioteca's Hemeroteca, where I spent countless hours poring over *La Patria degli Italiani*. I also want to thank Alejandro Fernández for kindly taking the time to share his research and sources with me.

Old Dominion University (ODU) generously supported the completion of this book through a Summer Research Fellowship Program Grant, which allowed me to visit the David M. Rubenstein Rare Book and Manuscript Library at Duke University and the Hagley Museum and Library in Wilmington, Delaware. I am indebted to the staff and archivists at these two libraries for their expertise and assistance. Over the last six years, ODU's Department of History has provided me financial help to attend conferences while serving as an intellectually spirited space for revising the manuscript. I wish to especially thank members of the Junior Faculty Writing Group—John Weber, Brett Bebber, Erin Jordan, Timothy Orr, Jelmer Vos, Anna Mirkova, and Megan Nutzman—for reading multiple iterations of manuscript chapters and providing smart feedback. I had the good fortune to find in this group and their families not only excellent and dedicated colleagues but giving and caring friends as well. I also want to thank Maura Hametz, who read and commented on grant applications and manuscript chapters, and our past department chair, Douglas Greene, and current chair, Austin Jersild, for their constant support of my project. During my third year at ODU, a National Endowment for the Humanities summer stipend gave me the precious time and support to reframe my manuscript and complete a key chapter, making it overall a much stronger work.

Finally, a postdoctoral fellowship at the Culinaria Research Centre in the Department of Historical and Cultural Studies at the University of Toronto Scarborough challenged me to push my intellectual comfort zone and see my project in a new light. Culinaria's dynamic and interdisciplinary group of faculty and students encouraged me to more fully flesh out links between mobile people and the movements in foods and culinary traditions that their migrations produce. It would be impossible to overstate my appreciation for the mentorship and friendship of Jeffery Pilcher, Donna Gabaccia, Dan Bender, and Jo Sharma. Whether shucking oysters or attending operas with Dan and Jo, or visiting Montreal, Niagara Falls, and Muskoka (thanks to the warm hospitality of Franca Iacovetta and Ian Radforth) with Donna and Jeffrey, my time in Toronto was one of the most intellectually and personally enriching moment of my academic career. I also want to extend my gratitude to Simone Cinotto, Mike Innis-Jiménez, and Glen Goodman, whose participation in the Migrant Marketplaces Workshop helped me sharpen my analysis and contemplate the larger significance of my project. The friends and colleagues I met in Toronto, especially Jean Duruz, Irina Mihalache, Camille Bégin, Sanchia DeSouza, Nick Tošaj, Lesley Davis, and Matt White, offered me support and much fun.

I could not have found a more enthusiastic and capable editor than Marika Christofides at the University of Illinois Press. I thank her along with Jill R. Hughes and all those at the press who made this book possible. Since both of my readers have revealed their names, I am delighted to provide my gratitude to Linda Reeder and Simone Cinotto for the tremendous amount of time and consideration they put into their readers' reports and for the many ways they helped me refine the manuscript. Jeffrey Pilcher also gave me key feedback during one of the project's most crucial phases. My ODU colleague David Shields very graciously resized and adjusted the images that appear in the book. I claim any mistakes or shortcomings as my own.

This book owes a great deal to dear family and friends. While a student at Minnesota, I was lucky to acquire lifelong friends in Sonia Cancian, Melanie Huska, Trent Olson, Johanna Leinonen, and Pascal Gauthier. They have been pillars of support through academic and personal highs and lows, and their friendship is the only thing better than the book to come out of this work. Since I moved to Norfolk, Emily Moore, John Weber, Lisa Horth, Sonia Yaco, Erin Jordan, and Elizabeth and

Michael Carhart have been especially supportive of my professional and personal endeavors, and for this I am very grateful. Kerri Bakker and Maggie Powers and their respective families provided much-needed diversion and complete encouragement over the last ten years. My academic and life journey would not have been possible without the constant love and companionship of Katie Zemlick, whose friendship means the world to me. Between writing, researching, and teaching, trips home to Kalamazoo, Michigan, to spend time with family and friends, especially my grandmother Linda Markham and my godparents, Rene and Joan Adrian, revitalized my soul and reminded me of who I am. Finally, I want to thank my parents, Leo and Cathy Zanoni, and my brother, Dominic Zanoni, for their unconditional love and unwavering support. It was no surprise to me (or to them) that I opted to write on food, for it has been at the heart of our beautiful family culture as it has evolved over the years in Kalamazoo and in Bucerias, Mexico. It is their love and goofiness that keeps me going. I dedicate this book to them.

Migrant Marketplaces

Introduction

In November and December 1925, Cella's, a New York importer and seller of Italian food and wine, published a set of consecutive advertisements in the Italian-language newspaper *Il Progresso Italo-Americano*. Each advertisement presented a snapshot of Italian people and imports in the United States for 1895, 1910, and 1925 and did so in the changing figure of a stylishly dressed woman. The first advertisement, "1895," highlighted the masses of Italian newcomers at the turn of the twentieth century who joined established Italian communities in cities like San Francisco and New York (see fig. 1). "Italian-American trade follows hand in hand intensifying from our immigration," the advertisement read.[1] While the publicity makes no textual reference to women or female consumers, it is dominated visually by an illustration of a well-dressed Victorian lady in a long, corseted dress with puffy sleeves and with an elaborate bonnet fixed to her head. Two weeks later Cella's "1925" advertisement presented an image of a lean woman in a shorter, tubular tunic, playfully fidgeting with a long pearl necklace (see fig. 2). "Italians are now a living and active part of American greatness and civilization," the advertisement began, a civilization, it implied, that Cella's helped create by providing its customers with olive oil, canned tomatoes, wine, and other foods from Italy for over seventy years.[2]

As the series of Cella's advertisements suggest, migrants frequently discussed and represented global connections between trade and migration in Italian-language newspapers like *Il Progresso*. In major cities of the Americas, where transatlantic commercial and migratory flows converged, Italians formed their consumer experiences and identities based on an awareness and exploitation of the inherently global nature of migration and trade. They sold, bought, and consumed in

Figure 1. Cella's Inc., *Il Progresso Italo-Americano* (New York), November 29, 1925.

"migrant marketplaces," urban spaces defined by material and imagined transnational links between mobile people and mobile goods. As migrant marketplaces, cities like New York were global and gendered sites where male and female sellers and buyers interacted with products from their home and host countries in ways that shaped migrants' consumer identities and practices, the consumer cultures in which they were enmeshed, and wider transatlantic commodity networks.

The history of Italian migration to North and South America illuminates the historical formation of migrant marketplaces. Italians were one of the most mobile ethnic groups during the age of mass proletarian migration, and in the late nineteenth and early twentieth centuries their transnational labor paths opened and sustained global networks of trade in Italian products.[3] During these decades the United States and Argentina were the two most popular overseas destinations for Italian people and trade goods. Between 1880 and the beginning of World War II over four million Italians migrated to the United States and over two million mi-

Figure 2. Cella's Inc., *Il Progresso Italo-Americano* (New York), December 13, 1925.

grated to Argentina.[4] On average, during those same years, the United States and
Argentina received annually around 80 percent of all Italian products exported to
the western hemisphere.[5] The port cities of New York and Buenos Aires served
as principal gateways for Italian migrants and trade goods entering the Americas
through the North and South Atlantic. Merchants and business leaders connected
migrants in New York and Buenos Aires to Italy and to each other by facilitating the
flow of Italian goods, especially foodstuff, including cheeses, macaroni, fruit, wine,
and olive oil, but also nonedible items, such as clothing, textiles, and industrial
products. In migrant marketplaces of those two cities, Italian migrants constructed
changing and competing connections between gender, nationality, and ethnicity
through the consumption of these global imports. *Migrant Marketplaces* argues that
the formation of Italian migrants' consumer habits and identities was transnational and gendered, connected to food goods and to ideas about masculinity and
femininity circulating in the Atlantic economy.

Theoretical and Methodological Framing

The migrant marketplace framework encourages a global perspective that encompasses transnational and comparative approaches to the study of people and products on the move. It builds on a rich interdisciplinary body of literature by scholars who have used transnational perspectives to study cross-border migratory activities and identities.[6] Historians of Italian migration have offered some of the best examples of work that explores enduring cultural, political, economic, and familial links between migrants and the regions and towns from which they left.[7] Scholars of Italian migrants, like scholars who study other transnational people, have employed the term "migrant" because it moderates the role of the nation-state in defining the experiences of people as they enter (*im*migrant) or exit (*em*igrant) borders, while highlighting the multidirectional and circular movements that characterize the complex reality of migrants' lives.[8] The term "trans-local" perhaps best describes the ties Italians' maintained with their home country; as part of migration chains that linked one small *paese,* or village, in Italy to a community of its villagers abroad, Italy's disparate peoples identified with the traditions and values of their local *paesi* rather than with an idea of Italy as a unified nation or with other people from their homeland as "Italians."[9]

Despite labor migrants' tenuous connections or loyalty to the Italian government, Italy and other nation-states exerted real power over migrants' lives by managing and inhibiting transnational movements. As Mark Choate has shown, the Italian government sought to control the transnational movements of both Italians and products and incorporate them into nation- and empire-building projects.[10] Indeed, the migration and diplomatic practices of Italy, the United States, and Argentina often interacted and sometimes clashed with migrants' short- and long-term strategies. Furthermore, it is the varying geopolitical priorities of nation-states that generate variances in the way migration unfolds across the globe. The comparative and diasporic approaches used by both Samuel Baily and Donna Gabaccia show that while migrants' experiences and movements transcended national borders, Italian people and trade goods also remained entrenched in the economies, cultures, politics, and migration histories of particular nation-states.[11] By incorporating both transnational and comparative approaches, this study of New York and Buenos Aires as migrant marketplaces reveals how transnational linkages and nation-specific processes transformed migrant consumption. Furthermore, by treating North and South America as a single analytical site within the field of transnational history, it also exposes north-south hemispheric connections between migrant marketplaces in the Americas, connections fostered in large part through the United States' increasing presence in South America over the twentieth century.[12]

Until fairly recently scholars have examined migrants and the products they consumed separately, and within the context of a single nation, detached from other migrant-receiving countries and from the larger Atlantic world of commerce and migration.[13] Historians of global migration have focused mainly on the movement of people rather than on commerce, while economic histories of global trade have largely overlooked the role of labor migrants in shaping commercial flows.[14] And yet connections between the circulations of migrants and consumer markets were vitally important to the history of globalization in the late nineteenth and early twentieth centuries. By exploring Italian migrants as facilitators of global goods, and by drawing from historical work on both migration and trade, migrant marketplaces characterize everyday exchanges between migrant buyers and sellers as critical to the history of global integration.

The migrant marketplace paradigm also provides a methodological meeting ground where questions about ethnic enclaves and entrepreneurship that have traditionally interested social scientists can be combined with insights about identity, consumption, and representation from consumer theorists, cultural historians, and food studies scholars. Social scientists interested in ethnic enclaves have focused mainly on evaluating whether and to what extent these enclaves create opportunities for social and economic mobility and for eventual incorporation into the host society.[15] This literature has largely overlooked migrants' consumer experiences and how consumption produces practices, meanings, and ideologies that structure ethnic enclaves while connecting them to expanded national and global spaces.[16] Conversely, cultural historians and food studies scholars, while attentive to discourses that structure identity formation and culinary practices in migrant communities, have been less interested in exploring how meanings about and connections between gender, race, ethnicity, and nationhood evolved through and shaped commodity paths.[17]

One field in particular that has separated and continues to separate scholars interested in migrant economies is gender. An interdisciplinary group of scholars has analyzed the gendered nature of human mobility and how men and women experience the migration process in different ways.[18] Gender, however, does not figure centrally in dominant narratives of the history of migrant entrepreneurship or global capitalism, even though ideas about masculinity and femininity intersected with nationalism, race, and ethnicity to influence the formation of migratory and commodity links.[19] Migrant marketplaces redirect attention to consumption and retailing to incorporate scholarship on gender, identity, and consumerism into histories of enclave markets. This scholarship has identified consumption as a key factor in the development of capitalism and in the formation of national identities, especially for white middle-class women in Western societies.[20] However, aside from a small number of histories focused on migrant consumption—some of which have employed both gendered and transnational approaches—many questions remain about how consumption affected the creation of identities and practices among mobile people.[21] That Cella's utilized images of a demure Victorian lady and a flirty flapper from the Roaring Twenties to illustrate change over time in bonds between Italian migrants and food imports points to the role gender played in the establishment of migrant marketplaces. It also illuminates migrant women's evolving position in a larger, global consumer society that increasingly viewed consumption and femininity as interconnected.[22] In New York and Buenos Aires gender functioned concomitantly with ethnicity and nationality to organize transnational connections between trade and migration and to affect the consumer experiences of Italians living abroad.

In the migrant marketplaces analyzed for this book, food served as the central commodity around which migrants defined and redefined themselves as gendered and global consumers. While this study at times brings in discussion of nonfood items, such as textiles, another lucrative Italian export, it concentrates predominantly on foodstuff. Food-related enterprises, both small and large—import houses, grocery stores, restaurants, and industrial food enterprises—were the most numerous of the Italian-owned and -operated businesses in migrant marketplaces. Cheese; canned tomatoes; bottled wines, spirits, and olive oils; packaged pastas and tobacco products; and agricultural staples, including lemons, hazelnuts, and rice, counted among Italy's most profitable exports during the late nineteenth and early twentieth centuries. Quantitative trade and migration data make clear that migrant sellers, buyers, and consumers moved food merchandise across the Atlantic, and they did so to very specific locations. It is no surprise that the Italian government distinguished the "direct influence" that migrants abroad exerted on Italy's food export market through their desire for familiar homeland tastes from the "indirect influence" they had over other Italian exports, which were consumed by migrants and nonmigrants alike.[23] The Italian government therefore meticulously tracked the flow of foodstuff across the Atlantic and confidently celebrated the powerful influence that migrant consumer demand wielded over Italy's food trade.

Italian politicians and economists proved correct in assuming that perhaps more than any other commodity, foods intimately tethered migrants abroad to their premigration lives. Homeland foods went way beyond offering material nourishment to migrant bodies. Food products, rituals, and traditions also served as symbols through which migrants formed and maintained collective but changing transnational identities and practices, as well as power relations based on gender, race, class, and nationality.[24] Hasia Diner, Simone Cinotto, Donna Gabaccia, and Vito Teti, among others, have described the transformation of Italian migrants' diets over the early twentieth century and with it meanings about what *eating* Italian, and subsequently *being* Italian, signified to migrants, their children, and members of their host countries. Poor Italian day laborers, peasants, and *contadini* (farmers), accustomed to monotonous diets based on regional staples back home, found in the Americas new and more abundant ingredients, which they incorporated into traditional dishes, creating new, more varied and nutritious iterations of what they had eaten in Italy. Migrants clung to regional gastronomic customs from back home, seeing in them symbols of comfort and stability as well as the material means for solidifying familial bonds and communal identities. And yet Italian migrants, even poor laborers, used a profusion of cheap, new, and more abundant ingredients, especially meat, to build on their traditional foodways.[25] Indeed, in both the United States and Argentina, two of the world's largest meat-producing countries, cheap access to beef, pork, and other animal proteins slowly transfigured Italian food cultures abroad.[26] Moreover, migrant remittances and food products sent back to families in Italy, as well as new food knowledge and experiences carried home by returning Italians, altered the diets of Italians in Italy, evidencing how mass migration produced changes in food cultures on both sides of the Atlantic.

Focusing on food especially accentuates the gendered nature of consumption in migrant marketplaces, since Italian women, like women in other ethnic groups, held the responsibility of food production, purchase, and preparation both before and after migration. Gendered conceptions of labor division charged women abroad with combining the old and the new, a task that became particularly critical, and sometimes contentious, as mothers navigated the changing tastes and desires of their U.S.- and Argentine-born children, whose palates were less moored to the regional foodways of their parents.[27] Women's public and private interactions with foodstuff not only influenced the local economies of migrant communities and the wider global networks in which they were situated, but they also conveyed the complicated and often contested values and meanings associated with foods and eating.

The Archive of Migrant Marketplaces

Migrant newspapers like *Il Progresso*, in which the Cella's advertisements appeared, make a thorough discussion of migrant marketplaces possible because they ground migratory and commercial flows in specific cities, reminding us that global processes take shape in and depend on localized social relations. A 1916 advertisement for imported Turin-based Cinzano vermouth, appearing in Buenos Aires' *La Patria degli Italiani*, features a rotund, bespectacled gentleman lounging in a comfy armchair, reading the very paper in which the ad appears (see fig. 3). This intertextual strategy employed by Cinzano functions to wed the product and its consumers specifically to readers of *La Patria*. Cinzano's brand recognition derives just as

Figure 3. Ad for Cinzano vermouth, *La Patria degli Italiani* (Buenos Aires), September 20, 1916, 63. Courtesy of the Hemeroteca, Biblioteca Nacional de la República Argentina.

much from its placement in and reference to Italian-language readers in Buenos Aires as it does from the actual vermouth, which is conspicuously absent from the publicity; *La Patria* itself, as well as its readership, becomes an integral part of the Cinzano brand. Furthermore, Cinzano vermouth did not appear in Italian-language newspapers in the United States, which instead advertised other popular brands such as Cora and Martini & Rossi, demonstrating how the two diasporas supported particular transatlantic commodity paths.

The Cinzano ad demonstrates the vital role the foreign-language press played in shaping migrants' evolving consumer habits and identities.[28] Migrant print culture offers indispensable source material for the study of migrant consumption from a transnational perspective because it reveals material and discursive links between people and products and between home and host country. Italian-language newspapers include statistical data such as weekly ship manifests, information on national and international markets, and demographic details about migrant communities, institutions, and businesses. But they also contain discourse and representations about consumption in the form of op-ed pieces, articles, advertisements, and photographs. This quantitative and qualitative evidence together demonstrates that Italians did not encounter goods merely as economic actors but also as cultural mediators who produced new spaces and meanings about gender, ethnicity, race, and nationhood in their host countries. They depict migrants not only as laborers but also as serious consumers who took advantage of new opportunities to buy and consume Italian as well as Argentine and U.S. foods. And they paint a picture of migrant marketplaces that anchored Italians and products to specific locales while extending beyond them to connect migrants to Italians in Italy and to nonmigrants and other migrant groups in the United States and Argentina. Migrant print culture reveals migrant marketplaces as simultaneously bordered and permeable spaces, bounded by the communities and cities in which they developed, while equally influenced by urban, national, and transnational webs of commerce and migration.

This study focuses specifically on the Italian-language commercial press and on publications of Italian chambers of commerce in New York and Buenos Aires. I concentrate mainly on the two most popular Italian-language commercial newspapers during the late nineteenth and early twentieth centuries: *La Patria degli Italiani*, published in Buenos Aires, and *Il Progresso Italo-Americano*, published in New York.[29] Similarly, I focus primarily on monthly bulletins published by the Italian chambers of commerce in New York and Buenos Aires: *Rivista Commerciale* (New York) and *Bollettino Mensile* (Buenos Aires).[30] Members of Italian chambers of commerce shared with the owners and editors of *Il Progresso* and *La Patria* elite, or *prominenti*, status as middle-class community leaders. Their publications reflected a shared desire to unify Italy's migrants around a common *Italianità*, a sense of Italianness, which they believed would strengthen migrants' loyalty to their homeland and Italy's global repute. Migrant newspapers and business publications served as the principal channels through which Italian government representatives and merchants, as well as migrant business owners and retailers, made economic and cultural demands on migrants. But they were also sites where migrant entrepreneurs in New York and Buenos Aires, as well as U.S. and Argentine businesses, championed their own products and used them to fashion alternative identities, habits, and geographies for Italians abroad. In their desire to craft national identities around commodities, and to profit from these linkages, prominenti journalists and merchants struggled to manage working-class migrants' spending and consuming practices.

The experiences of migrants in New York and Buenos Aires, both of them urban, commercial, and global cities, hardly reflects the experiences of Italians migrating to and living in other parts of the United States and Argentina. However, as the two cities with the largest number and concentration of Italian migrants in the western hemisphere, Buenos Aires and New York were the principal nodes in a triangular geography connecting Italy, Argentina, and the United States. They were the cities where the large majority of Italian migrants and trade goods entered their respective countries and therefore served as locations where linkages between people and products had an especially visible and powerful impact. While there are limits to what migrant marketplaces of these two cities can reveal about migrants in other cities of the Americas, as the most important access points, the connections between migration and trade and the meanings and practices these connections produced influenced Italian communities in other urban areas. Moreover, *Il Progresso*, *La Patria*, and the chamber of commerce bulletins circulated way beyond New York and Buenos Aires, and they regularly discussed larger U.S. and Argentine economic, political, and cultural issues, as well as communities of Italians throughout their respective countries. As such, they provide insight about migrant consumption in other cities where Italians settled in large numbers, such as San Francisco, Boston, and Chicago in the United States, and Rosario and Mendoza in Argentina.

The Structure of the Book

While the Cella's advertisements joined migrant consumption to women during the 1920s, it was laboring migrant men who first dominated migrant marketplaces in New York and Buenos Aires. The first three chapters of this book concentrate on the late nineteenth and early twentieth centuries, a period characterized by the Italian government's official support of Italians abroad and by relatively low restrictions on mobile people and goods entering North and South America. The voracious demand for mainly unskilled manual labor in the United States and Argentina produced heavily male, temporary migrations from Italy to the Americas before World War I. Chapter 1 establishes a foundation for the study by mapping links between Italian migration and trade between Italy, the United States, and Argentina, and by discussing endeavors by Italian elites to exploit these links as part of larger nation- and empire-building projects. It compares trade and migration statistics to representations of Italian people and products in political tracks and export iconography to show how those elites masculinized migrant marketplaces at the turn of the twentieth century.

Chapters 2 and 3 consider differences and similarities in the migrant marketplaces of New York and Buenos Aires before World War I. Chapter 2 explains how assumed similarities between Italians and Argentines—racial, familial, linguistic, and cultural—generated shared consumer experiences in Buenos Aires for migrants and nonmigrants in relation to Italian foods in ways that they did not in New York. Furthermore, differences in the ethno-racial, economic, and political landscapes of the United States and Argentina allowed Italians in Buenos Aires, but not in New York, to assert their ability to influence Argentine trade policies in order to access homeland foods. Chapter 3 compares the development of *tipo italiano* products—Italian-style goods manufactured by entrepreneurs in the United States, Argentina, and Europe—to explore changing meanings of nationality and authenticity applied to foodstuff. Because they were cheaper than Italian imports, tipo italiano goods better met the financial goals of Italian male consumers in transnational family economies who strove to save money to send home. Italians in New York took advantage of their host country's robust industrial capacities and protectionist economic policies to manufacture domestically made wines, cheeses, and pastas on a mass scale. Together, these chapters show that while Italian elites sometimes treated migrant marketplaces in New York and Buenos Aires as interchangeable, and as inextricably linked to Italy, differences in how the United States and Argentina were integrated into the global economy produced varying opportunities and challenges for migrant sellers, buyers, and producers.

Chapter 4 describes World War I as a major turning point in the history of New York and Buenos Aires as gendered and transnational migrant marketplaces. It shows how migrants' involvement in wartime campaigns to buy Italian products abroad and to redirect trade to Italy normalized consumption as a "duty" to the homeland among a people who were more used to saving than spending and largely unfamiliar with nationalist feelings toward a unified Italy. In both countries, temporary gender-balanced and even female-dominant migrations brought on by the war, along with a growing, more gender-balanced second generation, made women central to these campaigns. During the interwar years, migrant marketplaces in the two countries feminized as advertisements for and discussion about Italian products, as well as the national identities they generated, became increasingly moored to women.

The final two chapters turn to the interwar years, when a worldwide depression, intensifying restrictions against mobile people and products, and rising nationalisms globally politicized migrant consumption in especially controversial ways. It is also a time when U.S. businesses' increased interest in Argentina as a site for consumer goods and investment changed the global geography of migrant marketplace connections. Chapter 5 shows that during and after World War I, U.S. food conglomerates began actively employing the Italian-language press to target Italian consumers in both New York and Buenos Aires as distinct consumer groups within their respective national marketplaces. As explored in chapter 6, ties between Italian consumers in these two major cities became particularly visible and politicized with the rise of fascism in Italy. Ironically, as Prime Minister Benito Mussolini tried to divorce Italian women from U.S.-style consumerism at home, migrant print culture employed links between women and consumer goods to generate Italian identities abroad. And yet while migrant consumerism emerged out of transnational ties to Italy during World War I, migrant marketplaces became less grounded within the national boundaries of either Italy or migrants' host countries by the interwar years. By the late 1930s, Italy, the United States, and Argentina all competed for the attention of Italian consumers, especially female migrants, who after the war became ubiquitous in advertisements for both Italian imports and domestically produced goods.

The book's epilogue ponders the fate of Italian transnational migrant marketplaces after World War II. Migrant marketplaces in New York and Buenos Aires came to exist increasingly in the imaginary and in commodified form rather than in the actual embodied movements of Italians and trade goods from Italy. Nevertheless imagined and deterritorialized ties between mobile Italians and foods continued to play a vital role in the performance of ethnicity for the descendants of Italians and in the consumption of Italianità for nonmigrants in the United States and Argentina.

CHAPTER 1

Manly Markets in
le due Americhe, 1880–1914

In 1888 a manual published in Genoa for Argentina-bound Italians included a dictionary of useful phrases in Italian with their Spanish translations. Among the many practical phrases designed to help Italian newcomers in Buenos Aires was "I do not like this wine. Bring me a bottle of Barbera or Barolo."[1] In providing Italians abroad with the vocabulary necessary for demanding homeland goods, the guide promoted a vision of the ideal migrant consumer, one whose preferences for the Piedmont wines traveling with him across the Atlantic heralded a new, imperial age for the Italian nation. While there is no way to know if and how migrants used the guide to manage their purchases, trade and migration data suggest that Italians stimulated international trade in Italian wine: Italians made up 63 percent of all migrants arriving in Argentina during the 1880s, and in the five years after the guide appeared, Italy rose from third in the quantity of wine exported to Argentina—behind France and Spain—to first place, surpassing its European competitors.[2] The guide anticipated that migrants' purchases of wine and other Italian edible goods would bind mass migration to mass exportation in building a *più grande Italia*, a "Greater Italy," a nation constructed in large part abroad by its peoples scattered across the globe.

Like the authors of the manual, Italians in positions of political and economic power during the late nineteenth and early twentieth centuries sought to capitalize on the consumer activities of the more than 16 million Italians who left Italy from 1870 to 1915. Italian politicians, economists, businessmen, and merchants desired to intensify transatlantic links between migrants and exports and to incorporate them into nation- and empire-building projects. Debates over migrant consumption focused particularly on the United States and Argentina in *le due Americhe* (the two Americas), which, as the two most popular overseas destinations for both Italian people and products, held the greatest promise as commercial colonies abroad. Elites enthusiastically pointed to trade and migration data as proof of migrants' potential to bolster Italian commercial flows and the economic power and prestige of Italy more generally. Unlike its European neighbors, Italy's future lay not so much in formal imperial rule, but in its unrelenting export of men and markets. Italian leaders assured themselves and the world that Italian-style expansion via men and markets was just as manly as state-dominated colonial pursuits in Africa and East Asia led by England, France, and Germany. Under the guidance of Italian merchant princes, and with the blessing of notable Italian men, past and present, transatlantic labor migrants transformed into armies of commercial explorers, warriors, and defenders of *la patria*.

Emerging from elite representations of Italian people and products abroad before World War I is a gendered geography of migration and trade that masculinized migrant marketplaces and the transatlantic spaces in which they were embedded. While labor migrations made up of predominantly male workers opened and sustained commercial routes for Italian goods to the Americas, these idealized man-centered depictions of overseas nation and empire building reflected the aspirations of Italian leaders rather than the more troublesome social problems and diplomatic realities facing the newly united nation-state. Furthermore, gendered representations of migration, trade, and consumption overlooked how labor migrants' commitments to families left behind restrained their purchases of Italian exports.

Elite Fantasies of Italian Migration and Empire Building in the Americas

In 1900 Italian economist and future president Luigi Einaudi popularized elite dreams of Italian empire building through emigration and trade with the publication of *The Merchant Prince: A Study in Colonial Expansion*. Einaudi described and praised what he called Italian "merchant princes"—migrant entrepreneurs who transplanted the country's capital, products, and culture to Italian communities or "colonies" abroad. These merchant princes, he predicted, would supply a growing class of Italian consumers overseas who, "used to consuming objects from their homeland, prefer among all, the wines and oils of their country because they better satisfy their tastes, and Italian furniture and clothes because they have the form and color that they like."[3] Einaudi focused especially on Enrico Dell'Acqua as a contemporary merchant prince, whose exportation of almost 50 million lire worth of yarn and textiles to South America won him high honors at the "Italians Abroad" section of Italy's 1898 National Exposition in Turin.[4]

Einaudi was one of a number of leaders at the turn of the twentieth century who debated the most effective means of dealing with the country's colossal emigration. In an effort to enhance Italy's international reputation as a great imperial power, Italian politicians and economists implemented a program for its citizens abroad that envisioned migration and colonialism as inextricably connected. Italian nationalists such as Prime Minister Francesco Crispi imagined state-sponsored migrant settler colonies in East Africa—where Italy had established its first African colony, Eritrea, in 1890—as the foundation for a reborn Roman empire characterized by conquest and territorial gain. However, after Ethiopia's defeat of Italian forces in 1895 frustrated nationalists' plans for expansion in Africa, liberals transferred the conception of colonies as demographic settlements to the Americas. Einaudi and his liberal supporters conceived of migrants in North and South America as modern incarnations of Genovese and Venetian medieval merchant princes whose peaceful commercial pursuits created voluntary emigrant colonies tied patriotically and economically to Italy. Migrant entrepreneurs in free settlements of the Americas, liberals agreed, represented a more sophisticated and less violent form of colonialism than expensive European conquests in Africa and Asia where commerce forcibly followed the flag.[5] While Italian political leaders—liberals and nationalists alike—continued to pursue African settlement, liberals contended that Italian commercial expansion was best achieved when Italian commodities followed the footpaths of Italian migrants to the Americas.[6] Advocating for a type of free-market imperialism in which migrants like Dell'Acqua formed Italian colonies abroad to absorb Italian goods and disseminate Italian culture, Einaudi hoped to transfer the glory and wealth of medieval Italy to the western hemisphere.

A year after the publication of Einaudi's manifesto, liberals in the Italian government passed the country's historic 1901 emigration law, the first piece of national legislation, globally, to actively guide, protect, and profit from its migrants. Its passage signaled a shift in many lawmakers' attitudes; rather than conceive of migration negatively, as an embarrassing hemorrhage of labor power, capital, and military resources, the law's architects depicted migrants as political and economic instruments to enlarge and secure international prestige and colonial terrain, and to assist Italy domestically through remittances and new markets for Italian exports.[7] The law was one of many progressive steps taken by the Italian government during the Giolittian era, named after Giovanni Giolitti, who as prime minister oversaw much of Italy's arduous path toward industrialization from 1900 to World War I.[8] Unlike Western European countries such as Germany, France, and Britain,

Italy's colossal migration, combined with a weak industrial sector and fledgling consumer market, necessitated a worldview that envisioned migrant consumption in the Americas as central to economic growth at home as well as to imperial pursuits abroad.

The wide-sweeping 1901 law established the Emigration Commissariat, an independent agency under the foreign ministry to oversee emigration, and the *Bollettino dell'emigrazione* (hereafter *Bollettino*), a monthly, voluminous bulletin dedicated to the documentation and study of Italian emigration.[9] National and local commissariat agents reporting on transatlantic migration in the *Bollettino* ruminated positively on ties between Italian migrants and trade. As a 1904 report confidently stated, "The links between emigration and commercial flows are many. Our emigrants take with them their lifestyles and their tastes, and they continue to consume products from Italy, which sustains an active exportation of our market." By requesting Italian articles of popular consumption, especially foodstuff such as wine, olive oil, pasta, cheese, and tobacco products, migrants exerted a powerful influence over trade routes. Pointing to large annual increases in the amount of such items to the United States, the report urged Italian businesses to continue catering to migrant demand. "And it is our hope that our exporters follow, with a shrewd eye, migration, and know how to respond to the needs of our colonies abroad."[10] The 1901 law, the Emigration Commissariat, and the *Bollettino* represented new mechanisms through which migrants' links to Italy's export markets were described, measured, and analyzed.

While the commissariat and its *Bollettino* gave Italy unprecedented bureaucratic authority to connect emigrants and exports, it was not the first governmental attempt to capitalize commercially on its citizens' purchases abroad. In the late nineteenth century, the Italian Ministry of Foreign Affairs set up chambers of commerce abroad in order to promote trade between Italy and the rest of the world. The chambers, run by prominenti migrants, worked assiduously to arm Italian merchants with the resources and information needed to boost Italy's export market. From major cities outside of Italy, they produced for their members regular publications that discussed trade and migration legislation, local and international consumer prices, current and potential markets for Italian exports, emigrant communities and their commercial activities, and national and international trade exhibitions and meetings.[11] And they looked specifically to migrant consumers as outlets for Italy's exports. As Mark Choate explains, Italian chambers of commerce as well as nongovernmental and private organizations that were focused on expansion, such as the Dante Alighieri Society and the Italian Colonial Institute, promoted a cohesive Italianità, a national identity that would use Italian exports, language, religion, and ethnicity to unite Italy's heterogeneous migrants.[12]

In the late nineteenth century, as increasing numbers of government officials, economists, and businessmen turned away from Africa and toward the Americas, they focused specifically on the United States and Argentina as the two most partic-ularly fruitful locations for Italy's overseas colonies. In 1893 lawyer and emigration promoter Guglielmo Godio described the United States and Argentina as the pro-totype countries of *le due Americhe* (the two Americas) in *America and Its First Factors: Colonization and Emigration*. He justified his focus on those two countries because of their "greater vitality and pronounced character, for the vastness of territory and for the splendor of destinies," and because they represented what he articulated to be distinctly different histories based on contrasting modes of European coloniza-tion by the Spanish and English.[13] Presaging Einaudi's study and the 1901 law by almost a decade, Godio wrote his comparative history of colonization in order to encourage the Italian government to protect its migrants and the export markets they developed. He anticipated that if and when Italy woke up to the importance of its emigrants, "Oh! Then, yes, the other exporting nations would have reason to be alarmed."[14]

Almost twenty years later Aldo Visconti, then a student at the Institute of Ad-vanced Studies in Trade in Turin, also focused on the United States and Argen-tina as the two destinations that were exceptionally suitable for wedding men and markets. Visconti began his 1912 thesis, *Emigration and Exportation: A Study on the Relationship between Italian Emigration and Exportation to the United States of North America and the Republic of Argentina*, arguing that Italian overseas migration made these two countries the most important outlets for Italian exports. While geographical proximity accounted for Italian commercial success in France, Germany, Austria, and Switzerland, migrant demand for Italian products explained why outside of Europe, Italian exports did so well in these two faraway countries.[15] By constantly comparing and linking le due Americhe, Italian trade experts like Visconti sug-gested that the commercial achievements of one of the two countries reinforced the potential of the other. Economist Luigi Fontana-Russo, who also devoted his 1906 article "The Emigration of Men and the Exportation of Goods" to the United States and Argentina, introduced his section on Argentina by writing, "That which we said about the U.S. finds ample confirmation in our relationship with Argen-tina."[16] Italy's commercial triumph in the United States and Argentina promised future achievement in other countries of le due Americhe with growing Italian populations, especially Brazil, Uruguay, and Canada. Elites envisioned the United States and Argentina as the principal gates through which Italian migrants and exports would disseminate throughout the western hemisphere.

While Italian leaders disagreed over whether the United States or Argentina offered more promise for linking migrants and exports, together they provided optimal locations for building a Greater Italy from the commercial activities of Italians abroad. Trade and migration statistics bolstered their faith that Italian commerce followed the footpaths of migrants across the Atlantic. Figure 4, based on Italian government migration and trade data, represents the percentage of all

Italian people and products headed to the western hemisphere that arrived in the United States and Argentina from 1880 to 1913. This figure shows that the dynamics of export trade did closely follow the dynamics of migration in both countries. With few exceptions, when Argentina's and the United States' share of arriving migrants increased, so too did both countries' share of exports. In part this is because people and goods traveled together on the same ships across the Atlantic; transporting both people and goods made passenger traffic a more profitable business for steamship companies.[17] During downswings in the global economy, when fewer Italians migrated and when higher numbers of migrants repatriated back to Italy, exports decreased as the primary consumers of Italian products either dwindled or returned home. A focus on the two dotted lines representing exports also shows that migration to the United States and Argentina fluctuated considerably more than did exports to these two destinations; unlike migrants, who often returned home or made multiple trips back and forth across the Atlantic, Italian products did not repatriate.

Representing Italian migration and trade to the United States and Argentina as percentages of the total made to the western hemisphere also reveals the importance of focusing on both countries within a single analytical frame, since they constituted the two overseas countries that received the largest number of Italian people and goods. Almost every year between 1880 and 1913, the United States and Argentina received over 80 percent of all Italian goods and 60 percent of all Italian migrants headed to the Americas. Increases in the percentage of exports and migrants in one country almost always meant a decrease in exports and migrants to the other. This is reflected in the way the lines for people and products to the two countries appear in figure 4 as almost mirror opposites of each other. The figure also shows how booms and busts in the global economy influenced migration patterns to the Americas, and in turn how migrants' decisions affected national and international labor and trade markets.[18] A major depression in Argentina in 1890 and 1891 created a temporary break in what was until then a steady increase in Italian migration to the country. Italians responded to the Argentine downturn by reorienting their voyagers northward, where unskilled labor was in high demand; after 1890 more Italians migrated to the United States than to Argentina until the early 1920s, when restrictive immigration legislation in the United States made Argentina once again the preferred destination. Similarly, when the Panic of 1907 dealt a heavy blow to the U.S. economy, Argentina's share of Italian exports increased slightly the following year, as did Argentina's share of Italian migrants, who most likely responded to unemployment in the United States by turning their journey southward. Clearly, a symbiotic relationship existed between Italian people and products on the move, and between Argentina and the United States as the two most important receiving countries in the western hemisphere.

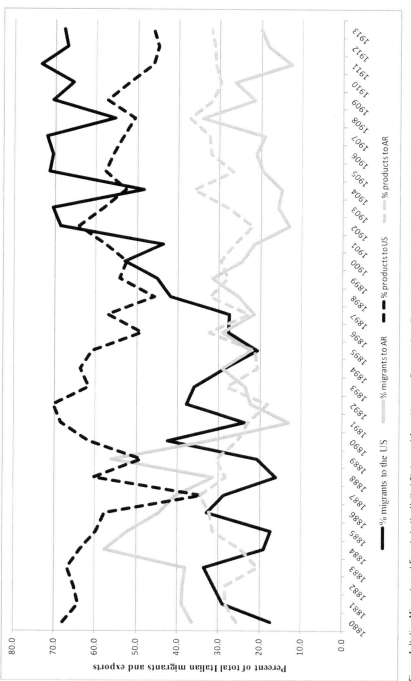

Figure 4. Italian Migrants and Exports to the United States and Argentina as a Percent of the Total of the Western Hemisphere, 1880–1913. Sources: Exports—*Movimento commerciale del Regno d'Italia* (1880–1902) and *Annuario statistico italiano* (1903–1913), publishers vary. Migrants.—Commissariato Generale dell'Emigrazione, *Annuario statistico della emigrazione italiana dal 1876 al 1925* (Rome: Edizione del Commissariato Generale dell'Emigrazione, 1926).

Although elites lauded the beneficial effects of migration on all Italian exports, they noted the particularly strong correlation between the joined increases of migrants and food exports. The *Bollettino* differentiated between exports like cotton and silk, over which migrants exerted an "indirect influence"—indirect because migrants and nonmigrants alike consumed such goods—and foodstuff, over which migrants' "direct influence" stimulated commercial flows.[19] The United States and Argentina absorbed by far the bulk of Italy's western hemisphere–bound food exports and, for certain food items, the majority of Italy's total global exports. In 1907 the United States and Argentina together received over 90 percent of all Italians headed to the Americas; that same year the United States was the leading importer worldwide of Italian cheese, pasta, canned tomatoes, hazelnuts, lemons, and citrus extract, and Argentina was Italy's number one global importer of olive oil, wine, rice, vermouth, and spirits. Argentina and the United States together imported 75 percent of Italy's global exports of vermouth, 76 percent of Italy's bottled spirits, 60 percent of Italian tomato preserves, 44 percent of Italy's exported oil olive, 37 percent of Italian cheese, and 34 percent of Italian wine.[20] Demand for these foods and beverages by emigrants drove the transformation of Italian agricultural processing and industrial food production at home. By 1911 food manufacturing was one of Italy's most industrialized sectors, with 14 percent of industrial workers employed in the industry, third behind textiles (22.9 percent) and engineering (16.7 percent).[21]

Trade and migration statistics demonstrate that migrants moved merchandise, especially the products of Italy's developing commercial food and agricultural industries, and to very specific transatlantic locations with large Italian populations. Quantitative data justified elites' initiatives and attached measurable monetary values to the consequences of migrants' consumer decisions. That both countries exhibited such strong correlations between migration and trade, elites surmised, confirmed that such links were not a coincidence but rather a particular ability or asset of the Italian nation-state. Italian merchants and business leaders abroad agreed; given the large number of Italians and exports in the United States and Argentina, it is no surprise that Italian chambers of commerce in these two countries became powerful migrant institutions. In Argentina the government founded the Italian Chamber of Commerce and Arts in Buenos Aires in 1884; the following year an Italian chamber of commerce opened in Rosario, capital of the province of Santa Fe. In the United States the Italian government created the San Francisco Italian Chamber of Commerce in 1885; two years later, the Ministry of Foreign Affairs established a sister chamber in New York City.[22] These chambers drew attention to migrant consumption in order to depict the United States and Argentina as exceptional among Italy's diasporas. "Only in *le due Americhe*," wrote Guido Rossati, head of Italy's wine department in New York, in an article appearing in the New York Italian Chamber of Commerce's bulletin, "has the exportation of national wine continued to progress," while exports to European countries decreased. This trend, Rossati concluded, proved "the truth of the axiom that trade follows emigration."[23]

Prominent Italians like Rossati built their dreams about trade and migration in le due Americhe on a number of assumptions. Economists, politicians, and academics confidently proclaimed that Italian migrants served as loyal consumers. As the image of the Barbera-drinking migrant in the 1888 guide suggests, emigration promoters largely presumed that Italians abroad would request Italian imports and reject domestically made alternatives. "The emigrant is the most loyal consumer of merchandise from the homeland," Fontana-Russo asserted. "He demands it even from far away, he pays more because it is better, he gives to homeland production a force that it did not have previously."[24] A publication prepared for the "Italians Abroad" display at the 1906 Milan International exhibition described Italians' diets in the United States as unsusceptible to Americanizing influences: "All that arrives on his table must be imported from Italy, or, when his finances do not allow it, made in the Italian way by producers in the colony."[25] Elites described migrants in the Americas as having discriminating tastes: they refused to drink weak wines made from grape varieties grown in the Americas and avoided bitter U.S. American and Argentine beers; they turned up their noses at cooking oils diluted with peanut and cotton oil, preferring the more costly but pure olive oils from Genoa and Lucca; and they purchased dried pasta manufactured in Italy from 100 percent *semola,* or semolina, flour.[26] At the first Congress of Italian Ethnography in Rome, in 1911, Italian social reformer Amy Bernardy, in her discourse on Little Italies abroad, noted, "Above all, it is in the kitchen where [migrants] conserve Italian habits, because it is very difficult for an adult emigrant to adopt, after many years, the cooking of the [host] country." "The Genoese do not renounce the *taglierini* with pesto, and the Neapolitans remain loyal to *maccheroni*," she reported.[27]

While Italian food exports offered the most obvious evidence of migrants' direct influence on Italian commerce, preference for Italian exports extended beyond edibles. Fontana-Russo, reflecting with wonder on the remarkable success of Italian textiles in the United States, noted that such growth could only be explained by migration. In an industrially sophisticated country like the United States, with a well-developed textile and clothing industry, Fontana-Russo surmised, "The fact has to be the emigrant, who does not change his taste and therefore abandon homeland goods; rather he spreads them among the people with whom he has contact."[28] Migrants' consumer tastes and habits were immutable and unchanging, preserving their character as they moved across the Atlantic and materialized in migrants' everyday lives. Visconti agreed, suggesting that Italian commerce to le due Americhe defied the natural laws of economic competition; commercial flows responded to a different, more powerful pull: the insatiable demand for homeland goods by Italians in migrant marketplaces like New York and Buenos Aires.[29]

In their optimistic predictions, the ultimate focus of much elite discourse was not on le due Americhe but on la patria. Liberal migrant supporters were particularly confident that migrants' purchases abroad would eventually turn Italians in Italy into consumers. They surmised that profits from the sale of exports, combined with migrant remittances, would stimulate Italian industry, developing a broader domestic consumer base and emancipating Italian consumers from foreign imports. Godio argued that migrants' achievements were best gauged by observing the changing spending patterns of *reduci* (returnees) and the enlarged consumer opportunities of Italians left behind by their migrant relatives.[30] While insisting that migrants maintained traditional consumer habits in the Americas, officials also conceded that reduci brought home new desires and dietary standards. A 1908 *Bollettino* report on meal service aboard Italian passenger ships proposed a new food regime with increased meat rations to better harmonize with the "habits and tastes" of migrants. Migrants who returned with savings "demand steak, and when they cannot get it, they buy it at an exaggerated price from first-class service, and in this habit acquired in the new world is the explanation of their success."[31] Success, as evidenced by the report, involved eating steak and other foods that were unavailable to most poor farmers and day laborers before migration. Migration and Italian commercial expansion, acknowledged a group of Italian businessmen in South America, increased the standard of living among Italian peasants, indicated by access to better-quality clothing, furniture, tobacco products, and especially nutritional and varied foods. As a consequence of migration and trade, they boasted, "Nearly every farmer [in southern Italy] is permitted the luxury of eating at cafes and taverns."[32] Italian government officials also directly attributed the improved diets of Italians to better economic conditions created by migration and trade.[33] Reporting from Basilicata and Calabria for the commissariat, Adolfo Rossi similarly commented that while at one time day laborers were content with a modest meal, "now they want instead food of the highest quality."[34]

These confident projections countered vocal opponents of emigration, who pointed to the perceived moral, economic, and social consequences of mass male migration. Rather than drain the country of workers and the consumer base necessary for stimulating domestic industries, liberal leaders predicted that migrants would not only activate consumption at home but would also eventually discourage emigration altogether.[35] Migrant demand for Italian goods triggered higher production levels and the expansion of Italian industry; the resulting economic development created better-paying jobs, which allowed workers to buy more Italian products, while curbing migration. "Therefore, an almost invisible force, born under the impulse of emigration, begins to work against emigration," Fontana-Russo predicted.[36] Until Italian transnational migration decreased, it would continue to encourage the growth of Italian consumer societies on both sides of the Atlantic.[37]

Finally, Italian elites assumed that migrant consumers would serve as conduits for enlarging consumption among non-Italians in the United States and Argentina. Luigi Einaudi confidently stated in 1900 that Italians abroad were "the most enthusiastic apostles of our worldwide expansion."[38] A decade later Visconti agreed; migrants, he acknowledged, "represent a bond between two countries," who, by demanding products from their homeland, "are truly the most effective and precious agents of international trade."[39] Italian chambers of commerce abroad often reported on the diffusion of Italian exports in the larger Argentine and U.S. marketplaces. Migrants not only served a "direct commercial function" by favoring Italian products, but "they become without even having realized, free traveling salesmen of trade for the Italian export houses." Their consumer habits and Italians' increasing interactions with non-Italians, especially around foodstuff, developed among the people of their host countries the desire to buy Italian olive oil, pasta, and wine.[40] The Italian Chamber of Commerce in New York gloated, "It can be asserted, in fact, without exaggeration, that the Italian emigrant to the U.S. was and is still the major salesman for projecting Italian articles among American consumers."[41] Italian migrants would serve as the primary medium for introducing U.S. Americans and Argentines to new consumer practices around Italian products, practices that advanced Italy's economy and global repute.[42]

Although their inflated confidence in Italy's transatlantic transfers of men and markets would wane over the course of the twentieth century, prominent Italian men had high expectations for Italian consumers in migrant marketplaces abroad. They surveyed the growing amount of government data from both sides of the Atlantic confirming that Italian mass migration fostered overseas trade routes and revolutionized agricultural and industrial production at home. In the face of broader economic and social insecurities confronting the newly unified nation-state, such high expectations would be met only by turning Italy's male labor migrants in le due Americhe into merchant princes and commercial warriors for the homeland. In the hands of Italian business, economic, and political leaders, the exportation of men and markets became manly markets.

Gendered Representations of Migrants, Markets, and the Italian Nation

An 1895 trademark for Lombardy-based Enrico Candiani, one of the largest manufacturers and exporters of Italian cotton fabrics, captured Italian elites' hopes of capitalizing on Italian migration, trade, and consumption (see fig. 5). Candiani's trademark features a crowned image of a classically robed "Italia," female embodiment of the Italian nation, being propelled eastward across the Atlantic on her chariot of exports by small *golondrine* (birds of passage), the Italian word given to the thousands of male laborers who migrated abroad during the winter harvest

Figure 5. Trademark for Enrico Candiani cotton fabrics, Archivio Centrale dello Stato, Ministero dell'Industria, del Commercio e dell'Artigianato, Ufficio Italiano Brevetti e Marchi, fasc. 3098, 1895. Courtesy of the Archivio Centrale dello Stato.

seasons and then returned to Italy.[43] While Italia floats celestially in the heavens, she is being pulled along well-established routes linking cities of departure and arrival for both emigrant golondrine and the exports they convey.[44] A commercial expression of Einaudi's axiom "the flow of commerce must follow the flow of Italian emigration," the image produced and then compressed space and time to articulate a triangular geography between Italy, North America, and South America.[45]

Candiani represents one of many turn-of-the-century businesses that employed a complex repertoire of symbols and images in their packaging and marketing to project gendered meanings about the Italian nation and about the global commercial and migratory links connecting Italy, the United States, and Argentina.

That Italia holds the reins attached to the golondrine's wings suggests she exerts a degree of influence over migration and trade; yet she is also being escorted by male migrants, who carefully guide both her and the purchase orders nestled in their beaks. As detailed above, in political and economic tracts, government reports, and publications issued by chambers of commerce in the United States and Argentina, Italian elites detailed and promoted the formal commercial routes created and maintained by migrants. These discussions echoed trademarks for Italian exports such as Candiani's in projecting gendered messages equating exports, migrants, and consumption with Italian notable male personas, industrial power, imperial pursuits, and military dominance.

In accordance with Italian laws governing trademark registration and protection, Italian businesses registered their trademarks with Italy's Office of Intellectual Property within the Ministry of Agriculture, Industry, and Commerce.[46] Italian firms typically commissioned trademark art from Italian lithography and printing companies, who in turn hired artists to create the graphics. While little is known about the relationship between Italian manufacturers and the printing establishments that produced the images, before the advent of professional advertising firms after World War I, Italian business owners played an active role in the creation of trademarks.[47] The varied messages about Italian migration, trade, and consumption in visual representations did not mirror economic and political realities, nor can they provide insight into how migrants as viewers and consumers in New York and Buenos Aires responded to such depictions. However, exploring commercial and discursive depictions of Italian migration and trade reveals recurring patterns in the gendered ways Italian elites presented people and products to themselves, to migrants, and to other nations as they struggled to secure Italy's location in the competitive world of commerce, diplomacy, and imperialism. Rather than reflect consumers' actual interests, these trademarks projected their creators' gendered worldview, a worldview colored by their own positioning as elites who saw their personal goals and the progress of the Italian nation intimately tied to manly markets of migration and trade.

Trademarks for Italian exports before World War I often used images of well-known men to associate the production and consumption of products abroad with Italian notables. Chief among these were the monarchs of the House of Savoy, the dynastic family who ruled the Kingdom of Italy from 1861 to the end of World War II. Garofalo's pasta factory, founded by Alfonso Garofalo, was one of the many small pasta factories in Italy that began exporting macaroni during the late nineteenth and early twentieth centuries.[48] In the Garofalo trademark, Italia balances a shield with the red and white crest of the House of Savoy while overlooking the smoke-billowing Garofalo pasta factory in Gragnano near Naples (see fig. 6). At the bottom of the trademark is a series of medallions featuring almost identical-looking busts of whiskered Italian leaders. One pictures Vittorio Emanuele II,

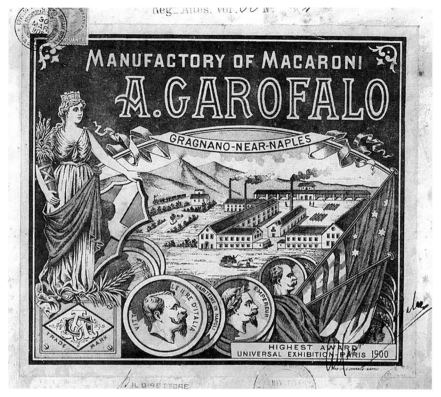

Figure 6. Trademark for A. Garofalo macaroni, Archivio Centrale dello Stato, Ministero dell'Industria, del Commercio e dell'Artigianato, Ufficio Italiano Brevetti e Marchi, fasc. 6976, 1905. Courtesy of the Archivio Centrale dello Stato.

the first king of a united Italy and emblem of the Risorgimento, the Italian unification movement; the other two are less recognizable figures—although one labeled "empereur" suggests links between Garofalo and the legacy of imperial Rome, while the medallion of the merchant marine in Naples ties the product to Italy's contemporary commercial fleet. Although the trade and migration routes in this trademark are less explicit than Candiani's westward-bound golondrine, the intertwined U.S. and Italian flags in the lower left-hand corner, along with the writing in English, symbolize transatlantic routes between producers in Italy and consumers in the United States. The regular use of monarchs and other royal imagery gave trademarks like Candiani's an official quality by featuring Italian kings as product endorsers.

Internationally renowned spiritual and cultural leaders joined Italian kings as masculine bearers of Italian civilization. A number of trademarks for exported Italian food products used the image and name of Italian poet and statesman Dante Alighieri. A 1907 trademark for Dante-brand olive oil, produced by Giacomo Costa, one of the largest exporters of olive oil to the United States, featured an illustration of the laurel wreath–crowned Dante in profile.[49] Depictions of the winged and often nude Mercury, the Roman god of commerce and protector of merchants, appeared regularly in trademarks for Italian exports.[50] These real and mythological male luminaries meshed well with liberals' articulation of Italian migration and colonization as a commercial conquest secured not only through trade, emigration, and consumption but also through the spread of Italianità embodied in Italian history, civilization, and culture.

Trademarks also employed well-known Italian migrants, especially Giuseppe Garibaldi and Christopher Columbus, to market their wares for consumers in the Americas. Numerous companies, especially those exporting to Argentina and other parts of Latin America, featured nationalist leader Giuseppe Garibaldi in the packaging of their products. Before fighting in the Risorgimento, Garibaldi led the Italian Volunteer Legion, which fought to defend Uruguay against forces led by Juan Manuel de Rosas, the dictatorial governor of the Province of Buenos Aires, and the ex-Uruguayan president Manuel Oribe.[51] An 1899 trademark for Milan-based Brancaleoni and Company, a manufacturer and exporter of *fernet* (an *amaro,* or bitter herbal liqueur), linked Garibaldi to Vittorio Emanuele II, the principal architects and symbols of Italy's independence struggles (see fig. 7).

Explorer and merchant Christopher Columbus, the most famous Italian transatlantic voyager, served as a key emblem of Italian commercial and demographic expansion in trademarks.[52] As the original medieval merchant prince and real-life Mercury, depictions of Columbus served as symbolic devices for communicating a new era of Italian economic and cultural imperialism made possible by Italy's abundant migration. These commercial images of Columbus complemented the writings of Italian elites who constantly compared Italian merchants and migrants

Figure 7. Trademark for Fernet Brancaleoni *amaro*, Archivio Centrale dello Stato, Ministero dell'Industria, del Commercio e dell'Artigianato, Ufficio Italiano Brevetti e Marchi, fasc. 4217, 1899. Courtesy of the Archivio Centrale dello Stato.

in the Americas to medieval and early modern explorers. In 1904 Edoardo Pantano, a member of parliament and future head of the Ministry of Agriculture, Industry, and Commerce, described Italian merchants abroad as "enterprising and determined, with an astute intuition, with unceasingly activity, unsettled and bold wanderers, reminiscent of our medieval merchants."[53] A *Bollettino* report that same year equated early twentieth-century Italian exporters with the "audacious merchants from Lombardy, Genoa, Naples and Sicily, that, in the Middle Ages, traveled all the known world." These contemporary Columbuses, the report concluded, "contributed to maintaining the tastes of our emigrants for national products, enjoying therefore an impulse to Italian exports."[54]

Other trademarks glorified modern incarnations of Columbus-like men who moved abroad. Silvano Venchi, founder of the S. Venchi & Company in Turin, began his chocolate and confectionary factory in 1878; by 1906 Venchi manufactured more than two thousand different types of candy, chocolate, and caramel and exported his confections within Europe, as well as to North and South America.[55] The company's 1901 trademark for *Stella Polare Cioccolato* (Polar Star Chocolate) shows a snowy scene from the North Pole, where men hoist the Italian flag at a campsite surrounded by boxes of Venchi chocolate (see fig. 8). Suspended above the landscape are pictures of Luigi Amedeo of Savoy-Aosta (the Duke of Abruzzi), grandson of Vittorio Emanuele II, and Captain Umberto Cagni of the Italian Royal Navy. The Duke of Abruzzi, an Italian prince and explorer, became internationally esteemed for his Arctic explorations, especially in Alaska, and later served as an Italian admiral during World War I. The trademark reenacts his expedition with

Figure 8. Trademark for S. Venchi & Company chocolate, Archivio Centrale dello Stato, Ministero dell'Industria, del Commercio e dell'Artigianato, Ufficio Italiano Brevetti e Marchi, fasc. 5126, 1901. Courtesy of the Archivio Centrale dello Stato.

Cagni to the North Pole in 1899 on the steam whaler *Stella Polare*, from which the brand derived its name.[56] The duke, a turn-of-the- century Columbus and real-life prince affiliated with geographical surveys, heroic adventures, and the Italian military, represented the model migrant envisioned by elites.

Export iconography featuring male notables also functioned as geographic and diplomatic bridges between Italy and le due Americhe. Trademarks paired famous Italian male persona, including Italian royalty, not only with generic U.S. and Argentine emblems, such as national flags and the Statue of Liberty, but also with specific male political leaders from these two countries. Such trademarks condensed transatlantic space for buyers and sellers to create imaginary forums where well-known men from Italy, the United States, and Argentina were joined together through Italian oils, liquors, pastas, chocolates, tobacco, and other exports. Moreover, these trademarks presented a set of layered yet interlocking pasts—from ancient Rome to the Age of Exploration, to colonial and early U.S. and Argentine history—in ways that inserted Italy into the political traditions and foundational myths of le due Americhe. A 1902 trademark for Fernet-Branca from Milan, one of the most popular Italian exported bitters in North and South America, for example, included illustrations of the busts of both Christopher Columbus and George Washington.[57] A 1907 trademark for Emanuele Gianolio's America-brand Italian vinegar, manufactured in Genoa, exhibited a most bewildering mixture of manly geopolitical links (see fig. 9). The trademark displayed Columbus flanked by major nineteenth-century Latin American revolutionary figures, including Simón Bolívar, Domingo Faustino Sarmiento, José de San Martín, and José María Paz y Haedo.[58] A robed woman embellishes the leaders' portraits with laurel wreaths while attending to a *fasce*, the ancient Roman symbol of the Roman Republic representing strength and unity.[59] Trademarks made transnational and transhistorical links between nations by associating Italian migrant explorers and military heroes with Argentine and U.S. American male political figures. Appearing on cans, boxes, and bottles, commercial iconography eliminated real temporal and geographical divides in gendered ways to create and display relationships between prominent men who personified their countries' past, present, and future accomplishments.

Trademarks, like the one for Garofalo's pasta (see fig. 6), regularly paired portraits of Italian monarchs and other male notables to illustrations of factories in order to associate Italy's industrial successes and export capacities with male political leaders. Expressions of progress and abundance tied to industrial production became standards in advertising during the late nineteenth century when Western industrialists began emphasizing their material and cultural contributions to modern civilization by relying on popular representations of productivity and efficiency, including factories, locomotives, railroads, and electricity.[60] Italian companies' glorification of science, technology, and modernity in factory-centered trademarks

Figure 9. Trademark for Emanuele Gianolio vinegar, Archivio Centrale dello
Stato, Ministero dell'Industria, del Commercio e dell'Artigianato, Ufficio Italiano
Brevetti e Marchi, fasc. 8593, 1907. Courtesy of the Archivio Centrale dello Stato.

for exports tapped into global discourses of progress to present an Italian nation
on par with other major manufacturing powers.

Trademarks frequently coupled effigies of well-known male personas and industrial settings with images of the medals companies received at Italian and international expositions.[61] A 1909 trademark for Martini & Rossi vermouth displayed
its steam-powered, sprawling factory in Turin under text reading "awarded with
40 medals," including "a gold medal at the Paris World's Fair in 1876" (see fig. 10).
The factory is flanked not only by two figures—one, a crowned female, her arm
resting on a map of Europe, and the other a male, wearing a headdress and indig-

Figure 10. Trademark for Martini & Rossi vermouth, Archivio Centrale dello Stato, Ministero dell'Industria, del Commercio e dell'Artigianato, Ufficio Italiano Brevetti e Marchi, fasc. 9408, 1909. Courtesy of the Archivio Centrale dello Stato.

enous attire, next to a map of the Americas—but also by a colorful assemblage of national flags and gold exposition medals.

Italy's dependency on Italians abroad as economic resources required international platforms for intensifying migrants' loyalty to their homeland. As the Martini & Rossi trademark illustrates, international exhibitions offered Italian companies' optimal sites for such international nation- and empire-building ventures around men and markets. Italy hosted the world's fair twice in the early twentieth century: the Milan International in 1906 and the Turin International in 1911. Like other nations, Italy used these occasions to affirm its place among expansionist countries; however, in contrast to other participating powers, Italy employed fairs to showcase migrants as the instruments upon which Italy's industrialization, modernization, and empire building would be built. Prominenti

Italian migrants with political and business clout participated in such venues, staging special exhibitions of "Italians Abroad," sponsored by the Commissariat of Emigration, to highlight their contributions to Italy.[62] During both expos the commissariat awarded a number of medals specifically to Italians who established new markets for Italian products abroad and who built Italian industries outside of Italy that employed Italian workers. The Italian Chamber of Commerce in Buenos Aries, reporting on its support for the "Italians Abroad" display at the Turin International, professed, "There is probably not a region of the world where the Italian—encouraged by a thousands of years of migratory traditions—has not reached everywhere demonstrating his admirable habit of colonization, his brilliance in industrial work, in scientific investigation, in artistic creation, the robustness of his arm like the quick agility of his mind."[63] Trademarks that tied Italian royalty to award-winning exports affiliated Italy with other industrialized exporting nations, who also incorporated exhibition medals into their marketing campaigns, while projecting positive messages about migrants and the transnational markets they nourished.

The ubiquity of industrial scenes also suggests the gendering of work and space accompanying Western industrialization throughout much of Europe, a process that slowly, unevenly, and not without challenge legitimized factories as male-dominated public establishments built on the wage work of men and the rational, scientific, and mass production methods fostered by their owners and overseers. While there was much regional variation in how industrialization, urbanization, and land reform affected divisions of labor between men and women throughout Italy, the transition from a subsistence economy to a market economy redefined wage earning in industry and agriculture as male, even in regions where women continued to work as agriculturalists.[64] Unsurprisingly, then, when women did appear in trademarks for exports, they were used exclusively as either allegorical representations of the Italian nation or as representatives of an idealized peasant past untouched by the forces of global capitalism. Artists and advertisers have frequently employed women to represent traditional values, a romanticized homeland, nature's abundance, or a sense of timelessness.[65] Such trademarks for industrialized food goods featuring female agriculturalists offered a sharp contrast to the modernist and masculine images of factories and industry common during this period. A 1904 trademark for Sasso-brand olive oil, one of the most common brands of olive oil exported abroad, especially to Argentina, included a barefooted woman in classical robes collecting olives (see fig. 11). Trademarks for olive oil, canned tomatoes, pasta, and dairy products displayed women as rural harvesters dressed in traditional folk clothing contentedly collecting fruit or wheat, or milking cows in rural bucolic settings.[66]

Ironically, while such trademarks referenced a gendered trope that associated women's labor with a precapitalist past, the commercialized agricultural and pro-

Figure 11. Trademark for P. Sasso e Figli olive oil, Archivio Centrale dello Stato, Ministero dell'Industria, del Commercio e dell'Artigianato, Ufficio Italiano Brevetti e Marchi, fasc. 6676, 1904. Courtesy of the Archivio Centrale dello Stato.

duction methods that produced the foods destined for mass consumption and export were part of larger gendered processes in which women were gradually disappearing from the agricultural work force. Turn-of-the-century trademarks for exports projected gendered versions of Italian productivity: one tied women to agricultural labor portrayed as exclusively female and outside the industrialized economy; the other associated Italian men with migration, industrial development and commercial expansion, and well-known political leaders.

Italian monarchs, medieval and contemporary merchant princes, and Italian cultural figures mirrored liberals' vision of migrant colonialism modeled after the medieval maritime republics—one that was built on trade, consumption, culture, and industrial productivity—as an alternative to nationalists' more costly and bloody forms of expansion based on the rejuvenation of an ancient Roman empire.[67] And yet while liberals often publicly eschewed this alternative model, references to and images of war-making and conquest, and to ancient Rome, also appeared frequently in trademarks. National leader and general Giuseppe Garibaldi surfaced often in export iconography, as did more generic images of Italian soldiers or combat troops, such as in a 1906 trademark for the Società per l'esportazione e per l'industria Italo-Americana (Society for Exportation and for Italian-American Industry) (see fig. 12). The society included a partnership between a number of successful textile firms led by Enrico Dell'Acqua, the "merchant prince" protagonist of Einaudi's 1900 manifesto.[68] The trademark portrays an armored Italian cavalryman bent forward on lookout with a flag in hand. Ready for action, he gazes at a railroad and steamship, the modern transportation marvels that made mass migration and exportation possible. The word *vedetta* (sentry) underneath him characterizes the scout as a commercial warrior and protector of the society's economic activities.

A number of trademarks also utilized the specific military figure of the Italian Bersagliere, a special corps of the Italian Royal Army that fought for Italian unification and is still in existence today. Images of *bersaglieri*, identified by their unique helmets adorned with black capercaillie feathers, associated Italian migrants and exports with virile and belligerent methods of national defense while harking back to the grandeur of the Risorgimento past. A trademark for canned olive oil produced by Raffaello and Pietro Fortuna from Lucca showcased a war

Figure 12. Trademark for the Societá per l'esportazione e per l'industria Italo-Americana, Archivio Centrale dello Stato, Ministero dell'Industria, del Commercio e dell'Artigianato, Ufficio Italiano Brevetti e Marchi, fasc. 7730, 1906. Courtesy of the Archivio Centrale dello Stato.

monument commemorating fallen soldiers in Lucca's Piazza XX Settembre, a city square dedicated to the final event culminating in the unification in Italy: bersaglieri's capture of Rome from the Papal States in 1870.[69] Depictions of armed men and military leaders like Garibaldi tied Italian migration and commerce to a nation-state in competition with other industrialized countries looking to expand their markets and influence abroad.

These commercial images matched the masculinist and militarist language used by Italian business and economic elites to describe Italy and Italian migrants in commercial battles with other imperial powers. Italian businessmen who published the trade guide *Italy in Latin America* proudly described Italian migrants in Argentina—"our faraway brothers"—as an "army of our children always enlarged with new troops. They are, in our mind, the pioneers of a radiant civilization rising on the horizon."[70] Einaudi himself dubbed Dell'Acqua and other exporters as "army generals," and "captains of industry" and referred to Italian migrants as "a disciplined army which moves as one, under the leadership of captains and generals in the conquest of a continent." Emigrants, Einaudi and others asserted, were vital assets in competing against other nations in "a fertile economic and social war."[71] Similarly, economist Visconti discussed the entry of Italian people and products in the Americas as an "invasion" and the competition between Italian and other countries for markets as "battles" won victoriously because of Italy's migrants.[72] In 1900 the Italian Chamber of Commerce in Buenos Aires described its "noble scope" of helping Italian trade surmount difficulties in the "fight" against its "rivals."[73] This martial language matched trademarks featuring commercial explorers and warriors who moved abroad to open up trade routes and guide and protect them.

Multiple temporal landscapes—ancient imperial Rome, medieval Italian city-states, and the more recent Risorgimento—and the different versions of migrant colonialism they implied served as gendered symbolic tools for Italian elites. Together they projected confident and positive messages about Italy as a great European power and expansionist country, one that could vie with other industrialized nations by relying on the past glories and future successes of its male migrants. This was especially critical as Italy, "the least of the great powers" of Europe, battled to maintain international standing. As Richard Bosworth argues, Italian liberal policy makers constantly fought to maintain prestige among the country's European neighbors, even though they lacked the resources to do so.[74] While commercial imagery and elite discourse emphasized Italy's role in fueling the commercial revolution of the Middle Ages and Early Modern period, in reality the Italy of the late nineteenth and early twentieth centuries had a less glorious reputation as a major supplier of cheap human labor and agricultural staples. Critics pointed to Italy's failed colonial adventures in East Africa, its regional divisiveness and inability to control peasant revolts and socialist uprisings, and its inadequate spending on military and naval buildup as proof of its junior status in the league of European power players. Despite the real economic and social achievements during the Giolittian era, on the eve of World War I Italy continued to lag industrially behind Western Europe; the country suffered from a frail banking system, depended heavily on foreign investment, and relied almost entirely on Great Britain for its coal supply, which slowed Italy's industrial and transportation revolutions.[75] Only the colossal remittances sent home by migrants abroad kept Italy's balance of payments positive during the country's volatile industrial takeoff.[76] Furthermore, while migration helped make Italian exports as a percentage of the country's total gross domestic product double from 6.2 percent in 1861 to 12.4 percent on the eve of the war, Italy's European competitors far exceeded Italy in terms of total exports, despite economists' boastful language about Italy's commercial capacities.[77]

Italian elites' martial discourse and commercial imagery also offered a rebuttal to the feminized language often used by Italy's European competitors to describe the country's people, culture, and economy. Leaders of Germany, Britain, France, and Austria, as well as some Italian politicians, viewed Italy with condescension, addressing the nation and its people using gendered stereotypes, as, for example, a "Young Italy" of emasculated and powerless children, or an effeminate country with *un popolo donna* (a womanly people).[78] Demographically speaking, mass male emigration did leave Italians as un popolo donna. The nation came to be depicted as such by Italian and non-Italian writers, both critics and supporters of emigration, who all agreed that emigration was strictly a masculine undertaking. Professor Giovanni Lorenzoni described migration from southern Italy and Sicily not only as a modernizing process in which the old farmer is carried into the "larger world of modern industrial life," but also as a process through which he "becomes a real man" by finding wealth and being adequately compensated for his work.[79] As Linda Reeder writes, "In the cultural imaginings, the migrant and the act of migration became identified with characteristic male traits (strength, virility, and action), whereas the people who chose to remain behind became imbued with female qualities (weakness, passivity and dependence)."[80]

Elites countered such feminized portrayals of the Italian nation by depicting Greater Italy as being built largely from without, through manly markets of migrants and exports. By linking male migrants' purchases abroad to Italian nation and empire building, Italy differed from its Western European neighbors in its gendering of consumption. While Britain, France, and Germany made increasing distinctions between the masculine realm of production and politics and the feminine sphere of consumption and family life, Italian leaders tied consumption to its male citizens abroad.[81] Italy was not entirely immune to trends that were prevalent in industrialized countries such as Britain, France, and the United States, where middle-class ideals of gendered divisions of space framed women's emergence as consumers and connected them to the nation-state.[82] However, this model was not wholly practical for migrant-sending nations like Italy that relied largely on consumerism and remittances of male migrants outside its borders to construct a più grande Italia. Instead, Italian leaders focused much of their effort on the country's migrants abroad, rather than middle-class women at home, as the major consumers of Italian goods and as one of the principal protagonists in fueling Italian global expansion.

Migrant consumption was instrumental in carrying out and legitimizing male-centered Italian-style expansionism. Elite representations of manly markets showcasing famed political leaders, factories and industry, merchant princes, and war and imperial Rome all employed images and discourses of masculinity to associate the consumption of Italian commodities overseas with manliness and to assert Italy's position as a great power in the Atlantic political economy.

Reluctant Consumers in Transnational Family Economies

Man-centered descriptions factored so centrally into elite depictions of migratory and commercial links between Italy, the United States, and Argentina in part because Italian worldwide migration during the age of mass migration remained overwhelmingly male, particularly for the years before World War I.[83] Male-predominant migrations were even more pronounced in transatlantic flows to le due Americhe. The percentage of migrating males from Italy hovered between 69 and 78 percent for the United States and 65 to 74 percent for Argentina from 1880 to 1914.[84] However, male migrants did not travel undetached from homeland social and familial networks nor from the obligations those networks carried. Italian political and economic leaders, in their fervor to make manly markets abroad, overlooked how transnational familial and economic arrangements influenced Italians' consumer decisions in migrant marketplaces.

Italian liberals' plans for linking Italian people and products would function effectively and profitably only if migrants conceived of themselves as consumers rather than primarily as producers and savers. However, while migration introduced Italians at home and abroad to novel consumer items, practices, and values, transatlantic migration did not fully nor abruptly wrench them out of the subsistence economies and cultures in which they were embedded. In nineteenth-century Italy, subsistence production was organized around family economies in which all members contributed to the survival and well-being of the household and where little cash was earned or used. With some regional variation, most rural families in Italy made almost everything they ate and wore, with women largely in charge of producing food, clothing, and other consumable goods.[85]

In the late nineteenth century new taxes instituted by the Italian state, the intensification of commercial agriculture, and failed land reform schemes forced more

Italians, especially in the south, to search for wage work. Family economies in rural areas adjusted by sending men to earn money as laborers on faraway estates or northern cities, leaving women and dependent children to cultivate food, without wages, in fields closer to their villages.[86] Women also continued to produce cloth for family consumption and as part of a disappearing cottage industry in the south, where cheap cloth imported from abroad was slower to affect women's traditional work.[87] As it became increasingly impossible for peasants to work and possess their own land, they responded to the growing needs of a global economy by sending mainly male family members abroad for agricultural, mining, construction, and factory work and by extending family economies across national borders. These transnational family economies functioned best when male wages earned outside of Italy combined with the continuation of mainly female semi-subsistence production of food and clothing at home. The ultimate, but usually unrealized, goal for most rural families was to use male wage earnings abroad to buy land back home and live off the profits from rents, thereby freeing themselves from low-status, dependent, and poorly paid wage work. Excessive consumption in the Americas, therefore, was not conducive to the success of transnational family economies. Migrant laborers tended to spend the higher salaries they earned abroad not in their host countries, where the cost of living was high, but back in Italy, where the cost of living was low and where their earnings helped strengthen families' position in the extant social and economic hierarchy of rural Italy.[88] Despite elites' dreams, migrants sought to enhance the material security of their individual families in transnational economies, with little regard for the economic well-being and reputation of the Italian nation itself.

Indeed, most migrants lived thriftily in le due Americhe, choosing to forgo rich diets, high rents, expensive consumer items, and entertainment in order to create the savings necessary for the purchase or improvement of their homes and social standing in Italy. Male labor migrants in New York and Buenos Aires saved money by boarding with families or relatives from their home villages or in larger boardinghouses. Lodging expenses included laundry and meal service, often provided by an Italian woman, or sometimes by the *padrone*, the gang boss. There, workers were served mainly inexpensive meals consisting of pasta dishes; bean-, vegetable-, and rice-based soups and stews; and coffee with bread, which they supplemented with cheese, wine, produce, and other foods purchased from street vendors, restaurants, and grocery stores or raised in small urban gardens.[89] And although living thriftily may have prevented Italians from buying imports regularly, they reported eating well because food was more abundant and cheaper than back home. After migrating to Argentina in 1902, Oreste Sola wrote to his parents in Biella, Piedmont, about the plentiful "fruit of all sorts," especially peaches and pears, which, he wrote in amazement, were so numerous "they are used to fatten pigs." Argentina's booming cattle industry gave migrants like himself consistent access to meat. "We eat steaks and grilled meat every meal just like eating potatoes at home." He assured his mother, "Here I lack for nothing. We are in America, and so everything is available here."[90]

While overseas Italian migration remained predominantly male, women in Italy played central roles in transnational family economies, financing and sustaining the migration networks in which Italians operated. Women and their work were not unaffected by the larger forces that sent their husbands, fathers, and brothers abroad. These commercial images of subsistence economies—recalling unchanging and idealized worlds of peasant families, hermetically sealed off from the larger forces of capitalism—belied the world of motion affecting Italian women. Linda Reeder has demonstrated how mass male emigration transformed Italian women's lives, especially for Sicily's *vedove bianche* (white widows), women left behind by their migrant husbands. In contrast to contemporary depictions of such women as abandoned victims of male migration, and, in their removal from male supervision, as threatening to traditional familial arrangements and gender roles, most migrants' wives actively invested in migration as a strategy to enhance their family's standing. Women's handling of remittances—invested in house improvements, property, small businesses, educational opportunities, and dowries—represented an extension of rural women's long-standing role as managers of family resources and money.[91] Within a larger patriarchal system limiting female autonomy and authority, women held substantial power over the domestic sphere, including control of the family's finances. Mass male migration enlarged these responsibilities for women, who employed them to make demands on the state on behalf of their families and faraway husbands even while traditional power relations between men and women remained relatively unchanged.

Male migrant frugality allowed for an increasing number of rural women and men in Italy, as well as returnees, to help build and participate in the country's embryonic consumer society. Using remittances sent home, migrants' wives purchased food, furnishings, and cloth for their homes, families, and businesses. Like their male relatives abroad, women left behind also moved merchandise from faraway places into small agro-towns and rural villages throughout Italy in ways that transformed Italy's relationship to global markets. When elites acknowledged rural women's increased consumer activities, they usually did so with skepticism and disapproval, identifying women's spending power with immorality, indolence, and a reversal of proper class hierarchies. Reporting for the government from Molise and Abruzzi in southern and central Italy, Cesare Jarach noted that emigration produced "*la donnicciuola*" (a gossipy woman) who "comes to the city on Sundays to shop, handling with great arrogance large bank notes."[92] One affluent *molisano* (a person from the southern Italian region of Molise) lamented to a government investigator that "the wives of the Americans arrive at the marketplace and buy up all the fresh fish newly arrived from Termoli, regardless of price."[93] Another official proclaimed that in the southeast region of Campania, "remittances from abroad allowed women to live in complete idleness."[94] Such reports portray women as inimical to Italy's economic and cultural progress as a nation. However, they also disclose how rural women, like male migrants, shaped and were shaped by the forces of globalization—migration, trade, industrialization, imperialism, and commodity culture—which in turn transformed Italy and Italy's relationship to the world.[95] While it would be unwise to exaggerate the diffusion of consumerist activities in early twentieth-century Italy, especially in rural areas of the south, by putting cash in the hands of women, male migration linked Italy's inchoate consumer society just as much to women left behind as it did to male migrants abroad.[96]

Despite elites' predictions of a più grande Italia built on migration, trade, and consumption abroad, male labor migrants remained reluctant consumers. Mass migration stimulated Italy's export market, especially in foodstuff, while helping to slowly revolutionize Italy's agricultural, manufacturing, and commercial sectors. However, to the frustration of Italian leaders, migrants' transnational familial strategies constrained migrant spending on Italian products in le due Americhe and on the consumer goods of their host countries. While trademark iconography relegated women to a precapitalist, rural, and idealized past, the same modernizing forces motivating global migration and exportation altered women's activities in transnational family economies and ultimately hindered elites' plans to construct manly markets abroad.

*   *   *

In December 1912 the magazine of the Italian Colonial Institute, formed in the early twentieth century to promote Italian colonial expansion, published "Ten Commandments for Italians Abroad," iterations of which would appear in migrant papers across the Atlantic. Together, the final two commandments evidenced the unstated links between transnational family economies and the manly markets they fostered. The institute's ninth commandment instructed Italians abroad to sell, buy, and consume Italian products and to reject foreign goods. The tenth and final commandment read, "You shall marry only an Italian woman. Only with this and by this woman shall you be able to preserve in your children the blood, language, and feelings of your fathers and of your Italy."[97]

By exhorting Italian labor migrants to accept Italian wines and Italian wives exclusively, the country's leaders hinted at gendered and transnational connections between migration, trade, and consumption, connections represented in political tracts and export iconography. To elites, trademarks for Italian exports connoting industrial, commercial, and cultural superiority and national defense served as transnational and sometimes transhistorical commercial sites for linking Italian nation and empire building abroad almost entirely to men. Equating migrants and exports with warriors, weaponry, industrialization, and colonialism confronted stereotypes depicting Italy as a politically, economically, and militarily weak and effeminate nation. Rather than un popolo donna, migrant merchants and consumers, and the trade paths they sustained, would yield a nation of producers and consumers led by migrant "merchant princes" worthy of respect from other industrialized, imperial powers.

As the twentieth century progressed, however, the strong faith elites had in migrants as loyal consumers of Italian exports would begin to wane. In remarkable ways, consumer demand from overseas facilitated trade in Italian goods, especially foodstuff, to le due Americhe. But to elites' consternation, Italians in migrant marketplaces abroad often prioritized their own economic and familial interests over a commitment to constructing a Greater Italy through manly markets. The Italian Colonial Institute's tenth commandment, directing migrants to "marry only an Italian woman," stifled migrants' ability to follow the ninth commandment, to buy Italian exports, as migrant consumption detracted from laborers' obligations to wives and children back home. It would take a world war, a slow but significant reconfiguration of gender ratios in migrant marketplaces, and intensifying ties between women and consumption globally to turn migrants toward consumption for the homeland.

CHAPTER 2

Race and Trade Policies in Migrant Marketplaces, 1880–1914

U.S. tariff policy, claimed Italian food importer C. A. Mariani in 1911, turned Italian migrants toward crime. In an article for the *Rivista*, the New York Italian Chamber of Commerce's bulletin, Mariani articulated a link between the racialization of Italian migrants and the recently passed Payne-Aldrich Tariff Act. Higher tariffs on Italian imports imposed by the act prohibited migrants from purchasing the olive oils, tomato products, cheeses, and pasta that were so characteristic of Italians' "mode of living." By forcing migrants to consume U.S. foods exclusively, the U.S. tariff policy compelled the Italian "to get a new stomach, which is only within the power of the Almighty." Even more nefariously, it tempted consumers toward whiskey, a product unknown to Italians back home. Excessively high taxes on the Italian wines that migrants had consumed without drunkenness in Italy drove them to drink whiskey and other less expensive, domestically produced beverages, leading to moral decay. "Many crimes are charged to the Italian emigrant which should be charged to American whiskey and the American form and habit of drinking," Mariani concluded.[1] While nativists pointed to Italians' inborn tendencies toward criminal behavior to explain high crime rates and incidences of public intoxication, Mariani blamed prejudicial tariff policy.

A year before the publication of Mariani's article, an ad for Luigi Bosca and Sons' Piedmont wines in Buenos Aires' Italian-language daily *La Patria* boasted that despite elevated tariffs, consumers in Argentina preferred Bosca wines over all others. While in New York, Italians' wine-drinking habits linked migrants to criminal activity, the Bosca ad characterized Italian consumers as a civilizing force in Europe and in the Americas. Bosca wines produced in the "ancient" Piedmont winery were known "throughout the civilized world," the ad explained. Starting with references to Virgil's *Aeneid*, the publicity recalled Italy's illustrious history from the Roman Empire to the proclamation of Rome as capital of the Kingdom of Italy in 1871. Popular throughout Europe by the mid-nineteenth century, Bosca's sparking white wines reached South America with Ligurian migrants in the 1860s. The same high tariffs that in the United States seduced Italians toward a life of illegality and depravity proved innocuous to Italians in Argentina, where Bosca wines had conquered the Argentine and South American markets.[2]

As these examples attest, migration and exportation were not unrelated to each other and to notions about race and ethnicity; instead, they collided in migrant marketplaces to shape migrants' consumer options and host societies' perceptions of newcomers and their foodways. Italian elites' optimistic proclamations of transoceanic empire building through men and markets often disregarded on-the-ground differences in le due Americhe. But clearly these differences generated distinct experiences relating to Italian foodstuff for migrants in the United States and Argentina. In New York perceived racial dissimilarities between Italians and Anglo-Americans and the foods they ate inhibited migrants' ability to use consumption to forge ties with non-Italians. Instead, Italians used food to articulate differences between Italian and Anglo-American eaters. Migrant merchants, however, resisted racialization by arguing that anti-Italian prejudices could be bridged as more native-born U.S. Americans consumed high-quality Italian imports. In Buenos Aires, on the other hand, migrants used connections between migrants and markets to construct bonds between Italians and Argentines as members of the "Latin race," rooted in assumptions about the superiority of Italian and European civilization. The real and invented solidarities between the people and food products of Italy and Argentina opened up opportunities for shared consumer identities and experiences involving Italian imports between newcomers and natives in Argentina.

Imported foods in migrant marketplaces provided Italians an opportunity to shape racial boundaries but also to comment on their host countries' economic policies vis-à-vis Italy. In Buenos Aires, Italians used their prominent place in the larger Argentine social and economic landscape, and the often-discussed commonalities between Italians and Argentines, to position themselves as commanding players in the nation's economy with the power to affect Argentine trade legislation. Conversely, as only one of many foreign-born groups in a much more multicultural, industrially mature nation, Italians in New York struggled to portray themselves as having the ability to influence U.S. economic policy. However, they too denounced tariff increases as biased detriments to Italian consumption and inextricably linked to corporate trusts. In both countries, migrants employed connections between mobile people and mobile foods to represent themselves to their communities, to Italy, and to their host countries as formidable consumers in the larger transatlantic political economy.

Nation-Specific Differences Shaping Consumption in Migrant Marketplaces

Italian migrants' locations within the ethno-racial and socioeconomic structures of the United States and Argentina produced different foundations for assembling and challenging racial identities as food consumers and for arguing against protectionist legislation. Late nineteenth- and early twentieth-century industrial transformations in both countries relied enormously on a steady supply of international migrants as sources of cheap labor and, increasingly, as consumers. However, while more than twice as many Italians went to the United States than to Argentina from 1880 to 1914, they represented a much smaller percentage of the United States' total foreign-born population. Argentina, on the other hand, attracted Italians disproportionately as compared to other migrant groups.[3] Italians made up 59 percent of Argentina's total migrant population but only 6 percent of the United States' between 1881 and 1890. During the first decade of the twentieth century, Italians constituted 29 percent of the United States' total migrants, whereas in Argentina, Italians counted as 45 percent. In New York, which by 1910 had a larger population of Italians than any other city in the western hemisphere, Italians made up just 7.1 percent of the city's foreign-born residents, whereas in Buenos Aires, Italians more than tripled that, at 23 percent.[4]

Given Italians' demographic dominance in Argentina, it is no surprise that Italians made up the largest percentage of the foreign-born in almost every occupational category. Yet it was their overwhelming presence in Argentine commerce and industry—especially in the importing, manufacturing, and selling of foodstuff—that made Italians exceptionally powerful protagonists in shaping the country's consuming patterns and identities around imported edibles. This ascendancy can be traced back to Italians' earlier presence in Argentina; among the diverse group of exiles, laborers, and elites arriving in Argentina starting in the 1820s were groups of Ligurian merchants and commercial elites who by the 1870s counted among la Plata's leading families, setting the foundation for subsequent migrations and Italian commercial activity.[5] By 1901, when the Italian Ministry of Commerce requested a list of all Italian-owned commercial firms in Buenos Aires, Eduardo Bergamo, president of the Italian Chamber of Commerce in Buenos Aires, stated that the list could include only major Italian establishments. "Otherwise," he noted, "it would be materially impossible to make a complete list of the millions of Italian firms that specialize especially in the sale of edibles and drinks."[6]

Argentina's two-class system, which was divided into a small group of mainly native-born elites and a large non-elite population, explained the prominent location held by Italians in the nation's food trade. The Argentine upper class focused predominantly on law, politics, and land ownership—especially before 1912, when federal electoral reforms began to draw more migrants into politics—allowing the foreign-born and their children to almost completely control commerce and small-scale manufacturing, as both employers and employees. The Buenos Aires Chamber of Commerce reported in 1898 that Italians made up 58 percent of all merchants in Argentina, with the Spanish coming in second at 9 percent.[7] In 1914 the Argentine government found that foreigners, predominantly Italian, operated 70 percent of the country's total food-related commercial establishments. Of the 3,409 bodegas, retail shops, and warehouses enumerated in the 1910 Argentine census, Argentine nationals owned 28 percent, compared to the 47 percent owned by foreigners. Migrants also owned a large majority (71 percent) of the nation's liquor shops.[8] These Italians joined an emerging middle class of merchants, retailers, artisans, manufacturers, white-collar workers, and bureaucrats that grew as the country expanded economically and demographically.[9]

While small-scale migrant merchants and retailers predominated, not a few developed their businesses into some of the largest commercial houses in the country. One of the most prominent was Sicilian Francesco Jannello, former sea captain in the merchant marine and Italian Royal Navy. After overseeing the transatlantic shipping of Palermo-based winemaker I. V. Florio, Jannello opened a branch of the company in Buenos Aires in 1891 with exclusive rights to sell Florio products in the city. From his store on San Martín Street in the city's financial center, Jannello sold Florio wines and cognacs, especially the company's famed Marsala dessert wine. By 1910 Florio Marsala represented 90 percent of all imported Marsala in Argentina, helping make it more popular than similar dessert wines from France, Spain, and Greece. In 1908 Jannello also became the sole representative of another widely popular Italian liquor, Martini & Rossi vermouth, from Turin, and he aggressively marketed both imports in *La Patria*.[10] Jannello's association with the city's leading commercial and fraternal organizations provided him with access to the personal connections, ethnic networks, and financial resources necessary for building his successful import and retail business. He was vice president of the Italian Chamber of Commerce in Buenos Aires from 1901 to 1906 and was involved in several Italian fraternal organizations; he also served as a board member of the New Italian Bank from 1893 to 1907 and as an officer of the Bank of Italia y Río de la Plata, two important financial institutions in Argentina for migrant *prominenti*.[11]

In the United States, Italians faced a different set of economic, social, and cultural challenges that prevented migrants from competing as successfully as their counterparts in Argentina for control over their host country's commercial food sectors. In New York, Italians encountered a middle class that was already dominated by native-born whites and older migrant groups, such as second-generation Germans, Scandinavians, and Irish. With some regional variation, Italians settling in urban areas were relegated to the working class in construction and factory jobs with less opportunity for social and economic advancement than their counterparts in Buenos Aires.[12] Furthermore, while industrialization and the development of commercial agriculture in both countries required, above all, semi-skilled and unskilled workers, the United States' more rapid transformations starting in the late nineteenth century attracted larger numbers of seasonal and unskilled workers from Italy. Using census samples, Samuel Baily estimated that at the turn of the twentieth century almost twice as many Italians in Buenos Aires worked in nonmanual, white-collar positions—mainly commerce—than in unskilled jobs as laborers or servants. In New York the opposite pattern reigned: twice as many Italians worked in manual laborer positions than in white-collar work.[13]

With less opportunity for socioeconomic mobility in a country where they counted as only one of many migrant groups, Italians made up a smaller percentage of the United States' total number of food merchants and retailers. Although Italians came to dominate the importation of select bulk agricultural products, such as lemons, most Italian food businesses prospered chiefly within Italian urban enclaves like the Mulberry District south of Fourteenth Street and East Harlem on the Upper East Side in Manhattan, where they sold their wares to other Italians.[14] Observers characterized Italian neighbors in New York as isolated from the larger city, culturally, socially, and commercially. Social worker Louise Odencrantz wrote in her study of Italian families in New York, "They form small communities in themselves, almost independent of the life of the great city. Here the people may follow the customs and ways of their forefathers. They speak their own language, trade in stores kept by countrymen, and put their savings into Italian banks. . . . The stores all bear Italian names, the special bargains and souvenirs of the day are advertised in Italian, and they offer for sale the wines and olive oils, 'pasta,' and other favorite foods of the people."[15]

Despite Italians' smaller presence in the United States' import and food industry sector, as in Argentina, a number of migrant-owned businesses evolved into sizable firms. Among the biggest was L. Gandolfi and Company, established in New York City in 1883 by Luigi Gandolfi and Ettore Grassi, migrants from the northwest Italian province of Lombardy. The company offered consumers a wide variety of Italian foods at their store on West Broadway in the Mulberry District, such as cheeses from Milan and Parma, olive oil from Lucca, and dried pastas from Genoa and Naples. They also represented Fernet-Branca, a Milan-based maker of an amaro, and Florio-brand Marsala wine, making them Francesco Jannello's North American counterpart. Like other large and prosperous migrant importers, Gandolfi joined Italian associations; he was a longtime member of New York's Italian Chamber of Commerce and served as vice president of the Italian American Trust Company, a major credit-granting institution.[16]

Nation-based differences in the two countries' ethno-racial makeup and socioeconomic hierarchy produced different opportunities for interactions between Italians and non-Italians involving imports. While Italian merchants in both places dominated the Italian import market, those in Buenos Aires benefited from entering an economic and class sector that was not dominated by native-born Argentines. Italians' command of the nation's food-related commerce—as well as other industrial sectors—meant an expanded consumer base beyond Italians to include Argentines. In New York, Italian merchants, who made up a small percentage of the nation's total food-related establishments, had a less extensive market reach beyond Italian communities.

Commercial newspapers depict Buenos Aires' migrant marketplace as a permeable location where migrants and Argentines forged shared consumer experiences. Ads for Italian foodstuff had a more formidable presence in the Argentine Spanish-language dailies than in nationally circulated English-language newspapers such as the *New York Times*, where publicity for Italian goods did not show up with regularity until after 1920. As early as 1900, for example, both Buenos Aires' *La Prensa* and *La Nación*, the nation's chief Spanish dailies, included ads for Italian imports, especially liqueurs, wines, and sparkling water.[17] Italian retailers also ran ads for various Italian goods in *Caras y Caretas*, Argentina's popular Spanish-language magazine.[18] In 1910 a full-page ad for Cora-brand vermouth sold by José (Giuseppe) Peretti featured two well-dressed ladies lounging on a sofa, enjoying a glass of the vermouth, "the father of Turin vermouths."[19] By the early twentieth century, large department stores in major cities like Buenos Aires and Rosario, which attracted a wide range of consumers from different national backgrounds, sold Italian imports. An ad for La Gran Ciudad de Chicago in Rosario reminded readers of *La Patria* that it had "the most complete assortment of Italian articles."[20] These department stores, products and emblems of an emerging urban modernity built on merchant capital and mass consumerism, clearly sought out the pesos of Italians, the largest foreign-born consumer base in the country.[21] Conversely, department stores in New York, such as Wanamaker's and Macy's, which stocked a wider range of both domestically produced goods and imports, advertised regularly in New York's *Il Progresso* only after World War I. Spanish-language dailies also more regularly covered stories related to Italy and to the Italian migrant community than did English-language newspapers in New York. As early as 1900, Argentina's *La Nación* ran a regular column titled "Vida Italiana" (Italian Life) that focused on Italian politics and economics and on Italian migrant fraternal organizations.[22] Ads and articles in the Spanish-language press reveal a shared print culture that promoted the consumption of Italian imports among Italian and Spanish speakers in Buenos Aires.

That the consumption of imports both reflected and helped forge connections between Italians and Argentines is also evidenced by the large number of ads in

Spanish for Italian foods that appeared in the Italian-language *La Patria*. From the late nineteenth century through the 1930s, a quarter to a half of these ads ran in Spanish rather than in Italian. Conversely, publicity for Italian imports in New York's Italian-language *Il Progresso* remained almost exclusively in Italian; ads for Italian foodstuff rarely appeared in English.[23] And while the bulk of articles in both *La Patria* and *Il Progresso* ran in Italian for the entire period of Italian mass migration, information dealing with international trade often appeared in Spanish in Buenos Aires but remained in Italian in New York. In *La Patria*, detailed *manifesti* (ship manifests) listing the importer, quantity, and often brand name of Italian imports entering Argentine port cities were consistently published in Spanish.[24] New York's *Il Progresso* issued a regular column called "The Commercial Bulletin" in Italian beginning in the early twentieth century, but the column included a very limited list of the local market prices of mainly food items. It evolved to include some international commerce, but only regularly after World War I, and they remained in Italian, not English.[25] The widespread use of Spanish in ads for and information about Italian imports in Argentina's Italian-language press indicates not only linguistic similarities between the two romance languages but also a readership and consumer market that included both Italians and Argentines, increasing numbers of whom were the children of migrants. Linguistic anthropologists might argue that *La Patria* was a language contact zone, where Italians' dominance in commerce combined with linguistic similarities to produce a mixed-speech community of buyers and sellers who facilitated the consumption of Italian imports by both Argentines and Italians.[26]

Italians in Buenos Aires had a more formidable presence as food merchants, retailers, and consumers in Argentina than did New York–based Italians. The greater scale and scope of the U.S. economy and its more heterogeneous foreign-born population limited the reach and influence of Italian imports beyond Italian consumers in New York. Italians' prominent place in the nation's food sectors made Buenos Aires' migrant marketplace a more commercially fluid site that engaged both Italians and Argentines. These nation-specific differences provided migrant food purveyors and consumers with dissimilar foundations for constructing ideas about race.

Constructing Race in Migrant Marketplaces

In 1909 the Italian Chamber of Commerce in New York set out to debunk the myth that in South America, particularly Argentina, Italian migrants and the trade in Italian imports they opened had a more promising future than they did in the United States. As an example, the chamber argued that the derogatory term "gringo" used by Argentines to describe Italians equated to the pejorative terms "dago" and "guinea" used by U.S. Americans. According to the chamber, "The Argentine government, always hospitable and courteous, hides under the promise of fraternity." And yet, "It is useless to deceive oneself. Latin America follows *il programma yankee*" (the Yankee way). Argentina, the chamber concluded, "does not give a damn about their and our *Latinità*" (Latinity).[27]

Italian-language newspapers and business publications in Buenos Aires, however, suggest that migrants employed the "promise of fraternity," suggested in the concept Latinità, to account for their acceptance and commercial success in Argentina. In New York the "Yankee way" frustrated Italians' attempts to attract food consumers; in Buenos Aires, Italians argued that Latinità elevated Italian foods and their consumers as "forces of civilization."[28] They suggested that migration, the "foundation of commercial relations between Italy and Argentina"—a foundation built on the wine, oil, cheese, rice, and "hundreds of other Italian articles" in Buenos Aires—could not occur in the less racially hospitable United States, where the Italian "finds himself among people who seem to be another race, which has diverse customs."[29] Ideas about race shaped, and in turn were shaped by, migrant marketplaces of New York and Buenos Aires, making the two cities radically different locations for identity building and racial formation.

Global conversations about race circulating among academics, physicians, and politicians in the mid-nineteenth century shaped U.S. and Argentine attitudes and policies toward Italians and other migrant groups. During this period, racial thinkers in Europe and the Americas began categorizing humankind in a system of castes, with the assumption that racial characteristics were the most important societal indicators. Using allegedly scientific methods to study and typologize populations, eugenicists blamed social problems, such as illiteracy, poverty, and immorality, on the supposedly inherited and unchanging character traits of degenerate groups. Eugenicists joined Social Darwinists in applying evolutionary principles to society in order to justify actions and inactions by nation-states that were increasingly worried about protecting superior racial stocks and the higher levels of civilization they represented. Against the backdrop of massive global migrations, migrant-receiving nations responded anxiously by using pseudoscientific thinking to judge the racial fitness of the people entering and exiting their borders.[30]

Italians held an ambiguous position within these developing racial typologies. By the late nineteenth century, experts concurred that northern and southern Italians belonged to two biologically different and unequal races. Italian criminal anthropologists, led by Cesare Lombroso, characterized southern Italians—who by the early 1900s made up the majority of overseas migrants—as racially inferior and innate criminal types, "barbarians" associated with organized crime and immorality. He joined other racial scientists in blaming the social and economic

problems of the *mezzogiorno* (southern Italy) on southern Italians' inborn racialized traits. And yet there remained a number of academics who counted Italians
as part of the civilizing "Mediterranean race"; they were reluctant to completely
deny Europe's and modern Italy's links to the Roman Empire, the Renaissance,
and the Age of Exploration, although they increasingly connected that illustrious
history to northern Italians exclusively.[31]

Notwithstanding shared dialogues about race among elites in Europe and North
and South America, differences in the U.S. and Argentine racial landscapes affected
Italians' wherewithal to practice racial inclusion through imports and the foodways
they helped create.[32] In the viceroyalties of Peru and of the Río de la Plata, a system
of castes, a hierarchical socio-racial classification used by Spanish colonial elites
to categorize mixed-race people, privileged Catholic, Spanish-born *peninsulares* and
criollos, those born in the colonies of Spanish ancestry.[33] After Argentina declared
independence in 1816, a leading group of liberals—many of them wealthy criollo
landowners—strove to eradicate the nation's allegedly barbaric and savage elements. These elements included indigenous populations and the then significant
number of people with African ancestry, along with *caudillos* (rural warlords) and
their gaucho (cowboys of the pampas plains, often of mixed race) followers in the
countryside.[34] European modernity and civilization, professed future president and
leading liberal Domingo Faustino Sarmiento in his 1856 opus *Facundo*, offered the
best model for Argentine nation building.[35] His ideas manifested in the Argentine
Constitution of 1853, which included a special clause encouraging European migration. The government recruited and subsidized transatlantic migration, especially
through largely unsuccessful programs to settle migrant farmers in Argentina's
interior.[36] Europeans were encouraged to populate and civilize an allegedly empty
and wild pampas frontier and to "whiten" indigenous and mestizo populations
through intermarriage. European migrants helped Argentine elites to construct
Argentina as a "white" nation, especially in comparison to Latin America's other
most populous countries, Mexico and Brazil.[37]

While Argentine immigration policy did facilitate Italian migration, over time
eugenicist ideas, social and political agitation, and financial woes led Argentina's
liberal oligarchy to rethink whether Italians posed a possible threat to Argentine
national and racial identity.[38] The majority of Italian migrants were poor, uneducated, and assumed radicals. Elites' claims that left-leaning foreigners brought
about social unrest and working-class protest in cities like Buenos Aires resulted
in the passage of the Law of Residence in 1902, which allowed for the exclusion
and deportation of radical and criminal migrants, as well as the comparable 1910
Law of Social Defense, which banned anarchists from the country.[39] However,
Argentine elites also reluctantly admitted that their country depended on Italians
as the largest foreign-born group. As Nancy Leys Stepan argues, Argentines could

not deny the country's obvious Latin roots as a Hispanic nation, even while not fully accepting that Argentina's racial identity would be Latin rather than Anglo-Saxon.[40]

It would be the United States, not Argentina, that would eventually exclude migrants based on racial and class categories.[41] While Anglo-American founders in the United States shared with Argentine elites an antipathy toward native peoples and strategies for removal, the country's long history of slavery concretized naturalization laws and citizenship rights based on a black-white binary, a more rigid and less complex—if equally oppressive—racial system than that which developed in colonial Latin America.[42] Whiteness in itself, however, did not guarantee Italians entry or inclusion.[43] The popularity of Social Darwinism and eugenics, the end of slavery, U.S. and global imperialism, and the arrival of millions of non-Protestant migrants from Europe, Asia, and Latin America created a cauldron of racial anxieties directed at foreigners. In 1882, when the U.S. Congress passed the Chinese Exclusion Act, which barred all Chinese laborers from entering the country, it became the first major piece of legislation to exclude a group based on racial selection.[44] By the 1920s nationalism brought on by the war and the ensuing "red scare" made increasing numbers of migrants fall under suspicion as inferior and excludable. The resulting Immigration Act of 1924 used a racially prejudicial quota system to drastically reduce the number of Southern and Eastern Europeans from migrating to the United States.[45] Racial discrimination, combined with elites' continued fear of national degeneracy through miscegenation, encouraged Italians to locate ethnicity as the centerpiece of their hyphenated identities, particularly among the second generation. This was different from Argentina, where eugenicists viewed racial mixing, albeit of superior blood types, as conducive to positive nation building, an approach that encouraged Italians and their children over time to identify simply as Argentines.[46]

The distinct ethno-racial histories of the two countries provided different contexts within which Italians in New York and Buenos Aires constructed race through trade and consumer experiences. In Buenos Aires, Italians associated their foods with both Latinness and Europeanness to explain Argentina's acceptance of and demand for Italian imports and people. In New York, where migrants confronted a dominant culinary culture that largely disdained Italian foods, merchants maintained that racial divides between Italians and non-Italians could be overcome as more Anglo-Americans appreciated high-quality Italian imports. Food practices, food studies scholars have shown, serve as a central arena around which unequal social hierarchies based on race are inscribed materially in bodily practices.[47] Imported "foreign" foods, because they arrive from the outside, often at the behest of the "foreign" migrant consumer, held the potential not only to racialize migrants as unpatriotic, perpetual outsiders but also to challenge notions of national sovereignty and racial purity. Imported olive oil, wine, pasta, cigarettes, canned to-matoes, and other foodstuff offered migrants the wherewithal to produce, as well as object to, racialized linkages between race, nation, and commerce.

When Ernesto Nathan, mayor of Rome, stated in honor of Argentina's cen-tennial in 1910 that Argentina and Italy were "the two branches of the Latin race that populate the old and the new world, and shake hands across space because they move toward the conquest of human progress," he expressed a common as-sumption articulated by sellers and buyers of Italian imports in Buenos Aires.[48] In a country where skin color, socioeconomic status, and descent combined with a belief in the superiority of European culture to define race, Italians regularly re-ferred to migrants and goods arriving from Italy as important civilization builders in Argentina. Civilization and race collided in Italians' repeated use of Latinità, a category employed to construct shared racial similarities between Argentines and Italians and to explain and enforce ties between Italian migrants and exports.[49] *La Patria* called attention to "the intimate relations of interests and sentiments that unite Italy to the great country [Argentina] where its children find, as is said, a second homeland and carry in exchange a continuous, precious contribution of the vigorous and refined Latin blood, of commercial and industrial genius and of honest work."[50] Basilio Cittadini, editor of *La Patria*, cited Argentina's immigration and trade policies as proof that the country welcomed Italians and encouraged them "to assume its share of partnership in the common work of civilization, of intellectual and economic evaluation of the Nation as it marches toward its high-est destiny."[51] As both "children of the Latin race," Italians and Argentines worked harmoniously in migrant marketplaces toward the modernization of Argentina.[52] Migrant publications often claimed that Italians essentially *made* Argentina, hint-ing at the deep debt the country owed migrants for their commercial, economic, and cultural successes. In 1910 the Italian Chamber of Commerce in Buenos Aires reminded readers of its bulletin that the Italian community had been and continued to be "the principle factor of development and progress in this country in every area of human activity."[53]

Given the predominance of men in Italian migration to Argentina, it is no sur-prise that expressions of Latinità overwhelmingly characterized relationships be-tween Italians and Argentines as brotherly rather than as sisterly. Italian-language newspapers and business publications in Buenos Aires described Argentines and Italians as "brotherly people" and were filled with emotional declarations of fra-ternal love for their Argentine "brothers."[54] Argentines, too, often discussed rela-tions between themselves and Italian migrants using sentiments of fraternity. The Argentine daily *La Nación*, in an article titled "Genuine Fraternity," claimed, "Italian-Argentine brotherhood is not limited to the exchange of international politeness or to the pretense of official state sentiments; this fraternity is a fact; it is in the heart that loves as it is in the brain that thinks."[55] Migrants and Argentines alike constructed Latinità through the gendered language of brotherly affection.

While migrants used the narrative of Latinità to cement cultural and commercial bonds between host and home country, they sometimes asserted their superiority over Argentines within this larger transnational Latin family. They echoed Italian elites like Luigi Einaudi, who insisted that in South America "there lives a similar race to ours at a level of civilization not superior to ours and sometimes inferior to the level of Italians."[56] Latinità allowed Italians to position themselves and their trade goods as harbingers of European civilization, with the potential to *italianizzano* (Italianize) Argentines. In 1908 the Italian Chamber of Commerce in Buenos Aires ran an article that described Argentina as a "second Italy," while claiming that a third of the nation's population had Italian blood in their veins—proof that Argentina had been Italianized.[57] Italian judges awarding Italian importers and industrialists for their commercial success in Argentina noted that Italians there (as opposed to in the United States) transplanted Italian pasta, wine, sweets, and olive oil easily and with success because of Argentina's "weaker population," and because "almost all the industrial fields are formed by Italians."[58] Latinità and Italians' demographic and commercial authority made Buenos Aires' migrant marketplace a permeable site where both Italians and Argentines consumed Italian imports, and where through such interactions migrants Italianized a "civilization in formation."[59] Economist Aldo Visconti agreed that similarities between the two populations made this process inevitable and imperceptible: "Despite its contrary will, despite all its efforts to avoid and continue to avoid it, the Argentine population had too many similarities with the Italian element to not be in part Italianized."[60] He saw this occurring most visibly around the consumption of Italian food imports; Argentines, as well as Italians, desired Italian products; by simply following their own preferences, Argentines favored the growth of Italian imports without even realizing it.

The Italianization of Argentines also occurred through women's reproductive labor. Visconti pointed to the large number of Italian families formed by Italian men who called their wives over from Italy and by Italian men who married Argentine women. Citing statistics showing that Italians had the highest birthrates in the country, he wrote, "In this way, the Italian population in Argentina continues to increase in intensity, not only due to continued immigration, but because they rapidly multiply, so that in innumerous families one can verify the infiltration of the Italian element."[61] City-level demographic data confirms that Italian migrants did have many more children than Argentines in the late nineteenth and early twentieth centuries. In 1900, for example, 11,468 babies born in Buenos Aires were born to Italian parents, compared to only 3,926 born to Argentine parents. The racial and cultural similarities embodied in the concept of Latinità also produced families constituted through mixed marriages between Italian men and Argentine women. In that same year, 2,190 children were born to an Italian father and an Argentine mother, far more children than those resulting from unions between Argentine mothers and migrants from other countries, including Spain.[62] By 1917 half of Italian men and a quarter of Italian women in Buenos Aires married either Argentine or migrants from other countries.[63]

While migrant newspapers and business journals most often emphasized fraternity or brotherhood as the foundation for Italy and Argentina's common Latin origins, Latinità was not a wholly masculine construct. The presence of women—Italian women, their daughters born on Argentine soil, and to a lesser extent the Argentine wives of Italian men—provided further means for asserting Italians' demographic dominance and for constructing racial connections between Argentines and Italians. Italians in the United States were harder pressed to conscript women into race making in migrant marketplaces. Despite the heavily male migrations to both countries before World War I, Italian migration to Argentina was slightly more gender-balanced than it was to the United States. Furthermore, because Italians migrated earlier and over a longer period of time to Argentina, by 1910 a larger second generation and more gender-balanced community formed in Argentina than in the United States.[64] And while the Italian government admonished male migrants to marry Italian women exclusively, commonalities in familial arrangements, language, and religious beliefs between Italians and Argentines seem to have facilitated couplings, however limited, between Italian men and Argentine women in ways that did not occur in the United States, where Italians were slightly more likely to marry other Italians.[65] In theory these mixed marriages furthered the goals of Argentine liberals, who anticipated that intermarriage between Europeans and the country's local population, and the improved offspring such intermarriages produced, would push Argentines up the civilization hierarchy. Conversely, in the United States, racists' fears about the fecundity of migrant women and about the "racial suicide" of the Anglo-Saxon race discouraged unions between Anglos and Italians.[66]

The large numbers of Italian children born in Buenos Aires, as compared to other nationalities, and mixed marriages between Italians and Argentines provide clues as to how women's presence strengthened the consumption of Italian foods in Argentina. As food buyers and meal preparers, women who cooked for Italian husbands and children literally and figuratively reproduced Italian food preferences, habits, and rituals. In discussions of Latinità, Italian print culture employed family metaphors that included, at least discursively, feminine depictions of Italy and Argentina as a rhetorical strategy for strengthening connections between the two countries. Tapping into the well-established trope of the feminized nation-state, in 1898 the Italian Chamber of Commerce in Buenos Aires defined the "demonstration of fraternity" between Italians and Argentines as "that sentiment of fraternity and affection that comes from a commonality in origin, aspiration, needs, and that makes this young Nation more than a friend, but our sister."[67] As "sister nations" Italy and Argentina were portrayed as members of a Latin family that extended beyond the Atlantic and included the sons and daughters of Italians born in Argentina, children counted as "Argentine" by Argentina and "Italian" by Italy.

In arguing for Italians' superior positioning within the larger Latinità hierarchy that included Argentines, migrants echoed the often-stated attitudes espoused by Argentine elites who saw European migration as a vital ingredient in Argentine nation building. And yet Argentine liberals questioned whether "Latin" migrants—especially poor, seasonal, and working-class migrants from Italy—represented the quality European civilization necessary for Argentine modernization, turning instead to Northern and Western Europeans, "Anglo Saxons" and "Nordics" from countries like Britain and Germany. At the very moment when Argentine elites increasingly dissociated Italian migrants from desirable forms of Europeanness, Italians used imports to characterize themselves and their foodways as both Latin and European. Insisting on both the Latinity and Europeanness of Italians and their trade goods inserted Italians into Argentina's obvious Latin heritage, rooted in Spanish colonization, while concomitantly connecting them to a desirable European imaginary.

The simultaneous Europeanization and Italianization of Argentines occurred most visibly around the consumption of imported Italian foodstuff, especially beverages. In 1910 *La Patria* praised a speech by Italian criminologist Enrico Ferri in which Ferri called on the Italian government to favor exportation toward Argentina and other Latin American countries. "There are millions and millions, between edible and industrial products, that pour out of European countries toward these overlooked lands," boasted Ferri, "lands one can consider still virgin, barely marked with the invasive murmur of our civilization."[68] Ads conflated Italian imports with refined European-style consumer practices. Luigi Bosca assured consumers that his imported Italian wines were "known throughout all the civilized world."[69] Florio-brand Marsala, publicity proclaimed, was the only Marsala "allowed on the tables of the European courts."[70] The Italian Chamber of Commerce in Buenos Aires pointed to Italy's export market, especially the market for wine and liquors, as proof that Italy was winning in the global commercial battle to dominate the "young" Argentine market. Martini & Rossi vermouth was enjoyed "in Italy and beyond, in all the European nations, in those of new and old continents that deserve fame that comes from incomparable exquisiteness of the product." The vermouth's reputation as a refined European and Italian *aperitivo* accounted for continued consumer demand in Argentina. "It's evident that in all these stores, in the numerous bars, in the cafes, the exquisite product from the renowned Turin firm is demanded by all consumers of good taste."[71]

Discursive constructions of Latinità in articles and publicity served as powerful vehicles for assembling meanings about race through imports due to real similarities in the food cultures of Italians and Argentines. These similarities are rooted in Spain's long-standing presence in the Italian peninsula during the medieval and early modern period, including the Kingdom of the Two Sicilies of southern Italy and Sicily, which for most of the fifteenth century though 1860 fell under the control of rulers with ties to the Spanish crown. Furthermore, the Spanish empire forged connections with Ligurian merchants and explorers from Italy's northern city-states who sailed in service of Spain. Spanish colonization of the Americas and Spanish rule in Italy produced commonalities between Italian and Spanish foodways and eventually a receptive platform for Italian foods in Argentina. Already by the seventeenth century, Spain, as well as Portugal, introduced "New World" foods such as tomatoes, potatoes, corn, and chili peppers to Italy, eventually manifesting in typical regional dishes like polenta, potato gnocchi, and tomato- and pepper-based sauces, stews, and condiments by the time of mass Italian emigration.[72] These early modern Atlantic circulations of foods and culinary knowledge between Spain, Italy, and the viceroyalty of the Río de la Plata influenced the culinary repertoires of all three regions and, without homogenizing them, created overlap in their cuisines as they evolved over three centuries.[73] Notwithstanding local traditions that influenced the creation of Argentine cuisine—making it distinct from both Spanish and Italian foodways—Spanish expansion in the Mediterranean and Atlantic meant that Italians arriving in Argentina found food staples such as maize, tomatoes, and peppers that many would have recognized from back home.[74]

Latinity, Italian migrants' dominating presence in the country's food sectors, and Argentine elites' emulation of European culture produced an emerging Argentine national cuisine and public dining culture in which Italian foods, chefs, and consumers loomed large. As in the United States, Argentina's landed elites looked to France in particular as the epitome of culinary sophistication. As Argentine food scholar Rebekah Pite has noted, affluent *porteños* (Buenos Aires residents) expressed their racial superiority over the country's local indigenous and mestizo populations by having their cooks prepare French meals and by eating out in fine French and European restaurants. And while wealthy Argentines may not have considered Italian cuisine on par with French, neither was it linked to the foods consumed by Argentina's poorest, mixed-race eaters. A book published by the commission in charge of Argentina's centennial celebrations in 1910 associated Italian food with national progress and cosmopolitanism. The modest and monotonous meals of yerba mate (a caffeinated beverage made from the leaves of a tree in the holly family), empanadas, and the various meat and vegetable stews such as *locro, puchero,* and *carbonada* of the colonial era, the authors claimed, had been "almost completely banished for . . . the modern cosmopolitan cuisine," listing as examples the popularity of *tallarines* (*tagliatelle* pasta), *ravioles* (ravioli), and *milanesas* (a thin cut of beef, breaded and fried, that originated in Italy as *cotolette alla milanesa*). While the book's authors identified French cuisine with the eating patterns of the rich, they wrote, "The Italian cuisine, with its famous macaroni and risotto, is the one that has been popularized in all the homes of the middle classes, for the reason of the preponderant number of the immigrant element from the Italian peninsula." The poor, instead, ate Spanish food, and the "purely *criolla* cuisine has passed its final end, and more than a characteristic [of Argentine cuisine], is a memory."[75] While local criollo dishes like *asado* (grilled meat), the Hispanic pucheros, and the African-influenced squash dishes increasingly found inclusion in Argentine cookbooks and were enjoyed privately by Argentines, it would not be until the ascent of nationalist Juan Domingo Perón in 1946 that native criollo foods would take center stage as "national" dishes.[76] Clearly, Argentine elites' emulation of European culture produced a hierarchy of cuisines in which Italian foods were perceived as part of Argentina's "modern cosmopolitan cuisines," above local criollo, indigenous, and even Spanish traditions, an attitude that aided Italian migrants in depicting their imports as civilization makers.

Despite Argentines' Francophile tendencies, Italians helped construct an Argentine national cuisine as it developed, and Italian migrants worked hard, with much success, to associate Italian foods with fine European cuisine. Argentine food scholar Aníbal Arcondo found that Italian food writer Pellegrino Artusi's 1891 *La scienza in cucina e l'arte del mangiar bene* (Science in the kitchen and the art of good eating), considered the first Italian cookbook for the middle classes, was well received in Argentina; Arcondo cites a print run of forty million copies of its twelfth edition in 1908 as evidence of its wide diffusion.[77] Italians owned some of the most notable Argentine restaurants. In a special supplement in honor of Argentina's centennial in 1910, the Spanish-language daily *La Nación* featured the Londres Hotel in Buenos Aires and the Gran Hotel Italia in Rosario, as well as the hotels' restaurants, run by Italian chefs. *La Nación* wrote that the Genoese cooking at La Sonambula restaurant in the Londres Hotel attracted the nation's "gourmands," among which counted the country's most notable people in politics, finance, commerce, and ranching.[78] When Le Cordon Bleu, the prestigious French culinary institution, opened in Buenos Aires in 1914, it came under the leadership of an Italian migrant chef, Angel Baldi. And cookbook writer and food personality Doña Petrona, whom Pite shows was instrumental in building a national Argentine cuisine, was the granddaughter of an Italian migrant.[79] Italians as well as Argentines constructed Italian dishes as part of both a European cuisine and an emblem of Argentine social and cultural progress.

Similarities between Italians and Argentines and the foods they ate appeared more prominent when compared to the supposed racial and cultural incompatibilities between Italians and non-Italians in North America. Whereas Italians in Buenos Aires talked constantly about similarities that bound migrants to Argentines as "brotherly people," Italians in New York reflected regularly on the "total difference" between Italians and Anglo-Americans.[80] Noting the wide disconnect between Italians and U.S. Americans, E. Mayor des Planches, the Italian ambassador in New York, wrote that despite better economic opportunities in the United States, it was understandable that Italians preferred countries like Argentina in South America, which, he said, were "more alike in language, in race, in customs, climates, religious beliefs."[81] Buenos Aires' *La Patria* often referred to migrant and nonmigrant populations as sharing the same Latin roots; New York's *Il Progresso,* on the other hand, often described Italians and U.S. Americans as a "community of two people, of two races" and criticized Anglo-American attitudes toward Italian migrants.[82]

Despite apparent racial differences, Italian prominenti in New York, like their counterparts in Buenos Aires, argued for full membership in U.S. society based on a racial construction of Italianness that used the magnificence of ancient Rome and the Renaissance to link whiteness to Western civilization.[83] And yet the United States' more rigid racial hierarchy, in which working-class migrants from Italy, especially southern Italy, were seen as racially inferior to Anglo-Americans, as well as to migrants from Northern and Western Europe, made the country a less productive site for forging affinities between Italians and U.S. Americans through imported foods. Many U.S. political leaders viewed their country's heritage as tied to Anglo-Saxon and Nordic settlers in ways that marginalized the role that Spanish and French, or "Latin," colonization played in the nation's history. While U.S. Americans had been "crossing the boundaries of taste" since the colonial era by mixing regional staples and ethnic foods, as nativist and xenophobic sentiments increased, middle-class Anglo-American culinary nationalists reached back to an imagined New England cuisine with its roots in Northern Europe for what it meant to eat U.S. American.[84] In this context, self-proclaimed guardians of the country's culinary borders racialized working-class Italians though the foods they insisted on eating—foods that, in their eyes, were irrational, unsanitary, and lacking in nutrition.[85] In 1904 settlement house founder Robert A. Woods suggested that the foods eaten by Italian migrants ill-prepared them for life in the United States. "Their over-stimulating and innutritious diet," he wrote, "is precisely the opposite sort of feeding from that demanded by our exhilarating and taxing atmospheric conditions."[86] An 1888 *New York Times* article equated Italian and African American foodways and poked fun at the culture of poverty the writer saw informing Italian food culture, a culture in which Italians "dine royally upon four olives and a chuck of sunburnt bread" and rarely enjoy meat.[87] Italians like Baptist minister Antonio Mangano protested such characterizations. He argued in his 1912 thesis on Italians in New York, "The fact that an Irishman or a German gorges himself with a pound of steak at each meal does not make him superior to or a more desirable citizen than the Italian who is satisfied with a plate of macaroni or a plate of beans."[88]

This perception of migrants and their food traditions meant that Italian cuisine held a marginal place in the public restaurant cultures associated with fine dining that emerged in U.S. cities, even as Italian food entrepreneurs played a role in forging that culture. While Buenos Aires' La Sonambula advertised its cooking as Italian and associated it with desirable fancy European cuisine, its North American counterpart, Delmonico's in New York, founded by the Italian Swiss Giovanni Del-Monico, served mainly a version of French cuisine and trivialized foods associated with the founder's home region.[89] And while elite tastemakers in the late nineteenth century may have accepted a few, mainly French-style Italian dishes—like the macaroni *au Parmesan*, *à l'Italienne*, and *au gratin* served at fine-dining restaurants, such as at the Fifth Avenue Hotel in New York—the middle class remained largely disinterested in and even hostile to the foods eaten by Italian migrants.[90] This was quite different in Buenos Aires, where risotto and pasta had become associated with middle-class cuisine.

These perceived racial differences and Italians' less ubiquitous presence in the nation's food trade prevented migrants from exerting a commanding influence over consumers outside Italian quarters, making New York's migrant marketplace a less gastronomically porous site. And yet Italians insisted that racial divides could be bridged as more non-Italians came to develop a taste and appreciation for Italian foods. Given the small percentage of Italian consumers, as compared to the total number of U.S. consumers, it is no surprise that migrant merchants and retailers viewed their monetary success and prestige as wrapped up in tapping a consumer market beyond Italians. "Few of our products succeed in permanently penetrating the American consumer market," lamented wine merchant Emilio Perera. He admonished readers of New York's *Rivista* for boasting that Italy had conquered the U.S. market for foodstuff and drink when in reality "the consumption of our principal edible products and our wine in the United States . . . is almost exclusively due to our Italians."[91] While exerting a powerful pull on Italy's export market, working-class migrants who were bent on saving, merchants admitted reluctantly, did not always have the money to buy imported foods, which were usually more expensive than domestic alternatives. Given this reality, sellers turned their attention with fervor to non-Italian consumers. U.S. Americans' increasing interest in and admiration for Italian olive oil, canned tomatoes, wine, and other foods, merchants hoped, would challenge characterizations of Italian imports and their eaters as undesirable.

Challenging racial stereotypes related to Italian foods required, above all, educating U.S. consumers about Italian products and cuisine. Although growing numbers of U.S. Americans were coming to understand and value Italian foods—especially through their experiences in Italian restaurants in cities like New York and San Francisco, and during holiday tours through Italy—most U.S. Americans, migrant merchants declared, were largely ignorant about Italy's gastronomic contribution to high-class eating.[92] Merchants were particularly annoyed with U.S. consumers' lack of basic knowledge about some of Italy's most cherished exports, especially olive oil, which many U.S. Americans thought of as "a drink of fishermen" and for medicinal use only.[93] They had to be taught the various ways olive oil could be used in cooking and about its superiority to other cooking oils. Importer C. A. Mariani, for example, expressed frustration at a group of "American ladies, of unimpeachable standing in the community," who mistakenly thought "Virgin" referred to the brand name rather than the quality of the olive oil.[94] Similarly, consumers did not understand that the potentially toxic sulfate rind on gorgonzola cheese needed to be removed before consumption; their unfamiliarity with such high-quality imported cheese led to burdensome regulations that inhibited trade.[95] Equally irksome were narrow-minded "puritan" temperance advocates, who, blinded by prejudice and ignorance, likened wine to whiskey-based "patent medicines."[96] U.S. retailers needed instruction on how to handle, store, and display Italian imported foods so as not to damage, spoil, or misinform consumers about the items.[97]

In educating U.S. Americans about imports, Italians often denounced the palates of Anglo-American food consumers as bland, unrefined, and crude in a way that reversed the hierarchy of racialized tastes in which migrant food cultures stood toward the bottom. In an article on brined Sicilian olives—more bitter than the olives with which most U.S. consumers were familiar—Guido Rossati noted with disdain, "It is obvious that this article will never adapt to the American consumer, educated to dull and bland flavors."[98] In a piece on food fraud, the *Rivista* bemoaned that U.S. consumers with "defective palates" could not tell the difference between genuine olive oil and cottonseed oil.[99] The haughty tone with which the bulletin began an article directed at non-Italian retailers of Italian imports is emblematic of the way migrant merchants challenged taste hierarchies: "As a general rule, imported articles are superior to the domestic of the same class, as it stands to reason that superiority only can justify the higher price."[100] From this perspective, if high-quality imports did not bridge racial divides between Italians and non-Italians, the failure rested on U.S. Americans' unsophisticated palates rather than on Italian foods or their eaters. After the passage of the Pure Food and Drug Act of 1906, which set new, stringent standards in food production and inspection on imports, merchants regularly criticized U.S. Americans' nationalist attitudes for dampening the popularity of Italian imports.[101] In 1911 the *Rivista* pointed out that most edible goods that failed to pass federal inspection were domestically made rather than imported. Using Shakespearean English, the *Rivista* chided the U.S. consumer "who allowest thy patriotism to poke its nose into the business of thy taste" and who believed "the liking for imported articles is simply a fad."[102] In a follow-up article the paper again accused the U.S. consumer of thinking "he would commit a crime if he should happen to purchase something which is not a product of, or manufactured in, the United States."[103] Migrant sellers challenged the racialization of foreign foods as un-American, unclean, and unsafe by pointing out the popularity of quality Italian goods and their superiority over local items.

While migrants often portrayed non-Italian consumers as ignorant, they simultaneously worked hard to appeal to them. They believed that increased trade between the United States and Italy would "tone down the *angolosità* [Anglo-ness] and diminish the racial prejudice."[104] The *Rivista* debated how best to attract more U.S. buyers, especially after the passage of the Pure Food and Drug Act of 1906. Merchants increasingly insisted that tapping U.S. consumer markets required assuring consumers of their products' purity and that this was best accomplished by pressuring one another to import only the highest-quality foods, even if they were more expensive. Only the most excellent, uncontaminated, and safe edibles would preserve Italy's reputation and tempt U.S. Americans away from, for example, domestically manufactured canned tomatoes and toward Italian imported ones.[105] Another strategy for reaching U.S. eaters was to insist that manufacturers and retailers take great care in labeling their products; not only did packaging have to be aesthetically pleasing, but also, in order to comply with U.S. food regulations, labels had to clearly indicate the product's contents, origins, and weight.[106] New York's Italian Chamber of Commerce posited these specific strategies as part of larger organizational overhaul in the financing and regulation of the Italian-American trade. Without these reforms, Emilio Perera wrote, Italy's trade in food products and wine would remain "enslaved" by the consumer demands of Italian migrants exclusively.[107] To entice U.S. consumers and challenge the stigmatization of Italian foods, merchant *prominenti* at times distanced Italy's export market from the working-class migrant eaters who sustained it.

In their attempts to reach a non-Italian market, merchants often targeted middle-class Anglo-American women as holding the best potential for bridging racial differences between Italians and non-Italians. Chicago's Italian Chamber of Commerce noted that Italian Moscato champagne was "preferred to the dry and extra-dry French by ladies and gentlemen of the best societies."[108] Migrant trade promoters frequently talked about Italian imports in the context of U.S. homes and kitchens, where an item like Sicilian tomato paste "has become absolutely indispensable even in the American kitchen."[109] New York's *Rivista* insisted, "There is not an American family who does not consume canned tomatoes, and it is universally recognized that there does not exist better canned tomatoes than Italian ones."[110] As these affluent women came to value Italian foods and incorporate them into their family meals and dinner parties, a rising number of U.S. consumers would come to cherish Italian foods.

That merchant migrants focused their attention on Anglo-American female consumers made sense. They recognized the gendering of consumption, including food provisioning and preparation, as increasingly linked to white, middle-class women. Their interest in Anglo-American women also reflected the unbalanced gender ratios of Italian migration, as well as its working-class character. Italian women made up a small percentage of the total female population, and exogenous marriages between Italian men and American women were uncommon.[111] While Italian migration was heavily male in both the United States and Argentina, Anglo-Americans in the United States more often characterized the homosocial worlds created by male migration as abnormal, threatening to middle-class gender arrangements and to the increasingly consumerist function middle-class families served in the United States' industrializing society. Merchants, therefore, dissociated themselves from Italian women and the working-class consumption they represented, instead hoping to affiliate their wares with middle-class Anglo-American female consumers and their families.

Ironically, as merchants looked specifically to white, middle-class female buyers of Italian edibles to overcome gastronomic boundaries, Anglo-American women associated with Progressive Era reform organizations pressured migrant women to change their families' diets and eating rituals and to avoid imported goods, which were considered too expensive for working-class families. As one anonymous social worker reported after a visit to an Italian family, "Not yet Americanized, still eating Italian food."[112] A dietary study of working-class families in New York blamed the high rates of rickets among Italian children on Italian mothers, who bought small amounts of expensive imported cheese rather than cheaper and easily available milk.[113] Laboring hard to rationalize migrant foodways, social workers and domestic science practitioners viewed the migrant table as chaotic, filthy, and morally circumspect sites where the meals they served were unsanitary, repulsive, and nutrient-deficient.[114]

By the early twentieth century merchants measured their commercial success not so much by migrant consumption, but by the extent to which non-Italians in the United States purchased Italian imports. Merchants took migrant consumers for granted, believing that Italians "were used to using them [Italian products] since infancy and they know them and demand them," an assumption that migrant food entrepreneurs and their clientele in New York and elsewhere would challenge.[115] In 1909 the New York Italian Chamber of Commerce congratulated itself when Italian imports increased during a time of attenuating migration, attributing the sustained trade to U.S. consumers' increasing demand for Italian foods.[116] On the eve of World War I, migrant sellers were regularly depicting their products as popular items among U.S. consumers. As Emilio Longhi of Chicago's Italian Chamber of Commerce announced, "It is a well-established fact that the great American department stores, the large grocery stores, the clubs, the hotels now keep Italian goods; these are preferred, and the customers have the habit of calling for them, and one hears insistent requests for Italian macaroni, Sicilian tomato paste, Parmesan or Gorgonzola cheese, etc. not mentioning olive oil."[117] Like their counterparts in Buenos Aires, merchants in New York associated their products with the refined dinning and consumer practices of the well-to-do. But they did so with less success. Notwithstanding Longhi's confident projections, the continued stigmatization of Italy's migrants and food practices inhibited both the widespread diffusion of migrant culinary practices and migrants' ability to craft shared consumer experiences between Italians and Anglo-Americans in New York. As Simone Cinotto, Donna Gabaccia, and others have shown, although Italian food restaurants in New York's Little Italy attracted a small number of curious native-born Anglo-American eaters, many of whom went "slumming" in ethnic neighborhoods as a temporary exotic thrill, the widespread adoption of Italian foods by Anglo-Americans occurred only after World War II.[118]

While migrants in New York discussed differences between Italian and non-Italian consumers and their food preferences and debated how best to bridge such differences, in Buenos Aires there was a conspicuous absence of strategizing over how best to attract non-Italian eaters. Migrant newspapers and business publications in Buenos Aires certainly praised migrant demand for facilitating transatlantic commercial flows in Italian imports; however, Latinità, built in part on a transatlantic and Mediterranean history of food trade, made educating Argentines about foods from Italy seem unnecessary. Furthermore, as Argentina's dominant foreign-born group, and at the helm of the nation's food sectors, Italians' economic livelihood in Buenos Aires did not depend as much on reaching beyond the city's already gastronomically porous migrant marketplace, where migrants and Argentines both purchased and consumed Italian imported foods.

The ethno-racial landscapes of the United States and Argentina, and Italians' place within them, affected identity building and consumer experiences in migrant marketplaces. The selling and buying of Italian imported foodstuff offered a platform for migrants to shape racial understandings in ways they hoped would strengthen commercial and migratory ties between Italy and the Americas. In Argentina perceptions of a shared Latinità between Italians and Argentines allowed Italians to better position themselves and their products as harbingers of European progress than Italians in the United States. The people and foods of Argentina and

Italy shared affinities—both real and imagined—that shaped Italians' perceptions of themselves and the foods they consumed. These similarities made Buenos Aires a permeable migrant marketplace, one that was flexible enough to brand imports and their multinational eaters positively as Italian, Latin, and European. In New York, where migrants entered a more rigid system that viewed Italians and their foodways as racially inferior to Anglo-Americans, merchant migrants expressed contradictory impulses in their attempts to overcome racial prejudice; they both denounced and sought to change the palates of U.S. eaters while adjusting their trade to win them over. In the United States the low status of Italian foods and the working-class migrants who ate them, as well as the absence of a colonial past conceived of by U.S. elites as "Latin," produced a more culturally and gastronomically contained migrant marketplace, where less food exchange occurred between Italians and non-Italians.

"A chi giova? [Who benefits?]": Making Trade Policy from the Diaspora

In February 1902 the Italian Chamber of Commerce in Buenos Aires noted that imported Italian olive oil, cigars, cheeses, vermouths, fernet, and wine were flying off the shelves in response to an anticipated hike in the import duties on these items. Migrant consumers responded to the impending increase by stocking up on the lower-cost foods that had already cleared through customs.[119] Two years later, retailers could not get customers to buy Italian wine, even in December, a month that usually saw sharp increases in sales for the Christmas and New Year's holidays. The cause of the slowdown, the chamber noted, was the imminent abolition of a 10 percent internal tax on wine, which would make these items cheaper. The following month, after the tax went into effect, most retailers saw their stocks completely liquidated.[120] Tariffs and taxes were not disconnected from migrants' everyday consumer decisions and identity making in migrant marketplaces. In fact, the often-discussed links between men and markets relied squarely on the ability of products and people to move together across borders relatively unimpeded.

Most nations' migration policies began as an inextricable element of international trade strategies; until well into the twentieth century, political elites considered migration a part of diplomacy to be determined though commercial treaties with other countries rather than through governmental legislation. Diplomats assumed that the freedom to trade internationally rested on the rights of migrants to traverse national borders without restraints.[121] And yet as the labor-abundant economies of Europe interacted with the land-abundant countries of the western hemisphere, and as nations attempted to manage the often jarring effects of an emerging global economy on their nation-building ventures, the regulation of trade as well as migration become contentious issues.[122] Indeed, over the course of the late nineteenth and early twentieth centuries, "the tariff question" emerged as a key debate among lawmakers and economists in exporting and industrialized nations. Businesses also attempted to mitigate the fluctuating boom-and-bust cycles brought on by global integration through economic concentration. Created in part by high tariffs, business trusts became vehicles used by industrialists to control market share, labor costs, foreign competition, and access to raw materials. Such monopolist practices drew increasing criticism from labor unions, consumer advocates, and government officials concerned with capitalists' growing political influence.[123] A growing number of political economists argued that while free trade remained the ideal, temporary protection in the form of trade barriers and large-scale corporate trusts could advance industrial development at home and economic progress globally.

Migrants, both merchant prominenti and everyday consumers, expressed passionate interest in how debates over tariff legislation and monopolies among diplomats, economists, and politicians affected migrant marketplaces. "The tariff is always the order of the day," wrote New York's *Rivista* in 1911, noting the ubiquity with which the issue was discussed in political and trade journals.[124] Rather than passively react to top-down decisions, Italians abroad attempted to influence the commercial and economic policies of their host countries. "A chi giova il protezionismo?" (Who benefits from protectionism?), the title of an opinion piece in Buenos Aires' *La Patria*, became an important question for migrants, one that provided them with an opportunity to insert themselves into discussions about commerce and migration.[125] And while tariffs joined trusts in remaining the order of the day in New York and Buenos Aires, nation-specific differences emerged in migrants' arguments against economic protection. In Buenos Aires migrants perceived themselves as more powerfully affecting the policies of their host country and as exerting more control over their consumer options. Italians cited their dominant presence in Argentina and solidarities based on Latinità as affecting international trade policy. Italians in New York—a much smaller, segregated market in a more consumerist, protectionist country—struggled to portray themselves as influencing legislation, although they used anti-Italian prejudices to describe both tariffs and migration restriction as excessive and unjust. By objecting to tariffs on imported food items and to high-priced domestic foods protected by corporate trusts, Italians in both cities mobilized migrant marketplaces as transnational consumers.

Migrants in Buenos Aires exploited Argentina's dependency on Italian migration to protest against tariffs and trusts. *La Patria* regularly asserted that increased tariffs on imports discouraged transatlantic migration. Italians would stop migrating to Argentina and those in Argentina would repatriate if higher tariffs made the cost of living too high for migrant consumers. In articles discouraging Argentina from approving tariff hikes, Italians echoed the civilizing discourse inherent in expressions of Latinità by describing migrants as "the richest contingent to the vivacious force of this country" while touting the threat of migrant reduction and its impact on the Argentine economy.[126] They marshaled government statistics showing a decrease in Italian migration to lobby against tariff hikes. The Argentine government, *La Patria* argued, should do everything possible to reactivate migration, "without which all aspirations of greatness and prosperity for Argentina is a fallacious dream."[127] Noting the way Italian migration to the United States far outpaced migration to Argentina, the Italian Chamber of Commerce in Buenos Aries argued that the United States was threatening to "absorb, in a period not far away, almost the totality of our [Italy's] excess population."[128] The "big secret" behind migrants' repatriation or migration to the United States was the increasingly high cost of living, including food prices, a condition that protectionism exacerbated. According to the newspaper, if Argentina increased tariffs on imports in an effort to protect domestic industries, migrants would not be able to pay the higher, inflated cost of domestically made Argentine products.[129] Exaggerated protectionism, warned *La Patria* in 1900, made Argentina "uninhabitable for the working classes of Europe that emigrate in search of a good life."[130] Taxes were so high that migrants were "forced to consume items infinitesimally small but still very expensive." The exaggerated costs of food and other items in Argentina would no longer entice "the foreigner to abandon his homeland and transfer to a country that does not offer him an easy life and plentiful sources of nutrition."[131]

Migrants used tariff debates to position Italians as consumers of both imported products and Argentine goods. They argued that any attenuation in migration would have disastrous effects on the country's budding manufacturing economy. Argentine industries, *La Patria* stated, suffered from an "exaggerated confidence in the progress of the country"; protecting fledgling Argentine manufacturing with high tariffs did not make sense, because the country had not yet attracted a large enough consumer market capable of absorbing factory products.[132] The United States became protectionist and industrial, the paper reminded readers, only after an internal market expanded to absorb industrial goods. "First work to populate the country and then we will talk about protectionism," the paper concluded.[133] The Italian Chamber of Commerce agreed, pointing out that Argentina "does not export industrial products in the strict sense of the word, with very few exceptions; it is still not an industrial country . . . it relies almost exclusively on importations for its needs."[134] The anti-tariff stance taken by the Italian-language press and the Italian Chamber of Commerce became a pro-migration and pro-importation position that lauded Italians' consuming potential and importance in Argentine nation building. The future prosperity of the Argentine economy depended on an increase in Italian migrants as both a "consuming and producing population."[135]

It is difficult to assess migrants' actual impact on Argentine tariff legislation. Argentine elites' liberal economic policies, combined with the nation's slower path toward industrialization, are probably most responsible for the relatively low duties on imports before World War I. Tariff policy commanded much debate between government officials and industry supporters starting in the late nineteenth century when Argentine lawmakers first considered tariffs as a way to foster industrialization.[136] Migrants' anti-tariff stance paralleled the position of Argentina's powerful landed interests, which pressured the government to maintain low tariffs to preserve cheap access to imported machinery and to keep European markets open to Argentine wheat, beef, wool, and leather. Manufacturers, on the other hand, blamed the nation's weak industrial sector on the government's favoring of this agro-export sector and its unwillingness to pass higher tariffs. Nevertheless, industry did receive some protection: Argentina increased duties on imports starting in the 1870s, mainly in response to the 1873 and 1890 depressions, with tariff valuation set in law in 1905; furthermore, the Argentine legislature repeatedly revised the tariff schedule on select items in response to the concerns of wealthy industrialists.[137] Tariffs on many imported Italian goods actually decreased through the early twentieth century; duties on foodstuff and alcoholic beverages went down, although after the takeoff of the Argentine wine industry in the late nineteenth century, the country did increase tariffs on European wines.[138] In 1899 the Italian Chamber of Commerce in Buenos Aires expressed satisfaction with recent adjustments to the Argentine tariff schedule, which secured slight tariff reductions for most imported foods, including pasta, tomatoes, preserves, cheese, and spirits. By the early twentieth century, however, the chamber increasingly treated Argentine tariff hikes as one of the principal threats to Italian goods.[139] In 1905 *La Patria* lauded the work of the League of Commercial Defense, made up of a multinational group of merchants, whose lobbying resulted in the Argentine 1905 tariff legislation that maintained low tariffs for most Italian imports, tariffs "that burden importation making life more expensive for the worker."[140] Objections to tariffs tended to unite prominenti merchant importers and journalists with left- and radical-leaning socialists, two groups that were sometimes hostile toward each other but joined in their desire to keep the cost of consumer goods low for migrant laborers.

Migrants in Buenos Aires lobbied hard to maintain Argentine trade policies favorable to Italian imports. But they also attempted to influence the trade policies of their home country, arguing that Argentina's continued acceptance of both Italian products and migrants depended on Italy's willingness to accept Argentine goods. In 1908 the Italian Chamber of Commerce pointed out, "Italy enjoys special circumstances that allow trade between the two countries to prosper," referring to Italian migration.[141] The chamber regularly urged Italy to take Argentina seriously as a trade partner. It denounced Italy for favoring imports from Europe over those from Argentina and accused Italy of being a poor consumer of Argentine agricultural goods.[142] "Why," the chamber asked, "should Italian industry import items like wool and leather from European markets when it can get them from Argentina?" Illustrating migrants' awareness of ties between tariffs and immigration restriction as forms of protectionism, the chamber cited intensifying xenophobic sentiments against migrants in the United States to spur the Italian government to open its markets to Argentine products. "Perhaps the day is not far off in which European immigration will find serious obstacles to be welcomed where now it is directed," as has occurred in the United States, the chamber noted ominously. If Italy refused to strengthen its commercial relationship with Argentina, Argentina, in turn, might restrict Italian migration and most certainly enter into more favorable treaties with other countries.[143] The chamber employed increasingly discriminatory immigration policy in the United States as a threat in order to prod Italy into cultivating a stronger trade relationship with its migrants' host country.

Although Italians in Buenos Aires insinuated that some form of tariff protection seemed appropriate for its more industrialized and populated neighbors to the north, Italians in New York argued the opposite. And yet in their objection to tariffs, migrants in the United States faced a varied set of powerful obstacles. Compared to Argentina, the United States had a more diverse economy, a much stronger industrial base, a vibrant consumer society, and a larger and growing middle class of mainly native-born Anglo-Americans who consumed many of the manufactured goods produced in urban factories. In the United States industrialization sparked fierce debates between Democrats and Republicans over tariff reform, especially before 1912, when power over commercial diplomacy transferred from the partisan bickering of the legislative halls to the executive branch. Republicans and northern industrialists typically called for higher tariffs, in part to protect industries and U.S. labor against competition from abroad. Conversely, Democrats, mainly representing the South and Southwest, argued against tariffs out of fear that higher tariffs would encourage Europe to increase tariffs on U.S. agricultural staples and raw materials.[144] However, whereas in Argentina migrants' low tariff position paralleled the stance of influential leaders in the export sector, in the United States migrants campaigning against tariffs stood in conflict with a powerful coalition of pro-tariff Republican politicians who lobbied to keep import duties relatively high. Although the overall tariff values of the United States and Argentina were comparable, rates on some of Italy's most popular exports, such as canned tomatoes, bottled wines, cheeses, pasta, and olive oil, were slightly higher in the United States. Italians in New York were also disadvantaged demographically in their arguments against tariffs, as they made up a small percentage of the United States' total consumer market. Moreover, Italians in both countries naturalized in low numbers, and their delayed entry in U.S. and Argentine politics further diminished their impact on tariff debates. Even so, Argentina's disproportionate reliance on migrants from a smaller number of countries gave Italians in Argentina more real and perceived political clout. In the United States, where Italian imports faced major competition from domestically made goods, which received protection in the form of tariffs and corporate trusts, migrants had a weaker voice as transnational consumers in nationwide tariff debates.

Despite these obstacles, migrants in New York condemned tariffs and used such objections to characterize themselves as global food consumers. They echoed their Argentine counterparts in blaming tariffs for "promoting monarchial tendencies" by "concentrating large fortunes in a few hands."[145] *Il Progresso* described tariffs such as the "unfortunate" Payne-Aldrich Tariff Act of 1909 as "a direct cause for the high cost of living."[146] Migrant trade promoters viewed U.S. commercial and migration policies as nationalist manifestations of U.S. American "egoism." Luigi Solari, president of the Italian Chamber of Commerce in New York, argued that the United States—a country that was formed as a rebellion against the "political and economic impositions of Europe"—was now hypocritically espousing "egotistical principles of exaggerated financial, industrial and social protectionism." Solari rejected U.S. isolationism as conservative, an ultimately futile stance against the inevitable march of progress toward internationalism. "The epoch in which one state was able to nearly isolate itself from others and operate only and exclusively in their own interest is over," he concluded. "Men can get worked up about erecting barriers, but these fall to the powerful blow of civilization that advances."[147] In 1912 the *Rivista* again denounced U.S. movements, organizations, and legislation that promoted tariffs as "stick-to-a-protective-tariff-or-nation-will-bust associations, that rely on the tariff to keep the dreaded foreign food out of the land."[148]

Merchants in particular, who relied on migrants as consumers of their imported foodstuff, viewed tariff and migration policy as linked, and they associated both with xenophobic sentiment. Journalist and lawyer Gino Speranza, writing on Italian-U.S. diplomatic relations for the *Rivista* in 1905, announced, "the emigrant is a type of international citizen." Connecting migration and trade, he continued, "The movement of a population from one state to another is a question eminently international. It will be a delicate question, if one wants, but not impossible to deal with and resolve. If nations have been able to make international agreements regarding commerce, can they not make them with respect to men?" Not only did the United States need to treat migration as an international rather than domestic issue, but it also had to stop treating Italian migrants as if they were "a population of misers" and Italy as if it were a South American nation.[149] It seemed that an equitable diplomatic relationship between the United States and Italy rested on the racialization of South American countries like Argentina, whose low status within the hierarchy of nation-states apparently called for a less egalitarian and more unilateral approach to trade and migration policy. In his crusade against tariffs, olive oil importer C. A. Mariani offered the most poignant condemnation of tariffs as forms of racial discrimination against working-class migrant bodies. The United States, he wrote, reserved for Italians the most physically arduous work, work that required laborers to replenish their bodies with foods, and yet tariffs essentially starved migrants by denying them homeland olive oil, pasta, cured meat, and cheese, the only foods that truly replenished Italians physically and psychologically.[150]

Unlike their co-nationals in Argentina, migrants in the United States directed some of their most trenchant criticism not at tariffs, but at corporate food trusts. They treated trusts and tariffs as interrelated, since powerful corporations pressured Congress to raise tariffs while using monopolistic practices to increase prices on consumer items.[151] Armed with both trusts—the "imperialism of American capital"—and protectionism, U.S. capitalists "oppressed consumers who are forced to pay exaggerated prices," wrote *Il Progresso* in 1905.[152] The *Rivista* called trusts a form of "commercial feudalism" while arguing that all consumers, regardless of nationality, were "tired of everyday paying greater for life's comforts."[153] In 1910 *Il Progresso* criticized "meat king" Jonathan Armour, head of the Chicago-based Armour and Company meatpacking plant, for blaming the exorbitant cost of meat and other foods on migrant congestion in cities, where high demand elevated prices. Instead, the paper claimed it was "fat cat" industrialists like Armour who promoted that very congestion and accompanying price hikes by building factories in high-density urban areas.[154] *Il Progresso* attacked monopolies and tariffs as iniquitously connected and injurious to consumers, and regularly supported Democratic presidential candidates whose platforms called for low tariffs and corporate regulation.[155] Buenos Aires' *La Patria* also called attention to the power of trusts, especially in the bread and meat industries, trusts that "starve poor people" by raising the price of foods for working-class families.[156] In response to insufficient demand and periodic depressions, already by the late nineteenth century Argentine industrialists, including those in the food, cigarette, and beverage industries, moved toward economic concentration.[157] However, before World War I Argentina's weaker industrial base—the nation's inability to fully substitute imports with domestically made goods—largely shielded corporate trusts from severe criticism by Italians in Buenos Aires.

Italians in New York and Buenos Aires who denounced economic protectionism called attention to links between the movement of people and foodstuff produced by their transnational migrations. Migrants in both countries positioned themselves as global buyers of Italian imports and of U.S. and Argentine foods, buyers whose purchases affected and were affected by the trade policies of their host countries. In debates over economic protectionism, Italians in Buenos Aires used migrant print culture to more effectively position themselves as consumers with the ability to influence trade and migration patterns, in part because they made up a larger, more united consumer base in what was, perhaps ironically, a less consumerist society. Migrants in New York—facing more severe economic protectionism in the form of both trusts and tariffs, which they viewed as linked—connected and condemned restrictions on and prejudices against both Italian people and goods.

*   *   *

Differences in the migration histories and socioeconomic structures of the United States and Argentina helped Italians in Argentina generate shared food experiences for migrants and nonmigrants in Buenos Aires' permeable migrant marketplace. While merchants in the United States endeavored to extend the cultural, racial, and gastronomic borders of New York's migrant marketplace by attracting a non-Italian clientele, they struggled to engage U.S. Americans. Italian-language newspapers and business publications disclose New York and Buenos Aires as important sites where connections between migration and trade allowed migrants to produce and challenge racial identities linked to their consumer habits and food traditions. The consumption of imports also provided migrants with a tool for positioning themselves as powerful protagonists shaping their host nation's commercial networks and economic development. In this aspect as well, Italians' overwhelming presence in Argentina advantaged migrants in their efforts to secure access to homeland foods. Discussion about ties between trade and migration before World War I reminds us that migrants in le due Americhe labored and consumed at a time when nation-states increasingly expressed their wariness about the world through both tariff and immigration policy. As the next chapter explains, one way migrants responded to these restrictions was by producing *tipo italiano* (Italian-style) foods in the diasporas, which, while satisfying working-class migrant consumers with limited budgets, threatened Italy's profitable export market.

CHAPTER 3

Tipo Italiano: The Production and Sale of Italian-Style Goods, 1880–1914

A 1902 Italian consular report from Philadelphia echoed the pessimistic sentiments of an increasing number of Italian government officials stationed in major urban centers across the United States. Rather than advantaging Italian trade, transatlantic migration, the report concluded, was sometimes having the opposite effect: the report described a weakening of the Italian import market brought on by increased migration. Ambitious migrants in U.S. cities posed a danger to Italy's valuable food trade "in that they [migrants] successfully produce and counterfeit the articles most desired" by migrant consumers. The report noted that local pasta factories in and around Philadelphia, all owned by Italian migrants, "operate on a large scale, and with perfected American machines have expelled similar [imported] articles from Italy." The report concluded, ominously, that a similar process was occurring with "counterfeit" Italian olive oil and salami, which were slowly crowding out genuine Italian imports.[1]

The burgeoning number of pasta factories in Philadelphia and other cities with sizable Italian migrant communities was an outcome of the increasingly globalized world in which workers labored and consumed in countries thousands of miles away from their homelands. Italian elites in Italy and abroad were not alone in attempting to profit from manly markets of trade and migration. As Italian leaders predicted, migration created and sustained trade paths in Italian goods. But it also generated new opportunities for migrants who used their social and cultural capital to secure employees and consumers and to craft new food experiences in migrant marketplaces. As the products demanded most by migrants, food imports were also the most susceptible to substitution. By effectively competing with Italian imports, these substitution tipo italiano products—"Italian-style" goods like the domestically made pasta maligned in the consular report—created a rupture in Italian commodity chains linking Italian food producers in Italy to migrant consumers in le due Americhe. That transnational commodity chains, and not transnational family economies, fractured explains the success of tipo italiano industries. Because they were less expensive than imported products, Italian-style foods made abroad allowed hungry male migrants to maximize their savings to send home to family; driven by homeland obligations, migrant eaters experimented with new foods to save, survive, and create meaning in their everyday lives. The entrepreneurs, merchants, and retailers who supplied them successfully navigated their often awkward, liminal positions at the interstices of national and transnational economies, and migrant and nonmigrant consumers, all the while enlarging their own economic and social standing in diasporic communities.

Comparing the development and popularization of tipo italiano products in New York and Buenos Aires sheds light on the varied meanings of nationality, ethnicity, and authenticity in migrant marketplaces. While in both cities the presence of tipo italiano foods grew rapidly in the years before World War I, in New York migrant entrepreneurs took advantage of economic and market conditions that were distinct to the United States' more industrialized society in order to popularize a wider variety of lower-priced tipo italiano food products. Unlike importers who sought out Anglo-American consumers, specifically Anglo-American women, in hopes of raising the status of themselves and their foods, tipo italiano manufacturers used their ethnic backgrounds to secure a labor force of migrant workers who in turn bought the lower-priced foods they produced. Manufacturers used migrant print culture to publicize their domestically made pasta, wines, and other foods as transnational products in ways that reflected consumers' continuing links to homeland culinary traditions and their evolving identities and tastes in a society that continued to mark them as inferior.

While a vibrant tipo italiano industry of domestically made foodstuff also developed in Buenos Aires, the country's weaker industrial structures kept the industry limited in size and scope; before World War I, consumer demand outstripped local production and migrants continued to rely heavily on imports to satisfy their desire for homeland foods. In Argentina tipo italiano became associated not as much with domestically made alternatives, but with products originating outside Argentina, especially Spain and France. In their depictions of the competition between Italian imports and similar foods arriving from European competitors, Italians once again used masculinist language of war, diplomacy, and expansion to describe Argentina as a key overseas site upon which Italy would win "battles" by "conquering" markets. Migrants had used commerce and consumption to construct the people and products of Argentina and Italy as "Latin" while simultaneously depicting Italians and their foods as "European" to advantage Italian imports. These racial and cultural similarities, however, proved expansive enough to include the peoples and food cultures of other Mediterranean nations—Spain, France, and Greece—that were also understood by migrants as "Latin" and to the detriment of Italian trade.

Migrant Entrepreneurs and the Growth of *Tipo Italiano* Industries in *le due Americhe*

In 1909 the New York Italian Chamber of Commerce featured an article in their bulletin on the De Nobili Italian Cigar and Tobacco factory, established in 1905 by Prospero De Nobili, ex-deputy to the Italian Parliament. The three-story, thirty-thousand-square-foot building in Long Island City rivaled other modern enterprises in its manufacturing technology and managerial sophistication. De Nobili recruited technical and managerial personnel directly from tobacco-processing plants in Italy, plants owned and operated by the Italian government. And like the government-run plants in Italy, De Nobili received his tobacco from Kentucky and Turkey. Modestly paid, mainly female Italian migrants working for piece-rate pay composed the majority of its labor force; for every one hundred cigars these workers earned twenty-three cents, with a medium income of about twelve dollars a week. By 1910 the De Nobili factory produced close to 180 million *sigari di tipo italiano* (Italian-style cigars) daily and had begun to export its products to Argentina. The article underscored especially the centrality of migrant consumers to De Nobili's success. "It is the consumption of our innumerable emigrants," the article closed, "that, above all else, justifies and fortifies the factory."[2]

Migrant entrepreneurs like De Nobili creatively responded to the labor and consumer needs of Italians in migrant marketplaces of the Americas. Employing food knowledge, expertise, and workers from Italy, but using raw materials, industrial technologies, and distribution networks in the United States, large-scale manufacturers of tipo italiano foodstuff successfully competed with popular Italian commodities for migrants' attention. Celebratory articles about tipo italiano manufacturers by the migrant press and by Italian chambers of commerce also reveal the intermediary positions held by producers and sellers of Italian-style products as they navigated between the demands of Italian business and government leaders in Italy and the everyday necessities of working-class migrants. Publicly, Italian chambers of commerce often looked askance at products like De Nobili's sigari di tipo italiano for competing with Italian imports sold by migrant merchants and retailers, whose interests the chamber represented. And yet this public posturing against such items did not prevent merchants from simultaneously embracing their makers as intrepid "self-made men," as New York migrant

businessmen did during their annual banquet in 1907, where they called De Nobili's factory an "act of audaciousness," a model to be followed by other migrants
who were "still afraid of courageous initiatives."[3] In their praise for De Nobili and
other emerging tipo italiano industrialists, Italians linked men to migrant capitalism abroad, even while migrant food production and consumption in migrant
marketplaces was increasingly being done by women.

Starting in the late nineteenth century, an increasing number of Italian-style
goods such as the lauded De Nobili tobacco products and the disparaged Philadelphia pasta appeared in migrant marketplaces throughout the United States and
Argentina. Whether made abroad or imported as "Italian" from other European
nations, these products challenged Italian elites' faith in the "indissoluble chains"
between Italy and its diasporas abroad, those unbreakable links upon which dreams
of nation and empire building rested.[4] Notwithstanding importers' efforts to attract
U.S. and Argentine consumers—efforts met with more success in Buenos Aires'
permeable migrant marketplace—Italians remained the foremost consumers of
Italian foodstuff during the entire era of mass migration. Italian economists and
politicians predicted that if migrant consumers opted to buy cheaper, more easily
available tipo italiano foods, and developed a taste for and reliance on them, Italy's
commerce would weaken considerably. In 1912 economist Aldo Visconti pointed
to tipo italiano businesses as proof that the commercial advantages created by
transatlantic migrants could only be transitory. In his discussion of "on-site imitation" Italian foods in Argentina, for example, Visconti anticipated that in time
Italy's export market "would be damaged by our same emigrants that in Argentina
work and produce, or try to produce, our most sought after goods."[5] Migrants created a major demand for Italian goods, but once they improved their conditions
economically, they became Italy's biggest competitors.

Italian leaders clearly viewed consumption in migrant marketplaces with am-
bivalence. Ideally, ties between migration and commerce functioned to advantage Italian imports; however, if managed improperly, migrant consumers proved
damaging if not ruinous to Italian transatlantic trade. While regularly asserting
migrants' preference for genuine Italian cheese, canned tomatoes, and other foods
over alternatives, Italian elites also implied that migrants' consumer desires were
menacing in their unrestrained voraciousness; their needs were so robust that if
not met though imports, Italians would inevitably turn to substitutes. "When our
colonies grow and the demand for Italian products becomes more active," explained
the *Bollettino* in 1904, "industries for the production of articles demanded by Italian consumers appear abroad," pointing to the large number of pasta factories in
North America as proof.[6] Italian trade promoters worried about migrants whose
undisciplined food consumption prompted the growth of domestic food industries.
Given the perceived detriment such goods posed to links between emigrants and

exports, Italian consulates and Italian chambers of commerce regularly reported on the growth of Italian-style industries abroad and often expressed concern over their effects on Italy's export markets.[7]

To the consternation of Italian businesses, the markets for Italian imports and tipo italiano food products did indeed grow simultaneously in the United States and Argentina. The two countries possessed abundant fertile land for the cultivation of raw materials used in many of Italy's most profitable food exports. By the turn of the twentieth century, both countries were among the world's most productive breadbaskets and meat suppliers; the grains grown and cattle raised, slaughtered, and processed in the U.S. Midwest and mid-Atlantic regions and on the Argentine pampas fed their inhabitants and much of Europe.[8] The United States and Argentina both cultivated the *triticum* species of wheat, better known as durum wheat, which produced the semolina flour used in most commercial dried pasta in Italy by 1900.[9] Many of the fruits and vegetables in popular Italian exports, such as tomatoes (a "New World" crop) and grapes (an "Old World" crop), were cultivated in the Americas, allowing for the expansion of the canned tomato and wine industries. The burgeoning large-scale dairy, beef, and pork sectors provided migrants with the raw materials for producing various Italian cured meats, such as prosciutto and mortadella, and cheeses that approximated the asiago, gorgonzola, and parmigiano-reggiano varieties from Italy. Migrants building tipo italiano businesses relied on raw materials similar to those found back home, materials that could often be produced in larger quantities and at cheaper prices in their host countries, albeit with changes in quality and taste.

The availability of staple ingredients, access to manufacturing technologies, and minimal taxes, tariffs, and transportation costs combined to make most tipo italiano foods less expensive than imported versions. The lower price perhaps best explains the growing popularity of these foods, as they satisfied working-class migrants' simultaneous desires to nourish themselves with homeland foods and save money to send home. Producers and sellers of such foods better understood migrants' enduring ties to wives, parents, and children left behind than did importers and Italian business promoters who disregarded the way transnational family commitments restrained migrant consumption. Domestically made substitutes almost always cost less than imported products from the same food categories. For example, the *Bollettino* reported that in the United States in 1902, a pound of imported *maccheroni* cost nine to ten cents a pound, whereas "domestic" maccheroni cost six to seven cents; similarly, Italian rice cost eight to nine cents a pound while U.S. rice cost five to six cents.[10] In migrant newspapers, ads for Italian imports often linked their foods to the affluent consumption patterns of European aristocrats; manufacturers of tipo italiano goods made the modest cost of their products an integral part of their advertising campaigns while assuring consumers that their items equaled imports in excellence. The United States Macaroni Factory in Pennsylvania accentuated the "economic convenience that renders it [their pasta] more preferable to imported pasta" in a 1905 advertisement in New York's *Il Progresso*.[11] That same year Domingo Tomba advertised his Mendoza wine, "the best table wine in Argentina," in Buenos Aires' *La Patria* as having "modest prices convenient for all families."[12] Tipo italiano industrialists responded to the economic constraints of migrant consumers, who often made better buyers of cheaper products fabricated in the diaspora.

While the high demand for lower-cost tipo italiano products certainly encouraged the growth of such industries, so too did migrants' employment needs. A steady source of working-class laborers made tipo italiano businesses possible, because they supplied the industry's labor and consumer market. The informal small-scale domestic production of various foodstuff usually relied on the unpaid labor of migrant wives and daughters.[13] However, already by World War I, not a few of these food businesses had grown into great industrial establishments that employed large numbers of workers from Italy. As Simone Cinotto has shown, especially in his work on the California wine industry, Italian migrant entrepreneurs used their ethnic backgrounds as social and cultural capital to secure a cheap migrant workforce and to guarantee a loyal consumer base. Similarly in urban migrant marketplaces, manufacturers producing Italian-style foods exploited ethnic bonds between themselves and their employees to procure workers who frequently toiled under poor conditions for minimal pay.[14] Migrant print culture reported, often with pride, on the number of Italians employed at various larger-scale tipo italiano businesses.[15] Rocca, Terrarossa, and Company, founded by Giacomo Rocca from Genoa and based in Buenos Aires, became one of the country's leading meat-processing plants, where some 250 Italians slaughtered 50,000 cattle and pigs annually and produced Italian-style cured and smoked meats.[16] The California Fruit Canners' Association in Sacramento, founded by Ligurian migrant Marco Fontana in the late nineteenth century, employed hundreds of Italian migrants, many of them female, to can California-grown fruits and vegetables.[17]

Although Italian men outnumbered women in the United States and Argentina, tipo italiano food businesses employed increasing numbers of Italian women by the early twentieth century. By the end of the first decade, continuous migrations from Italy had pushed up the number of Italian women in the United States and Argentina; these new migrants, and a growing second generation of Italian women born abroad, provided migrant food industrialists with an expanded labor market. Female migrant workers made up a large portion of laborers in urban centers like New York and Buenos Aires, where some became active in radical labor movements.[18] In the Italian households of New York studied in 1911 by the U.S. Immigration Commission, 47 percent of family members were female; 38 percent of the 544

Italian-born women for which the commission collected data reported working in manufacturing.[19] Industrialists desired these female workers because they could pay them less than men and because they assumed women were more controllable and less radical employees. Textile work, tobacco production, confectionary manufacturing, and fruit and vegetable canning, as not entirely mechanized "light industries," were deemed especially appropriate for female workers. In Buenos Aires, for example, there were over twice as many foreign-born women employed in the city's bakeries than Argentine women in 1895; in the cigar business, migrant women more than tripled the number of Argentine women.[20] By 1919 Louise Odencrantz found that in the over three thousand Italian families in New York she studied, 91 percent of all daughters above the age of fourteen contributed to the family income, mainly through work outside the home in nearby factories, including those producing foodstuff such as candy and tobacco.[21] The Italian American Cigar Company in San Francisco, founded in 1910 by Edoardo Cerruti, employed sixty *sigaraie*, most of whom had worked previously in cigar factories in Lucca and Naples. These young, female cigar rollers helped turn ten tons of tobacco per month into Tuscan- and Napoli-style cigars.[22] Similarly in New York, Giovanni Battista Raffetto's confectionary factory, which specialized in candied fruit and chestnuts in syrup, hired almost exclusively female laborers.[23]

Such entrepreneurs built social and cultural capital in part through their understanding of how migration and industrial capitalism challenged the division of labor and gender and sexuality norms that had governed families in Italy; to a certain extent, these norms rested on patriarchal fears about the dangers of unwatched female sexuality. In order to meet their social and housing goals, rural peasant families who unified in cities abroad adjusted their work patterns. When possible, families kept mothers and wives at home, where they often "turned their kitchens into factories," mainly as garment finishers doing homework for subcontractors in the textile industry.[24] But other mothers and wives—and, increasingly children, both female and male—worked outside the home to earn wages in support of the family. As Jennifer Guglielmo and other historians have shown, Italian families relocating to New York greatly depended on the wages of these women for survival.[25] In textile manufacturing, owners and workers did not always share the same ethnic background, but this was more often the case in the tipo italiano food industries, where entrepreneurs commercialized culinary specialties that were specific to regions in Italy. The rising number of women in tipo italiano food establishments suggests that the Italian background of employers helped ease the movement of wives and daughters out of the home and into large Italian-style industrial pasta, canning, confectionary, and tobacco businesses. As Italian employers did with male laborers, they used their shared ethnic background, as well as the networks that female workers themselves cultivated, to obtain inexpensive feminized labor.

While access to cheap, increasingly feminized migrant labor contributed to manufacturers' success, so too did access to Italian working-class consumers in migrant marketplaces, consumers who were attracted to lower-cost alternatives. The *Bollettino* reported in 1906 that tipo italiano businesses "employ Italian workers and they find in our colonies their market." Such firms grew from migrants' "impressions of needs, real and imagined," hinting at the way exposure to new foods and expectations abroad moved migrants' eating concerns beyond material and nutritional considerations only.[26] Owners of tipo italiano factories carried production and consumption together across the Atlantic to migrant marketplaces; they domesticated what had once been a transatlantic commodity chain, but their success continued to depend on material and cultural connections to Italy.

As global spaces, migrant marketplaces also disclose how migration and trade together complicated attempts to mark foods by their national origins. Most emblematic of this were tipo italiano items made from imported Italian ingredients. In 1880 Angelo and Antonio Pini, Italian migrants from Lombardy, turned the modest liqueur factory started by their father into a four-thousand-square-meter factory in Buenos Aires while opening a second distillery in the northeastern province of Corrientes. Yet they produced their Pineral-brand aperitivo from imported Italian herbs, red and white wine, almonds, saffron, and orange and lemon rinds; the product itself, therefore, embodied the actual trade routes in Italian foods opened up by migrants like the Pini brothers and sustained by migrant demand for homeland flavors.[27] Similarly, the award-winning Canale-brand biscotti made by the Vedova Canale company, founded by the Canale family from Genoa in the late nineteenth century, employed ninety-five workers, mainly Italians or their children, in their Buenos Aires factory. While the flour, butter, eggs, and sugar used in the biscotti came from Argentina, every month the company used about one hundred kilograms of lemon extract imported from Sicily.[28] After migrating to New York from Liguria in the late nineteenth century, Giovanni Battista Raffetto began preserving imported Italian chestnuts and domestically produced fruits such as strawberries and peaches. Relying predominantly on female workers, his firm churned out its famous chestnuts in syrup and cognac using machines that he invented and patented in the United States and Europe.[29] The wine industry especially muddled demarcations between italiano and tipo italiano; early twentieth-century wine entrepreneurs in the United States and Argentina often blended heartier and stronger imported wines with the more "anemic" or light wines made in California and Mendoza.[30] At a time when merchants shipped much of their foodstuff in bulk rather than in small containers with clear labeling, the drawing of neat boundaries around what counted as tipo italiano and what counted as imported proved difficult. Migratory and commercial flows together produced hybrid foods and beverages that reflected in some ways the transnational lives of their migrant producers and consumers.

Despite their regular denouncement of tipo italiano products, importers and retailers affiliated with Italian chambers of commerce acknowledged these blurred lines between domestically made and imported foods and often sold both in order to maximize their profits and satisfy their customers. In 1900 New York importer Ernesto Petrucci advertised his store as a "grand deposit of California wines" while also lauding his Italian olive oils and pastas from Genoa and Naples.[31] In Chicago, Emilio Longhi sold numerous Italian-brand macaroni, olive oil, liquors, *antipasti*, and tuna, but also Cavalleria-brand cigarettes made in the United States with "tanning and tobacco identical to the Italian manufacturing."[32] Even Francesco Jannello, seller of the renowned Florio-brand Marsala and Malvasia wines in Buenos Aires, traded in Mendoza wines.[33] Selling both imported and domestic foodstuff allowed food purveyors to appeal to a wide variety of preferences among Italian migrants as well as their transforming palates. While *piedmontese* Oreste Sola lauded Mendoza wine as "one of the best products" in Argentina, his wife, Corinna, complained not only about Mendoza wine but all tipo italiano foods that crossed her path in Buenos Aires. In a letter home, she urged her in-laws back in Biella to drink a bottle of Piedmont wine, pointing out, "Here you can't do that since there isn't any good wine. Not only that, but in everything you don't get satisfaction, like cheese, salami; none of the food we eat here is satisfactory."[34]

Likewise, tipo italiano food and beverage manufacturers often sold imported food items along with their domestically made products. G. B. Matelli in Buenos Aires produced the noted Margherita-brand Argentine vermouth while simultaneously selling imported Barbera, Grignolino, and Moscato wines from Italy.[35] The Savarese V. and Brothers pasta factory in Brooklyn, which employed some one hundred workers, sold its domestically made pasta along with imported Italian cheeses, wine, and olive oil, and even pasta.[36] A number of tipo italiano food manufacturers began their businesses as importers, opening up enterprises only after acquiring a critical mass of capital, business contacts, and clientele. Upon migrating from Liguria, Antonio Cuneo opened a small grocery store in the Italian quarter of Mulberry Street in New York and soon began importing a variety of foods directly from Italy. After accumulating money and experience, he founded the Atlantic Macaroni Company in Long Island, which under his son's direction became one of the largest pasta producers in the United States.[37] Similarly, in 1891 importer Luigi Gandolfi opened a bottling plant in New York where his Italian employees treated imported wine to a natural fermentation process. Using better technology available to him in the United States, the sparkling wine he produced and sold under his own label, he claimed, rivaled the imported bubbly wines he continued to sell in his shop.[38] The global economic integration that allowed ingredients, technologies, and laborers to move back and forth across the Atlantic made ascribing national labels to both food products and the people that produced them increasingly difficult.

The evolution and growth of tipo italiano industries in the United States and Argentina reveals the intermediary position of migrant producers, importers, and sellers. Rather than serve as mouthpieces of the Italian government or Italian food businesses, migrants in food sectors knew that it was in their best interest economically to represent and sell both tipo italiano and Italian imports. They navigated their own sometimes delicate place as unofficial commercial attachés of the Italian government from their positions in Italian chambers of commerce and as migrant prominenti interested in bolstering their status among consumers. Migrant merchants showed off their expertise and Italy's dependence on them as commercial interlocutors by sending back advice and instruction to Italian manufacturers. In 1900 the Buenos Aires–based Italian Chamber of Commerce encouraged Italian businesses to send representatives to Argentina in order to better understand changing consumer needs. Their experience buying and selling in the Buenos Aires migrant marketplace revealed to them that Italian consumers' desires were not immutable and constant, concluding, "Our workers and our citizens acquire in America a marked tendency to change taste."[39] They chastised Italian exporters for making assumptions about migrant preferences rather than basing their decisions and advertising strategies on calculated investigations of specific consumer preferences. In response to a proposed Argentine law regulating the wine trade, the Buenos Aires *Bollettino Mensile* noted, "Our exporters must persuade themselves that in the modern battle of trade it is not the market that must conquer the product, but the product [that] must conquer the market."[40] Similarly, in 1910 the Italian Chamber of Commerce in Chicago advised the Italian Ministry of Agriculture, Industry, and Commerce to follow U.S. pasta manufacturers' lead by boxing Italian pasta intended for sale in the United States in small, rectangular, elegantly designed cartons. "It now lies with the manufacturers in Italy to follow their example," concluded the chamber.[41] By using the tipo italiano industry to show off their marketing and business acumen, migrants positioned themselves as indispensable trade experts with knowledge of both their host country and migrants' changing consumer expectations.

The growth of tipo italiano goods provides an example of migrant merchants' intermediary position between national and global markets of people and products and between Italian and non-Italian consumers. They understood and capitalized on migrants' long-term plans and economic restraints, as well as their desire for familiar, if transforming, tastes, by offering them a way to consume in an ethnically distinct manner but at a cheaper price. Migrant food entrepreneurs did not, as critics often stated, simply imitate or emulate Italian goods; rather, they appropriated links between migrants and exports and creatively and profitably responded to migrant laborers' economic exigencies, changing appetites, and shifting relationships to their host countries.

Comparing *Tipo Italiano* Industries in New York and Buenos Aires

Ernesto Bisi, with his United States Macaroni Factory, was one of many migrant entrepreneurs in the late nineteenth and early twentieth centuries who capitalized on the growing demand for foodstuff and for employment among Italian migrants and on the technologies and resources of his adopted country (see fig. 13). Associating his product with industrial modernity, Bisi advertised his pasta factory in Carnegie, Pennsylvania, as "the most perfect, the most modern and among the largest that exist in the United States."[42] He employed a diverse set of symbols that reflected the transnational worlds in which his migrant consumers lived. A large illustration of the Statue of Liberty tied the production and consumption of a popular tipo italiano food to a U.S. emblem of freedom and migration, as the statue stood near Ellis Island, the migrant processing station through which thousands of Italians arrived.[43] Bisi, like other migrant food entrepreneurs, used the migrant press to invent new meanings about the consumption of Italian foods.

Figure 13. Ad for United States Macaroni Factory, *Il Progresso Italo-Americano* (New York), December 1, 1905, 3.

The Bisi advertisement also suggests how nation-specific differences characterizing le due Americhe influenced the growth of tipo italiano goods in New York and Buenos Aires, creating various consumer experiences and notions of Italianness in migrant marketplaces. That Bisi produced pasta by the "carload" for buyers in the United States, Canada, and Mexico, and presented his establishment as an "Emporium of Foodstuff," hints at the industrial resources available to him in the United States but inaccessible to migrants in Argentine cities.[44] While pasta importers in both countries worried about the competition that entrepreneurs like Bisi represented, in the less industrially developed Argentina, importers fretted more about tipo italiano foods arriving from European countries such as Spain, France, and Greece.

Tipo italiano industries in the United States and Argentina relied equally on migrant laborers and consumers, as well as on the raw materials (fruits, vegetables, wheat, meat, and tobacco) used in Italy's most profitable food exports. However, migrant entrepreneurs like Bisi took advantage of the United States' more mature manufacturing sector and transportation and communication networks to produce, market, and distribute their domestically produced foods on a mass scale. The first tipo italiano businesses in both countries depended on mainly imported, hand-cranked machinery for making pasta, sausage, wine, and other products.[45] However, the industrial food system developed much earlier in the United States than in Argentina.[46] Already by the late 1800s food constituted the United States' foremost manufacturing sector; by the turn of the century food made up one-fifth of the country's total industrial output, and the nation moved from small-scale food production for local markets to large-scale, mechanized production for an increasingly integrated, national market. New fuel sources; advances in transportation, distribution, refrigeration, canning, and bottling; and novel advertising techniques helped birth several large-scale agribusinesses, milling giants, meatpacking plants, and industrial food corporations that fundamentally changed the eating patterns and food cultures of U.S. Americans. Italian migrants remained hesitant to embrace the standardized factory foods they encountered, even while industrialization offered migrants access to their traditional foods—coffee, sugar, wheat bread and pasta, and meat—available only to the well-off back home.[47] But the United States' sturdy industrial infrastructure also provided the resources used by migrant food entrepreneurs to begin manufacturing Italian industrialized food for co-nationals in migrant marketplaces. By the early twentieth century, the United States' more extensive manufacturing base meant that migrants with capital and connections had at their disposal much of the machinery used for processing, bottling, canning, and packaging industrialized, domestically made foods.[48]

Migrants in Argentina did not have access to the same range of industrial capabilities and technical resources to fabricate large quantities of tipo italiano goods.

Fernando Rocchi describes Argentina's industrial takeoff in the 1880s and 1890s when the state implemented policies, such as taking the peso off the gold standard and tariff protection, that helped jump-start the country's manufacturing sectors, including food processing. Migrant entrepreneurs began turning modest pasta and biscotti factories, vegetable and fruit canning operations, and wineries into larger establishments.[49] However, Argentina's food businesses remained slow to mechanize: handicraft labor continued to be used extensively in factory production; its restrictive banking system made it difficult for migrants to secure long-term loans; its small stock market limited businesses' wherewithal to raise capital; and many businesses continued to rely on imported machinery.[50] A book on Argentina at the 1906 International exhibition in Milan listed none of the 4,377 small-scale food manufacturers in the country (86 percent of which were owned by migrants) as mechanized.[51] Finally, Argentina's overall population was much smaller than that of the United States and had a difficult time supporting economies of scale.[52] As in the United States, migrants in Argentina used their family, kinship, and cultural ties to overcome financing, administration, and legal obstacles, and to capture consumer markets. However, with some important exceptions, such as wine, spirits, and meat production, tipo italiano industries in Argentina remained small-scale before World War I while still meeting the basic needs of migrant laborers for lower-cost foods. Argentina's weaker manufacturing economy forced consumers to rely heavily on imports until after World War I.

Tipo italiano migrant businesses in the United States also benefited from the country's more protective economic policies and food safety legislation, legislation criticized by *Il Progresso* and the Italian Chamber of Commerce in New York. Higher tariffs encouraged the establishment of tipo italiano industries, as they could produce locally made foods cheaper than imported versions, which even without high duties were disadvantaged by the higher costs of transportation and storage. Trade promoters in the United States lashed out at migrant entrepreneurs, claiming that as soon as migrants in the States established tipo italiano businesses, they lobbied politicians to increase tariffs on Italian products. New York's chamber explained that once migrants, especially those in California, cultivated agricultural produce typically imported from Italy, such as citrus fruits, hazelnuts, vegetables, and tomatoes, they "ask, implore, demand protective measures, which they almost always obtained for their products, to the harm of those imported."[53] While praising Italian winemakers in California for their hard work and determination, victories "that make us think of the Genoese and Venetians of the middle ages," New York's *Rivista* reminded readers, "Do not talk to them about solidarity with the mother country regarding custom taxes." "Free traders at home became protectionist in the United States!" the article continued. "Regarding all else they are Italian, the most Italian, ready to sacrifice for the mother country . . . but don't touch the protective duties!"[54] Conversely, Italians in Argentina before World War I enjoyed relatively low import duties on major Italian imported products, and with some success they lobbied their host government to keep tariffs on imported foods relatively low.[55] While tipo italiano migrant entrepreneurs in Buenos Aires may have campaigned for tariff hikes to benefit their locally produced foods, their voices were drowned out by the more unified stance of migrant merchants, consumers, and workers who wanted, above all, access to cheap foods and other consumer goods.

Importers of Italian foodstuff also saw U.S. food safety laws, especially the Pure Food and Drug Act of 1906, as unfairly handicapping imports while advantaging tipo italiano industries. U.S. government inspectors, claimed the New York chamber, applied laws regulating food purity more stringently to foreign foods arriving at major U.S. port cities than to products made in the country. Authorities applied these laws "in a manner that suggests almost that these provisions were imposed with hate on particular products," especially imported Italian cheeses.[56] The act, for example, did not apply to products that moved within states, and migrants in the United States could more easily defend their tipo italiano foods against sanctions than could Italian businesses located across the ocean.[57] Argentina also had strict sanitary requirements in place for imported food goods. Legislation called for food inspections, the labeling of food, and the prohibition of spoiled and toxic substances.[58] However, while described as a nuisance to international trade, migrants did not discuss food safety legislation as prejudicially disadvantaging imports.[59] Unlike migrants in the United States, those in Buenos Aires wasted little discussion on the deleterious effects of food sanitation measures on the import market, probably because imports faced less competition from these domestic industries.

A good example of how industrial, economic, and legal factors advantaged tipo italiano industries in the United States is the evolution of pasta production in le due Americhe. The great demand for pasta among the United States' large Italian population fueled the wildfire-like growth of pasta factories in cities throughout the country. By 1906 in greater New York alone there were more than a dozen large-scale pasta factories employing over two million workers.[60] The economist Aldo Visconti complained of the United States, "Wherever there are nuclei of Italians of some importance, factories of pasta develop and, having owners and training almost all Italian, function perfectly well from the beginning."[61] Even in small towns outside of the Northeast and Midwest, migrants established pasta factories. In 1903 the *Bollettino* reported that in Texas five pasta factories had appeared in cities with sizable Italian migrant populations, including two in Houston and one each in Galveston, San Antonio, and Fort Worth.[62] Ads in New York's *Il Progresso* further reveal the generous number of medium and large-scale pasta factories by the early twentieth century. By 1905 pasta manufactured by A. Castruccio and Sons in Brooklyn, New York, and B. Piccardo in Pittsburgh regularly publicized their *paste domestiche* in *Il Progresso*.[63] In cities with large migrant marketplaces like New York, San Francisco, and Philadelphia, a number of pasta factories grew into vast establishments with wide distribution networks. Rivaling Ernesto Bisi's United States Macaroni Factory was the Atlantic Macaroni Company, established and managed by Andrea Cuneo, which produced thirty thousand kilograms of pasta daily by 1906 in its Long Island factory.[64] A tax of five cents per kilogram on imported pasta helped stimulate this growth, as did the 1906 Pure Food and Drug Act, which, New York's chamber claimed, "produced grand inconvenience to importers," as imported pasta remained in dock warehouses, sometimes for days, waiting for inspection after being unloaded from ships.[65]

While producing obstacles for importers, such tipo italiano pasta factories in and around New York thrived because their modest prices allowed poor migrants to regularly enjoy the foods that back home had been reserved only for the wealthy. As a Sicilian woman living in New York said to Italian American educator Leonard Covello, "Who could afford to eat spaghetti more than once a week [in Italy]? In America no one starved, though a family earned no more than five or six dollars a week. . . . Don't you remember how our *paesani* here in America ate to their heart's delight till they were belching like pigs and how they dumped mountains of uneaten food out the window?"[66] The ability to overeat Italian-style spaghetti and even to waste food symbolized to working-class migrants a reversal of class hierarchies in Italy, a hierarchy that had kept their meals modest and unvaried.

In Argentina a substantial number of pasta factories, mainly operated by Italians, also flourished by the end of the first decade of the twentieth century. In 1910 the Argentine Industrial Census counted 177 pasta factories, employing a total of 794 people, 88 percent of which were owned by migrants.[67] While much fewer in number, several pasta factories in Buenos Aires evolved into major producers that supplied urban consumers and those in the interior provinces. Giovanni Casaretto opened his Buenos Aires pasta factory in 1895 after migrating from Liguria; just five years later, 50 Italian laborers produced more than 4,000,000 kilograms of pasta daily in his 1,200-square-meter, two-story factory.[68] Canessa, Pegassano, and Company also counted among the country's oldest and largest pasta manufacturers, making 15,000 kilograms of pasta a day by 1906 and employing over 150 workers, almost all Italian.[69] Despite these Italian pasta tycoons and despite the low tariffs on pasta imports, obstacles and limitations embedded in the country's weaker industrial base and consumer market kept most pasta establishments modest before World War I. Indeed, few ads for pasta factories appeared in *Il Progresso* during the first two decades of the twentieth century. In Argentina demand for pasta and other Italian-style food items outstripped local production capabilities, ensuring a continued market for imports. In 1909 the value of Italian pasta imports to the United States was almost 150 times greater than the value of such exports to

Argentina. Just six years later, in 1915, however, the value of pasta exports to the United States had decreased by almost half while imports to Argentina continued to rise. In 1916 Italy exported almost four times more pasta to Argentina than to the United States.[70]

The more rapid development of domestically made pasta in the United States mirrored the growth of other tipo italiano foodstuff in the two countries. Migrant newspapers help evidence this difference, as a wider array of products and a larger number of medium- to large-scale tipo italiano businesses appeared in New York's *Il Progresso* than in Buenos Aires' *La Patria*. Readers of *Il Progresso* viewed publicity for a growing variety of Italian-style wine, liquors, cigars and cigarettes, canned fruits and vegetables, candies, cheeses, and cured meats by the early twentieth century, whereas in *La Patria* ads for tipo italiano foods remained limited. Ads for such products in *Il Progresso* often referred directly to imports, assuring consumers that their products simulated and perhaps surpassed Italian foodstuff, certainly in price but also in quality. Ernesto Bisi assured readers that his pasta was "guaranteed to compete victoriously with imported pasta."[71] In 1900 the Italian Cigar & Tobacco Company of New York stressed that their Tuscan- and Neapolitan-style cigars were "fermented in perfect imitation of form and taste to the Italian ones."[72] In a country that, notwithstanding a revolution in the fields of marketing and mass advertising, depicted the mainstream U.S. consumer as homogenously white and middle-class, migrant entrepreneurs employed an assortment of national and regional symbols to appeal to migrant consumers' transnational livelihood, as well as their changing tastes.[73] Like Italian businessmen in their trademarks for popular Italian exports, they used the names and images of notable Italian men in their branding to emphasize the Italianità of the product. In New York the Razzetti Brothers manufactured and sold "Garibaldi" cigars, while the firm of G. B. Lobravico produced Dante-brand Italian-style cigars.[74] References to various national and regional identities and geographical indicators often merged together, as in the American Chianti Wine Company of Brocton, New York, which produced 12 million gallons of wine annually, wine made not from the Sangiovese grape, as the winery's name suggests, but from Concord grapes cultivated in New York State.[75] Likewise, De Nobili cigars, "vastly superior to any other Italian cigar manufactured abroad," used both U.S. and Italian flags in publicity for the company's products in ways that reflected the transnational experiences of De Nobili's workers and consumers.[76]

Entrepreneurs like De Nobili who called their goods "Italian" also pioneered in crafting national rather than solely regional identities for their products and clientele. Ads for imports in both New York's *Il Progresso* and Buenos Aires' *La Patria* tended to accentuate the local or regional origins of their products and brands— Ligurian olive oil, tomato sauce from Naples, Marsala from Sicily, Chianti wine, vermouth from Turin, candy from Siena, and so on. Ads for tipo italiano foods, however, more often referred to their foods simply as "Italian" while sometimes continuing to also employ regional labels. Despite Italian elites' hopes that their country's heterogeneous peoples scattered abroad would unite as "Italians," in part through the consumption of imports, migrants continued to identify with the regional traditions of their local paese well into the twentieth century.[77] And yet publicity for and articles about tipo italiano goods indicate that the formation of a more unified national identity among migrant consumers began already in the early twentieth century and around tipo italiano wine, cigars, pasta, and other foodstuff rather than in relation to Italian imports, which continued to be marketed as regional specialties. Migrants' identities became consolidated and commodified in migrant marketplaces through interaction with tipo italiano foods just as much as, if not more so, than through imported goods.[78]

As Simone Cinotto has found in his research on Italian foodways in New York during the interwar years, migrant food entrepreneurs constructed an Italianità based less on provenance or notions of authenticity as it relates to a primordial fixed past and more on the need to forge identities that helped migrants grapple symbolically and materially with their daily existence as low-wage transnational workers.[79] While tipo italiano producers viewed the mingling and merging of national and regional symbols as a key element in their marketing strategies, importers described such practices as consumer fraud. Indeed, the growing presence of tipo italiano foods in the United States unleashed a wave of concern over issues of food authenticity. The Italian Chamber of Commerce in New York regularly associated tipo italiano foods with "counterfeiting," "manipulation," "imitation," and "falsification," labels connoting poor sanitation, low quality, and illegality.[80] They took particular aim at manufacturers and vendors who mislabeled their products or used confusing and misleading wording. At the 1906 "Italians Abroad" exhibit, the jury awarded the Italian Swiss Colony in Sonoma County, California, for the quantity and quality of their wine but used the occasion to admonish the winery for using flasks similar to those in Tuscany and for including the word "Chianti" on their labels. This disingenuous practice made the California wine an "illegitimate" and "fearsome" competition to "true Chianti."[81] Similarly, Luigi Solari, president of the Italian New York chamber, rallied against tipo italiano businesses that used the adjective "yellow" to describe foods and oils that were just barely greenish in color and who pronounced their products "sweet" even when they were bitter. While he described these misleading words as "venial sins," he saved his most vociferous condemnation for store owners who passed off locally made products as Italian. Such retailers, he claimed, strategically employed phrases such as "oil from Lucca," "Romano cheese," and "Aged Chianti wine," instead of the more direct "guaranteed pure olive oil from Lucca or guaranteed real Romano." Fighting against this "parasitic commerce," Solari concluded, was one the most important obligations of Italian chambers of commerce abroad.[82]

Moreover, by focusing its efforts on imported foods, the New York chamber claimed that the 1906 Pure Food and Drug Act did not regulate the arenas where real harm was enacted on food consumers: U.S. factories, warehouses, and retail outlets, where migrants made Italian-style foods in unregulated, unhygienic tenement buildings, and where greedy merchants and grocers mixed genuine Italian goods that had just passed rigorous inspection with domestically made inferior ingredients. They counted as the worst offenders dishonest migrant importers and retailers who combined genuine Italian oils and wines with domestically made, inferior products such as cotton and peanut oil or grape varieties grown in California or New York. "It is an incontrovertible fact," an article in the *Rivista* claimed, "that many imported Italian products like olive oil, in passing from places of production to this grand consumer market, are often altered or artificial, adulterated or counterfeited."[83] To importers seeking to popularize their products among Anglo-Americans, who often held prejudicial stereotypes about Italian food, such fraud threatened the high reputation of Italian products that importers worked so hard to cultivate. This threat arrived just as the quality of many Italian food exports, especially wine and olive oil, was greatly improving. In Italy advances in knowledge about grape and olive cultivation, investment in new machinery and bottling facilities, the commercialization of monocrop wine and olive regions, and a growing national and international consumer market modernized these industries, as well as others such as pasta, tomato preserves, spirits, and cheese manufacturing.[84]

In response to the real and perceived dangers represented by tipo italiano foods, the Italian Chamber of Commerce in New York intensified its policing of boundaries between imported and domestically made items. Starting in 1909 the chamber began a program of inspection and certification in which importers and retailers could voluntarily submit to an examination by experts affiliated with the chamber—usually other reputable importers—in order to obtain a certificate of origin that guaranteed the authenticity and purity of imports.[85] In 1910 a short column in the *Rivista* called "Commercial Fraud" provided information on lawsuits against merchants accused of food falsification and imitation. In what amounted to a public shaming, the column included the details of the case: the name of the defendant, the products and brands in question, and the outcome. In October 1910 the *Rivista* reported that the Board of Food and Drug Inspection had arrested Giuseppe Rindone of New York for five accounts of counterfeiting, including the forging of Ferro China Bisleri and Fernet-Branca spirits. He was declared guilty, sentenced to five months in jail, and fined five hundred dollars.[86] In 1911 A. Fiore & Company in New York admitted to passing off their cottonseed oil as "olive oil," and the Board of Trade found that macaroni sold as imported by the Ceravolo Brothers of Philadelphia was actually made in the United States.[87] Migrants wielded U.S. food legislation that they felt had so damaged the import trade against co-nationals to monitor the boundaries of authenticity in migrant marketplaces. While merchants hoped this self-policing would ultimately boost the sales of imports, the tipo italiano industry continued to grow unabated. The chamber failed to understand that migrant consumers cared less about the origins of the foods they ate and more about whether those ingredients, meals, dishes, and brand names satisfied their everyday nutritional and cultural needs as transnational eaters.

Italian business leaders in Buenos Aires also reported on the growth of domestically made tipo italiano edibles and denounced migrants who passed off their goods as Italian imports, selling them under the most accredited Italian brand names.[88] However, the most perilous menace to Italian commerce in Argentina was perceived as originating abroad rather than from within Buenos Aires' migrant marketplace. Argentina's weaker industrial structure meant that the country's tipo italiano industries developed more slowly than in the United States. Migrant consumers in Buenos Aires continued to demand and rely on imported European goods; as such, Italian-style foods brought about less notice or sanction from trade promoters, who sometimes condescended to local industries attempting to compete with Italian imports. Despite the extraordinary increase in the local production of cheese in Argentina, the Buenos Aires Italian Chamber of Commerce declared, "Among all the countries that furnish this article to Argentina, Italy will probably be least disadvantaged by the 'indigenous industry,' because manufacturers here have not succeeded yet in imitating with success the Italian product."[89] Therefore, in Argentina merchants came to associate tipo italiano not so much with ambitious and sometimes unscrupulous migrant entrepreneurs, but with threatening competitors abroad in countries such as Spain, France, Greece, and Portugal.[90] It is no wonder that these countries competed aggressively with Italy for the pesos of Argentine consumers, and Italians specifically, as the largest consumer block in the country. While the Argentine population comprised only 8 percent of the total Latin American population, it constituted one-third of all imports into the region in the early twentieth century.[91] Argentina, more so than other Latin American countries, represented a lucrative site for European manufacturers, and to the exasperation of Italian importers, Italy's neighbors also reached out to migrant consumers there.

These countries, it turned out, produced similar products for export to Argentina, products so similar that they could substitute for Italian goods. Italian trade promoters obsessively compared the quality, quantity, price, and style of Italian imports from similar items arriving from other countries: Italian preserved fish, canned goods, wines, and spirits found fierce competition in comparable products from Spain and France; cheese from Switzerland rivaled Italian parmigiana-reggiano and gorgonzola in flavor and cost; Italy and Germany competed to be

Argentina's number one supplier of rice; and Portugal did the same with their exports of dried fruit.[92] In such discussions the Italian Chamber of Commerce in Buenos Aires portrayed Argentina as a critical site upon which modern commercial battles between European powers played out. Italian cheeses, noted a 1908 article in the *Bollettino Mensile*, competed with the products of "all other nations anxious to conquer a market such as this one, where all high-quality articles find easy and profitable positioning."[93] They echoed the language used by Italian business and political leaders in Italy by employing martial discourse to project masculinized images of combative commercial transactions in Buenos Aires. The "numerous armies" of Italian workers served as the "direct consumers of Italian products."[94] In this grand "fight for commercial expansion," Italy proved victorious, succeeding in the "conquest of the markets of *la Plata*"; importers sustained, "with praiseworthy tenacity, a loyal battle of competition with other similar products that are imported from Spain, France, Greece and other countries of the old world."[95]

In the United States merchants focused their ire on the Italian-style foods made, sold, and consumed in migrant marketplaces; in Argentina most of these foods were tipo italiano on arrival. In 1899 the Italian Chamber of Commerce in Buenos Aires wrote a letter to the Italian government expressing concern over Greek wine imported to Argentina under the guise of Italian wine. Imported Greek wine, entering as "Italian" but of inferior quality, the chamber claimed, was causing major damage to the good reputation of Italian wines.[96] Two years later the chamber noted with alarm the increase in the number of consumers buying "Swiss cigars imitated in Italian style" because they were less expensive than imported Italian tobacco products.[97] Merchants' constant fear of products arriving as "Italian" from Italy's competitors in Europe suggests that industrialists across the Atlantic were aware of Italian emigration patterns and looked to tap into Argentina's multinational consumer base.

Tipo italiano items arrived in the Americas not only from countries like Switzerland, Greece, and France but also from Italy itself. As the twentieth century progressed, Italians in Argentina became more vocal in their condemnation of dishonest Italian manufacturers and of the Italian government for not exercising sufficient vigilance on the producing end of Italian commodity chains. Merchants accused unscrupulous Italians of exporting foodstuff to Argentina that were labeled "Italian" but actually contained only a small portion of real Italian ingredients or none at all. These adulterated "Italian" products, migrant merchants complained, irrevocably damaged Italian commerce; fake "Italian" goods lost consumers' trust while destroying the good name of quality Italian foods.[98] In 1902 the Buenos Aires chamber began engaging in a series of heated discussions about Italian industrialists who had been exporting "guaranteed pure" Italian rice to Argentina that proved to be Italian rice mixed with Japanese and Burmese rice, allegedly of a much inferior quality. Two years later the chamber continued to rail against industrialists selling "Italian" rice made from a mix of Italian and Asian varieties, claiming that 75 percent of all Italian exporters involved themselves in this "work of demolition."[99] Similarly, in 1908 the chamber argued that most abuses to the Italian olive oil market were made in Italy, in places like Genoa, where companies mixed Italian oil with cottonseed oil from Tunisia and other places but then sent them abroad under the label "Lucca" or "Riviera."[100] Crooked Italian manufacturers in Italy proved just as threatening to Italian-Argentine trade and consumption in Buenos Aires' migrant marketplace as did industrialists in other European nations.

Rather than direct their policing efforts toward the production and sale of *tipo italiano* products internally, within migrant marketplaces—as did merchants in the United States—migrants in Argentina directed their patrolling outward, toward Italy and other European countries. In 1900 *La Patria* urged their Italian Chamber of Commerce to support a measure that would require every barrel of wine imported from Italy to contain a certificate of origin, the brand name, the number of the barrel, and its weight and quality as had been done with wines from the northwest Piedmont region of Asti. *La Patria* noted that better labeling and identification would help eliminate "falsification on a vast scale," which the paper estimated amounted to 75 percent of the "Italian" wine entering la Plata. Requiring certificates of origin for all Italian wines would "render a valuable service to our country and our numerous Italian communities in the Río de la Plata who long, in vain until now—with some honorable exceptions—to taste Italian wine that is truly authentically Italian."[101] This lobbying may have played a role in the Italian government's passage of a law later that year designed to combat fraud in the preparation and commerce of Italy's wine.[102] Merchants used the passage of this law to urge the Italian government to adopt similar measures for other goods in order "to protect the true Italian industry."[103] They pointed fingers at Italy, at times blaming the commercial policies of the government for the growth of *tipo italiano* industries both in European counties outside of Italy and locally in Argentina. In 1901 importer Francesco Jannello complained when the Italian government granted a monopoly to only one cigarette and cigar importer and retailer in Argentina. Jannello argued that the monopoly damaged Italy's tobacco market in Buenos Aires by keeping the price of Italian tobacco high, forcing migrants to buy alternatives arriving from other countries.[104] Four years later the chamber used the looming specter of *tipo italiano* businesses within Argentina to again denounce Italy for the monopoly; by inhibiting competition, concessionaries' high prices provoked not only the continued importation of Italian-style cigars arriving from countries like Switzerland but also the local production of *tipo italiano* cigars, evidenced by a new cigarette factory opening in Buenos Aires. In a spiteful letter to the Italian Ministry of Commerce, the chamber noted, "It had always seemed that the privilege of the concession on the importation of our tobacco would surely be a damage to the interests of Italian finance."[105]

U.S.-based merchants also made appeals to the Italian government to regulate and foster Italian trade, especially in the production of wine for export. In 1909 New York's *Rivista* published a report by Guido Rossati, head of the Italian Wine Station, to the Italian minister of Agriculture, Industry, and Commerce in which he blamed poorly enforced Italian laws governing wine production and labeling for compromising the reputation of Italian wine among U.S. consumers. Discussing abuses in the geographical designations, he wrote, "In this way Chianti has never seen its namesake region, Marsala has never seen the provinces of Trapani or Palermo, Barbera has never seen Piedmont."[106] He joined the members of the chamber in pressing the Italian government to pass new, more stringent and punitive laws regulating the classification and regulation of wine. However, aside from the wine trade, most of the chamber's efforts remained fixated on patrolling food authenticity within U.S. migrant marketplaces.

Tipo italiano Greek wine, Swiss cigars, Tunisian olive oil, and Japanese rice arriving from across the Atlantic reveal that in an industrially weak Argentina, threats to Italian commerce came less from savvy migrants in Buenos Aires and more from industrialists across the Atlantic. Of all European nations, Spain represented the most dangerous menace to Italian products in la Plata. In 1901 the chamber noted that Spanish food imports such as wine, oil, olives, and dried citrus fruits gave "serious competition to similar Italian goods" in Argentina.[107] Merchants felt threatened by the presence of Spanish migrants, the second-largest foreign-born group in Argentina, whose demands for homeland goods also sustained transnational commodity chains in Spanish foods. Migrants' boasts about the consumption of Italian imports often accompanied cautious warnings about what decreases in migration would do to Italian-Argentine trade. In 1901 the chamber assured themselves that Spain, a "population poor" country, could not provide the same large number of *braccia* (migrant manual workers) to justify the type of privileged trade status Argentina continued to have with Italy. And yet, the chamber concluded, Italy should not rest easily: Italy produced the same products as Spain and must turn to the same markets; therefore Spain remained a threat to Italian commerce in Argentina.[108] In 1908, the first year in over half a century when more Spanish migrants arrived in Argentina than Italians, Italians linked the dwindling of Italian migrants, who are "naturally consumers of the products of the mother country," to a slowdown of Italian commerce and a simultaneous increase in Spanish trade.[109] "As you see, the diminution is a continuous advantage to the Spanish element and can already be seen in many warehouses that sell edibles—in which the trade had been almost an Italian monopoly."[110]

A history of Spanish empire building in Latin America also led Italian migrants to view Spain as the most harmful threat to Italian commerce. Foods similar to Italian exports—olive oil, wine, and dried fruits, for example—proved popular not only because of the competing presence of Spanish migrants but also because of Spain's colonial legacy. Spanish products, the chamber claimed, "find complete acceptance by the public, maybe in homage to the ancient ties that unite the two nations as one community of race and interest, and for the bonds of tight solidarity that represent in one man all those who, abandoning Iberian soil, came in search of better fortune in this country of emigration."[111] The chamber worried about consumers in Latin American countries like Argentina who, "forgetting the bloody battles in support for their independence, respond with enthusiasm to the call of the country of origin."[112] Italian migrants used consumption to help construct Latinità, shared cultural and racial traits between Argentines and Italians that supported Argentina's continued acceptance of Italian people and goods and promoted Italian foods among Argentines. However, as a racialized category associated with ideas of European civilization, Latinità proved broad enough to also unite Argentines and Spaniards as "one community of race." That is, notions of Latinness were perceived as advantaging the foods and migrants of Europe's other "Latin" countries, especially Spain and France. Italian migrants built ties between Argentines and Italians based on a shared culinary tradition linked to Spanish colonization. The past Mediterranean orientations and imperial forays in Latin America by Spain, Portugal, and France, however, also affected the food traditions of these nations, of Latin American countries, and of Argentina's other "Latin" migrants."[113]

While migrant prominenti asserted that Italians' strong demographic presence, combined with cultural commonalities between Italians and Argentines, allowed for the "Italianizing" of Argentines, they simultaneously worried that such similarities could conversely "Argentize" or "Europeanize" Italian migrants. As increasing numbers of Italians remained in Argentina, learned Spanish, raised families, and bought tipo italiano products arriving from Spain, Greece, and France, Latinità threatened to absorb Italians into Argentine society and endanger Italian commerce.[114] The chamber did not always trust that Italian migrants could distinguish between the olive oils, canned goods, wines, and tobacco products arriving from Italy and those arriving from Europe's other "Latin" countries. In 1904 merchants in Buenos Aires urged the Union of Chambers of Commerce in Rome to hold an exhibition of Italian products in Buenos Aires. "The goal," their proposal read, "is to familiarize consumers, and to show that there is a difference between Italy's goods and those of other countries who have similar products."[115] While merchants often assumed that Italian migrants were duped into buying goods from other countries because of deceptive labeling, discussions also hinted at a larger fear that Italians, and especially their children, were voluntarily buying non-Italian products because of their lower prices and migrants' evolving tastes.

Italians dominated not only much of the international commerce in food imports but also the emerging domestically produced tipo italiano industries, which, while smaller and less varied than in the United States, made up an important, if not the most important, manufacturing sector in Argentina's slow process toward industrialization. Italians often pointed to migrants' contributions to Argentine industrial life, contributions so great that Argentine manufacturing was marked with an "Italian stamp." Pointing to Italians in the nation's food processing sectors—pasta, biscotti, bread, wine, beer, liquor, and the fishing and dairy industries—a 1906 report concluded, "How could Italians in Argentina demonstrate a more admirable industrial development?"[116] In their debates over whether to participate in the "Italians Abroad" exhibit for the 1911 International exposition in Turin, members of the Buenos Aires chamber haughtily asserted that even if the chamber opted out, Italians would be present at the fair because the items displayed at the Argentine pavilion were essentially the "product of the work of Italians."[117] Italian industrialists in Argentina, including producers of tipo italiano pasta, liqueurs, wines, chocolates, and cured meats, they complained, could not count in the "Italians Abroad" exhibit, "since their products are considered national [Argentine] production," revealing again ongoing fears that Italians and the "Italian" foods they made in Buenos Aires were being claimed by Argentina on the world stage.[118]

Italian migrants' predominance in both domestic and international food sectors meant that Argentines bought not only Italian imports but also domestically made Italian-style goods. And yet, unlike in the United States, where tipo italiano foods, like Italian imports, marked their consumers as distinct from Anglo-Americans, the commonalities upon which Latinità was forged in Argentina defied attempts to tie many local foodstuff to one nation exclusively. The production of *panettoni*— the cylindrical sweetened breads with candied fruits, originally from Milan and consumed especially during the Christmas holiday season—reveals how migration complicated efforts to affix particular national culinary traditions to foods in Buenos Aires. By the turn of the century, ads for the holiday bread, called *panettone* in Italian and *pan dulce* in Spanish, appeared regularly in both the Italian-language *La Patria* and in Argentina's Spanish daily *La Prensa*, suggesting that both migrants and Argentines ate the holiday treat. For example, Italian Chamber of Commerce member Carlo Gontaretti, owner of the Two Chinese bakery in Buenos Aires, advertised his *pan dulce a la Genovesa, Milanesa o Veneciana* (Genoa-style, Milan-style, or Venetian-style sweet bread) in *La Prensa* and *La Patria*.[119] Panettone's widespread consumption in Argentina signaled not only migrants' overwhelming presence in Argentina's bakeries but also culinary similarities that made the production and consumption of sweetened bread made from wheat flour a long-standing tradition in Italy, as well as in Spain and France.[120] In a country where cultural foundations— including culinary traditions—were linked, in perception more than in reality, to Northern Europe, it is no surprise that panettone, imported or domestically made, rarely appear for sale in New York's *Il Progresso* until the 1930s, nor was it absorbed into the U.S. culinary mainstream. While today "pan dulce" in other parts of Latin America, such as Mexico, refers to a larger category encompassing many variations of sweetened breads and pastries, such as *conchas*, *orejas*, and *roscas de reyes*, in Argentina "pan dulce" has come to refer mainly if not exclusively to the Italian panettone rather than to a wide variety of pastries.[121] By World War I what began as a food item imported from Italy and made in Argentina by Italians had been absorbed into Argentina's national cuisine.[122] Buenos Aires' migrant marketplaces allowed Argentines and Italians shared consumer experiences around Italian imports, tipo italiano goods arriving from European countries, and tipo italiano goods like panettone made in Argentina by migrant entrepreneurs.

Whereas Italian imports and tipo italiano goods arriving from Europe and made in Argentina helped migrants construct commonalities between Italians and Argentines, in the United States the consumption of these same goods racialized Italians as inferior, "foreign," and "illegal." This is evident most poignantly in the production and consumption of tipo italiano wine. Italians in both countries led the way in establishing the first large-scale wineries in the western hemisphere. Starting in the late nineteenth century, Italian migrants in the state of California, United States, and in the provinces of Mendoza and San Juan, Argentina, pioneered in experimenting with imported and domestically grown grape varietals, eventually dominating one of their host country's most essential agricultural sectors.[123] In 1882 Venetians Antonio and Domingo Tomba founded the Domingo Tomba winery in the central western province of Mendoza. The winery became the largest and most important in Argentina, producing 254,000 hectoliters (6.7 million gallons) of wine on their 59,000-square-meter establishment by 1909.[124] Their U.S. counterpart was the Italian Swiss Colony in Sonoma County, California, founded in 1881 by Ligurian merchant Andrea Sbarboro along with a group of migrant businessmen from Piedmont. By 1911, under the leadership of Pietro Carlo Rossi, the Italian Swiss Colony owned 5,000 acres of vineyards and a wine cellar with the capacity to hold some 4 million gallons of wine.[125]

As they had done with Italian imported goods, migrants in Argentina used tipo italiano wine as proof that migrants brought European civilization to Argentina's uncultivated frontier. Describing Argentina as a "young" nation, a publication of the Buenos Aires chamber cited the success of Domingo Tomba winery as evidence that "Argentina in every way wants to emulate the ancient nations." The book associated the winery with similar establishments in Italy and France in terms of size, modernity of machinery, and quality of product. Linking Argentine wine cultivation to ancient Rome, the book described the Tomba wine cellar as an "enormous warehouse, suspended by a long line from arches that look like those of a Roman aqueduct."[126] Ads in the Italian-language press also played up wine entrepreneurs' links to western civilization and the Age of Exploration. The *Bollettino Mensile* asserted that Italians saved Argentina's wine industry from Spanish colonists, who had abandoned colonial-era vineyards.[127] A full-page ad for the Domingo Tomba winery from 1910 featured an image of Christopher Columbus on a ship with two shirtless indigenous men genuflecting before him. The ad included a list of all the awards the winery had won at mainly European international exhibitions.[128] In the hands of Italian businessmen, tipo italiano wine entrepreneurs represented migrants' entrepreneurial genius and Italy's contributions to Argentine industrialization and progress. However, the long-standing tradition of wine production and consumption in Argentina also allowed Italian wine entrepreneurs to be subsumed in Argentine nation-building endeavors. In the Spanish-language *La Nación*'s special centennial issue, the paper included a feature story on the Domingo Tomba winery, in which the founders' nationality was not mentioned. Instead, the celebratory article, which described Tomba wine as "being admired as a true wonder in all the wine producing countries of Europe," praised the winery as representative of Argentine industrial and agricultural virtuosity.[129] While Italian prominenti called attention to the winery's Italian roots, the ads for the winery more frequently left out the nationality of its founders; publicity in Spanish and Italian by the early twentieth century described Tomba as "genuine product of the Mendoza winery, the best Argentine table wine."[130]

In the United States, conversely, migrants' desires to employ wine production as a covetable symbol of civilization ran up against racist attitudes toward Italians and a powerful temperance movement that increasingly saw migrants' drinking cultures as threatening to white, Protestant, middle-class notions of respectability.[131] As the stigmatization of wine and wine drinkers as "criminal," "immoral," and "foreign" intensified, migrants from France, Austria, and Germany exited California's nascent wine industry, leaving it in the hands of Italians. The racialization of migrant wine producers and consumers, therefore, presented business and market opportunities to savvy Italian entrepreneurs, who came to dominate wine production and distribution, mainly to Italian consumers in cities on the East Coast.[132]

While the U.S. temperance movement would not triumph nationwide until the passage of the Eighteenth Amendment in 1919, Italian winemakers battled prohibitionists intent on shutting down their operations by challenging unfavorable depictions of wine producers. In 1911 Ettore Patrizi, director of San Francisco's *L'Italia*, characterized the Italian Swiss Colony in Asti as an idyllic site of upstanding morality, where the company's one thousand workers found steady employment and lived healthy, virtuous lives. Patrizi described the colony as a "picturesque scene of happiness, of vitality and of health" with beautiful houses and a Catholic church "creating pleasant Italian settings with all the most beautiful and brilliant

characteristics of our race and our customs." He emphasized the modern ameni-
ties of the colony, which included an electric plant, a doctor, and a pharmacy, a
colony that painted a "most pleasing moral image."[133]But, unlike in Argentina,
Italian entrepreneurs struggled to position themselves and their businesses as
civilization builders. Merchants used the Prohibition movement to identify and
attack Anglo-American ignorance, suggesting that most U.S. Americans did not
understand wine as a high-class symbol of sophisticated European culture. They
expressed particular frustration and dismay when temperance advocates placed
wine in the category of "intoxicating drinks" and "patent medicines" made largely
of whiskey, accusing them of foolishly mixing "sacred and profane things."[134] Italians in Argentina also reflected on how wine consumption produced divide rather
than unity between Italians and U.S. American consumers. A 1910 article in Buenos
Aires' *La Patria* explained that while Argentines drank wine, U.S. Americans did not,
preferring beer; because U.S. Americans considered the drinking of wine "vulgar,"
the children of "Latin" migrants, especially Italians, adopted the habits of their host
country and abandoned wine to avoid stigmatization.[135] Argentine reformers in
the early twentieth century also expressed increasing concern over excessive alcohol consumption and linked the social problems it created to migrants. However,
wine's long-standing place as a regular component of the Argentine diet already
in colonial Argentina combined with Argentine leaders' continued gaze toward
Europe as a model civilization to work against successful campaigns to eliminate
the importation, production, and consumption of wine there.[136]

*　*　*

While entrepreneurs in both the United States and Argentina used their host nations' agricultural and industrial resources, as well as consumer and labor markets,
to jump-start tipo italiano food industries, migrants in the United States exploited
the country's mature manufacturing base and distribution networks to produce
and sell a richer variety of commercial foodstuff by the early twentieth century.
Migrants affiliated with the Italian Chamber of Commerce in New York feared
that the mushrooming tipo italiano sector would damage trade between Italy and
the United States, and they directed the bulk of their policing efforts within New
York's migrant marketplace. In Argentina tipo italiano came to be associated not
so much with migrant entrepreneurs in Argentina, but with unprincipled European
and even Italian industrialists across the Atlantic who produced and exported food
items that were similar in taste, appearance, and quality to Italian imports. As a
racial and cultural category, Latinità proved expansive enough to facilitate not only
the purchase of Italian imports and tipo italiano foods like panettone and wine but
also the consumption of foods arriving from other "Latin" countries in Europe,
such as Spain, France, and Greece. The growth of tipo italiano industries in the

United States and Argentina discloses how early twentieth-century globalization complicated rather than clarified attempts to attach national labels to global foods and notions of food authenticity.

Clever entrepreneurs on both sides of the Atlantic took advantage of ethnic connections and migrants' desire for homeland tastes in order to produce new foods as well as work and consuming experiences for migrants in New York and Buenos Aires. They capitalized on their intimate, firsthand knowledge of migrants' economic restraints, employment needs, and gastronomical traditions to satisfy migrant consumers, and often they did so in better ways than did importers. Even while increasing numbers of Italian women and their daughters labored in the food industries of the host countries, migrant prominenti only rarely acknowledged how women in Italy or abroad influenced migrants' consumer practices. This oversight continued despite the reality that male migrants' contributions to transnational economies—perhaps more than any other factor—explained the consumption of lower-cost tipo italiano foods. Migrants desired foods from home, but buying expensive imports regularly detracted from their ability to make good on familial responsibilities. Because consuming tipo italiano foods allowed migrants to more quickly achieve their long-term objectives, they frequently proved more meaningful than imported foods, whose sellers fixated on provenance and purity. As events in Europe mobilized women and men in migrant marketplaces in le due Americhe, however, provenance would come to matter as migrants organized around commodities in Italy and abroad as resources for their homeland in need.

"Pro Patria": Women and the Normalization of Migrant Consumption during World War I

It was no one's business how Italian migrants chose to spend their money, New York's *Rivista* asserted in 1911. After reminding readers of the extraordinary contributions Italians made to their host countries' economies as manual laborers, the author pointed to a double standard by which migrants' spending patterns were scrutinized: "When I go to buy a pack of cigarettes, do I ask the tobacco seller how he will spend the 20 cents that I gave him in exchange for the cigarettes? And then why would this same cigar seller, when he gives me a dollar in exchange for my day of work in his tobacco factory, ask me where and how I will spend that dollar, or insist I spend that dollar on a miss in America rather than send it to my family in Italy?"[1] Italian elites expected that laboring male migrants would sustain Italian trade paths through their robust, patriotic consumption of imported foods. Nevertheless, reports such as these suggested that the transnational family economies in which migrants were enmeshed constrained migrant consumption in the United States and Argentina. Indeed, by associating consumerism abroad with spending money on a "miss in America," the report represented male migrant consumption as paramount to marital and family disloyalty. Overseas migrants, with their wives, parents, and children at home, held tight to their hard-earned cash.

Italy's entrance into World War I in 1915, however, turned migrants in New York and Buenos Aires toward consumption and Italy simultaneously. In doing so, the war served as a turning point in the history of migrant marketplaces in le due Americhe. While prominenti migrants and Italian leaders had attempted to cultivate a national identity among its disparate peoples across the world, and partly in relation to Italian products, most labor migrants continued to identify with the local traditions and values of their small *paesi*. At the same time, labor migrants' enduring commitments to family back home curbed migrants' consumer behavior. World War I, however, created an unprecedented opportunity to mobilize migrants as Italians and as consumers around a common cause and enemy. Italian merchants, businessmen, and journalists in New York and Buenos Aires used the war to cultivate a more powerful commercial and cultural relationship between migrants and their homeland in need.

In May 1915, as soon as Italy entered the war to take back from Austria-Hungary its "unredeemed" territories in the Adriatic Sea region, migrants in the United States and Argentina organized national and local war committees to raise money, goods, and support for Italy. That same month, Italians in Buenos Aires formed a national Italian War Committee with numerous regionally based branches; similarly, in New York the Italian General Relief Committee ran a variety of "pro patria" fund-raising campaigns.[2] As the main publishing outlet for their communities' diverse wartime activities, the migrant press proved instrumental in intensifying bonds between migrants abroad and Italy. *Il Progresso* in New York and *La Patria* in Buenos Aires supported war committees' initiatives in almost daily reminders to assist Italy by, for example, donating money to the Italian Red Cross and investing in Italian government bonds. In addition to running their own fund-raising projects, the press meticulously reported on community events, regularly listing the names of donors and the exact amount of money donated by each individual, business, or fraternal organization. The regular logging of donors' names and contributions provided a public forum for attaching specific monetary values to the patriotic redirecting of migrants' resources to Italy. The newspapers reported on major fund-raising galas at important cultural venues such as the Teatro Colón and Teatro Verdi in Buenos Aires and Carnegie Hall and the Century Theatre in New York City as lavish, elegant affairs, during which migrants conspicuously consumed and donated for the homeland.[3]

These fund-raising campaigns, which characterized migrants' spending and saving as a *dovere* (duty) to Italy, politicized migrant marketplaces in ways that disclose consumption's role in national and ethnic identity formation. As members of transnational families linked to subsistence production in villages back home, migrants were reluctant consumers as well as reluctant "Italians," preferring to save their money and to preserve their local or regional identities. While tariff debates in the late nineteenth and early twentieth centuries first offered migrants a vehicle for presenting themselves as transnational consumers, it would be fund-raising campaigns during World War I that would ultimately turn migrants toward consumption. From Buenos Aires and New York, Italians vocalized their ability to affect the outcome of the war, in part by influencing the international commodity chains that they helped generate and sustain through a history of transoceanic migration. Furthermore, unlike turn-of-the-century tariff debates in which migrants focused on securing access to low-cost homeland foods for mainly male laborers, migrant women's active involvement in a variety of initiatives called public attention to their spending power. Wartime demographic changes increased the number of Italian women in migrant marketplaces; Italian wives and daughters who either relocated to join their families in the United States and Argentina or were born abroad increasingly represented the "miss in America." Against the background of these transformations, women's war work helped to normalize consumption among a people who were more used to saving than spending and to link migrant marketplaces to both women and Italianità. These fund-raising activities demonstrate a transition in the gendering of consumption abroad, a transition that feminized postwar consumption of Italian imported foods.

Nation-specific contexts that differentiated migrants' discussions about race and about tipo italiano industries in the United States and Argentina also produced varying wartime experiences in le due Americhe. In both countries the war feminized postwar consumption. In Buenos Aires, however, migrants' transnational wartime experiences were mainly in connection with Italian imports and portrayed consumers of non-Italian items as traitors to their homeland. In the United States, conversely, migrants consumed for Italy by buying imported foodstuff as well as foods produced in the United States, especially tipo italiano goods, but also U.S. industrialized products. In both countries the war revealed migrant marketplaces as global and gendered sites where migrants crafted complex transnational identities through saving, buying, and eating.

"Blessed Are the Socks": Migrant Women's Pro-Wool Campaigns

It would be women's public association with a nonedible agricultural staple in their host countries that first opened the door to the feminization of migrant consumption during the interwar years. Shortly after Italy entered the war, migrant women, working in women's committees or female sections of various male-led wartime organizations, collected wool, yarn, and fabric—or money to buy these items—and turned them into mittens, sweaters, socks, scarves, face masks, and other clothing items for Italian troops on the frigid Alpine battlefront.[4] As "workers of the wool," female migrant women were catapulted onto the pages of New York's *Il Progresso* and Buenos Aires' *La Patria* in an unparalleled manner.[5] The newspapers repeatedly encouraged women as they labored over wool with knitting needles and sewing machines, celebrating their work by quantifying the colossal number of woolen items produced and sent back to Italy. In 1916 the Italian steamship *Tomaso di Savoia* departed for Italy from Buenos Aires with 3,480 pairs of socks, 301 scarves,

273 face masks, 131 sweaters, 69 pairs of woolen underwear, 28 pairs of gloves, 4 body belts, and 4 shoe insoles.[6] In October 1915 *Il Progresso* reported on a fourth shipment of clothing, on the Italian ship *Verona*, which included over twenty cases containing 5,700 pairs of socks.[7] The various woolen items, *La Patria* reminded its readers, "provide our brave soldiers with clothing that makes the winter days less severe on the mountains conquered by the enemy."[8]

Shifts in global migration patterns and the demographic changes in migrant communities they produced provide an important backdrop for understanding how wartime campaigns such as the pro-wool drives affected the gendering of consumption and national identities. With the outbreak of the war, overseas Italian migration decreased significantly. In 1910 a little over 104,000 Italians arrived in Argentina and just over 262,000 arrived in the United States; by 1918, however, only 640 Italians migrated to Argentina and only 2,793 migrated to the United States.[9] Furthermore, 51,774 Italian men from South America and 103,259 from North America repatriated during World War I, most of whom were recalled by the Italian government to fight for their country.[10] Italy's mobilization of its male population for military service reduced the number of men headed overseas and created a temporary rupture in the country's traditionally male-dominant migrations. In 1905 only 29 percent of Italians arriving in the United States and 23 percent of Italians arriving in Argentina had been female.[11] In 1917, however, 65 percent of Italians migrating to the United States and 49 percent of Italians to Argentina were female.[12] While the more gender-balanced and even female-dominant migrations to the United States and Argentina ended immediately after the war, these demographic changes increased the presence of women in migrant marketplaces during the very moment in which the war heightened feelings of Italianness relating to buying, spending, and saving.

Against the backdrop of these wartime demographic changes, migrant women—either newly arrived during the war or left in the United States and Argentina by their male relatives who returned to Italy—as well as the sizable, more gender-balanced second generation born abroad, attracted unprecedented attention by manufacturers, retailers, and the press. It is not a coincidence that while a short Sunday column titled "Nel Regno della Donna" (In the woman's realm) ran sporadically in New York's *Il Progresso* during the early twentieth century, during the war the paper inaugurated a regular column called "La Moda" (Fashion), in which *Il Progresso* advertised and sold patterns for women's clothing.[13] The paper also increased the number of articles and columns devoted to cooking, hygiene, fashion, and the home. Similarly, while in Buenos Aires' *La Patria* a lengthy column titled "Vita Sociale" (Social life) started in the early twentieth century and included articles covering women's association meetings, fashion, food, and beauty advice, in 1915 the paper began an additional column specifically aimed at women called

"Cronache femminile" (Feminine chronicle).[14] The press's new concern for migrant women during World War I would help produce a postwar print culture that paid an increasing amount of attention to women as readers and as consumers.

In their coverage and support of the pro-wool campaigns, the migrant press in both countries positioned women's work as "the duty of Italian women," a patriotic obligation that linked together women of every social class and rank. The press implored its readers to unite across divides in migrant communities—especially class and regional divides—as Italians. In 1915 *Il Progresso* urged "every woman, whether patrician or plebeian, supreme artist or modest worker, wife or young girl," to collect money for the campaign at work, with friends, at association meetings, and at family gatherings.[15] Similarly, *La Patria* wrote that wool and clothing campaigns demonstrated "Italian women in marvelous solidarity who have provided an example that will remain in history."[16] The wool drive, the paper proclaimed, joined poor and rich migrant women in cities and rural areas, across class and geographical lines.[17]

The press also portrayed pro-wool campaigns as unifying Italian women generationally and transnationally. Pro-wool initiatives strengthened generational relationships between Italian migrant mothers and their Argentine- and U.S. American–born children. The press actively encouraged young girls to participate in the wool drive along with their mothers. In 1915 *La Patria* reported on a girls' school run by Colonia Italiana, a prominent Italian fraternal organization, where the school's young students had knit two hundred scarves for Italian soldiers. The paper applauded the students: "How much tenderness and how much patriotism do these little hearts express for the homeland of their parents!!!"[18] The pro-wool campaign provided migrants the chance to stake claims on a growing second generation, despite Argentine and U.S. citizenship laws that legally declared such children as their own. The press also depicted the wool initiatives as joining migrant women transnationally to their female counterparts in Italy. Italian women in Italy and migrant women abroad were "linked in this hour of our nation."[19] *La Patria* detailed the pro-wool campaign in Italy, noting that all women—old, young, middle-class, uneducated, and professional—had dedicated themselves to buying and producing clothing for Italian soldiers. "Blessed are the socks!" the article concluded. "May Italian women in Argentina follow their example!" The paper published patterns and sewing and knitting instructions for the fabrication of socks, mittens, and face coverings, handed down from Italian consulates in Argentina, and informed its readers that Italians back home were begging their *connazionali* (co-nationals) across the ocean to send wool clothing or raw wool.[20] Pro-wool initiatives, the Italian-language press in both countries suggested, bound together women in Italy, Buenos Aires, and New York in their specifically feminine obligations as both women and Italians.[21] The wool drives represent the first major event during which migrant communities in the Americas employed women to forge a sense of national identity that transcended regional, class, generational, and national boundaries.

These campaigns projected gendered images of national belonging by portraying entire migrant communities—men and women—uniting around female wool workers. As *La Patria* wrote, "The true greatness of a population—it has been written—is seen in the nobility of a woman's spirit." The article continued: "The true sentiments of a nation are often the sentiments inspired by mothers, by wives, by female friends."[22] These feminized versions of Italian patriotism and nationalism in coverage of women's wartime activities departed from earlier turn-of-the-century images linking Italianità abroad to masculine images of merchant explorers, commercial warriors, and male consumption. The press depicted migrant marketplaces as responding positively to women's work of raising and spending money on behalf of Italy. In 1915 *Il Progresso* singled out for special thanks Mrs. Elvira Giordano, Italian language instructor and member of a women's war committee, who collected over thirty dollars for the wool campaign from Italian business owners, including Italian banker Mr. Filomeno Pecoraro and Mr. Michele Gargiulo, proprietor of an Italian restaurant on Coney Island.[23]

The migrant press enthusiastically supported pro-wool drives in ways that recognized and lauded women's specific contributions to the war effort while using gendered language to describe them almost exclusively as mothers, wives, sisters, and daughters. In a September 1915 article, *Il Progresso* urged women to think of their "sons, brothers, spouses" in the wintery Alpine trenches. Understanding caregiving as an innate female trait, the paper told women, "We must dress them and keep them warm; we need to give them the comforts of materials that have always been made by mothers, by sisters and by wives. . . . The wool that they receive from us, worked by the gentle hands of our women, will have a little family tenderness in them, as if a perfume."[24] The press depicted sewing and knitting as part of a woman's natural skill set, a traditionally female task, which the war transformed from a somewhat trivial and even narcissistic feminine pastime into a serious and selfless activity. *La Patria* ended a 1915 article on the wool drive by reminding women, "Work for the soldier! Here is a hobby that no one will dare criticize, a pleasure that nobody will accuse of egoism, an ingenious female work of national unity, not suspected of invasion in the men's field."[25] Receiving wool clothing from women in New York and Buenos Aires allowed brothers, sons, and husbands in Italy to "turn their thoughts and hearts to that domestic nest where there was much warmth of affection, where there was domestic comfort."[26] The press assured readers that despite women's active involvement as "wool workers," gendered divisions in wartime contributions remained intact. It was precisely because the Italian community understood sewing as a fundamentally female occupation, often completed in the home, that the press could acclaim women's central role in one of the most popular war-related Italian diasporic movements without appearing to challenge traditional gender roles.

The pro-wool drive was one of a number of transnational initiatives declared appropriate for women by migrant communities. Given women's association with food provisioning and cooking, wartime fund-raisers that focused on the dietary needs of co-nationals in Italy and abroad also launched women into migrant print culture. Migrant women demonstrated their duty to Italy by raising money for and assembling *scaldaranci*, handmade torches made from old newspapers and wax, used by Italian soldiers to heat their meals and stay warm on the cold Alpine battlefront. "After a strenuous military operation," *La Patria* explained, "our soldier must eat a slice of preserved meat—excellent certainly, but cold—and quench his thirst with spring water—and often water from snow—that makes not only the teeth but the stomach run cold."[27] By the end of the war, migrants in Argentina sent almost 30 million scaldaranci back to Italy.[28] Similarly, in the United States the Italian-American Ladies Association in New York, made up of first- and second-generation Italians, called out to women to join their organization and help the National Association for the Soldiers' Scaldarancio in Milan, reminding readers that "all the Italian women of America" should contribute to the *scaldarancio* drive.[29]

The press further linked migrant women to food consumption through another campaign dubbed by the press as migrants' "greatest patriotic duty." That "duty of our colony" was to support *le famiglie dei richiamati*, migrant families whose male relatives had been recalled to fight in Italy.[30] "Wives whose husbands have responded to the call of the homeland, children whose fathers went to fight and fall with the fateful name of Italy on their lips," *La Patria* wrote, "are without resources and without bread. . . . The babies must eat, they want to eat."[31] Illustrative of the way the press used the war and women to privilege national identities over those that were regionally based, the paper described migrant "heroines" as calling out to their community, "We are *Italian* women and our relatives, who are leaving, are *Italian*, rushing home to sacrifice themselves for its glory. We are proud but the sacrifice that we made should not harm the children and the elderly relatives, not fit to work."[32] *Il Progresso* called on "Italians living in the Great America" to help "those for whom the horrors of the war . . . have tossed and will toss into the most horrible poverty."[33] In their coverage of such campaigns, the press represented women left behind in the United States and Argentina exclusively as dependents, despite the large numbers of first- and second-generation Italian women who worked for wages. The majority of reservists, *La Patria* pointed out, were manual laborers "without money and without resources who leave here their parents, their wives, their young children without protection and without means."[34]

At the same time, however, the press portrayed migrant women participating in initiatives designed to meet the needs of these families and interacting with wealthy donors, merchants, and Italian fraternal organizations in order to do so.

"In homes, in factories, at parties, above all demand from everyone a donation for the unfortunate; be propagandists for this sublime charity," *Il Progresso* urged women.[35] In addition to raising money, which was paid out to families in monthly installments, women working for the Italian War Committee in Argentina visited families to determine the exact requirements of each household; they documented their visits, using such information to distribute milk, food, medicine, clothing, and linens to women and children.[36] On September 20, 1915, as part of migrants' annual festivities honoring Italian unification in 1870, migrants ran a special campaign for families of Italian reservists when they put together and distributed *cesti-regali* (gift bags) for families of the recalled, containing mainly foodstuff. Each of the one thousand cesti-regali carried a combination of imported foods as well as items produced in Argentina, including wine, panettoni, cookies, candies, milk, oil, pasta, coffee, sugar, grain, flour, meat, fruit, vegetables, beans, rice, cheese, cigars, and cigarettes.[37] Leading migrant importers, retailers, and industrialists such as wine importer Giovanni Narice, retailer and representative of San Pellegrino sparkling water Giuseppe Ferro, and Italian food importer Lorenzo Leveratto donated the edible items for the gift bags. The collection and dispersal of the gift bags served "to strengthen the bonds of solidarity among Italians who feel in this solemn hour more alive than ever in their soul the saintly love for the common country."[38]

The gendered language used by migrant newspapers to portray such public, transnational campaigns as rooted in women's long-standing duties as wives, mothers, and daughters in the domestic sphere belied women's increasing visibility in migrant marketplaces that wartime exigencies facilitated. These various campaigns involved migrant women, as Italians, interacting with and pressuring male prominenti businessmen and journalists. During but especially immediately after the war, *Il Progresso* and *La Patria* began devoting more space to articles geared toward a female readership and to advertisements featuring female consumers. At the same time, while lauding women's public work for their homeland, newspapers tethered feminine wartime spending, saving, and wool initiatives to women's responsibilities as wives and mothers and to their transnational duties as "Italians" rather than to individual consumerist desires. In an attempt to manage migrant women's new, very public presence in migrant marketplaces brought on by the war, the Italian-language press sometimes reminded women and girls to resist cravings for clothing, toys, and other consumer items.[39] An anecdotal exchange between a father and his daughter published in *La Patria* served to curtail the spending of money, especially by women and girls, on anything other than war-related activities:

Father, will you give me a little money?

No, little one: I bought you a toy a little while ago and now is not the time to spend much . . . another time.

But I don't want to buy a toy.

And then why do you want the money?

Because the teacher told me that the Italian soldiers must spend winter on the Alps and that good Italians should buy wool to send to them winter clothes . . . you know it's cold in the snow.

Take it, little one: take the money and a kiss.[40]

Despite prescriptive messages instructing migrant women to direct their consumer energies toward the Italian nation-state exclusively, changes in migration patterns, combined with wartime fund-raising campaigns, placed women in a position to assist their home country while publicly participating as buyers in the burgeoning consumer cultures of their host societies. Ironically, "Wool, Wool, Wool" articles in *La Patria,* exhorting women to save their money for wool, often ran next to large ads for Buenos Aires department stores like Gath & Chaves and Harrods featuring fashionably dressed women buying fabrics of many textures and colors.[41] During the war the migrant press presented paradoxical images of migrant women as they emerged with more force as readers, spenders, and savers, first as wartime fund-raisers, but after the war as the major buyers and consumers of imported Italian products.

Diasporic Differences in Migrant Wartime Campaigns

In 1915 the Buenos Aires–based Italian Chamber of Commerce sponsored the adoption of a propaganda stamp on its publications stating, "Italians! Prefer always the Italian national industry if you have at heart the glory and the prosperity of the homeland."[42] The stamp appeared in *La Patria* alongside an appeal to "Italians residing abroad." Migrants held the responsibility to fortify their homeland against its enemies during the war by keeping Italian goods competitive on the international market. "It is the duty of Italians abroad to protect and defend the consumption of Italian products," the announcement stated, citing as an example Marchese Boccanegra–brand olive oil from Genoa, sold by Bernasconi and Company of Buenos Aires, one of the largest importers of Italian foods in Argentina. Italian migrants who gave preference to the foods of other countries, the paper stated, were simply bad patriots.[43] While New York's *Il Progresso* and Italian Chamber of Commerce also worried about the war's effect on Italy's export market and encouraged migrants to consume imported foods, they did not initiate an official propaganda campaign in support of Italian products. Instead, wartime consumption and identity formation in the United States became affiliated as much with U.S.-produced goods as it did with Italian foodstuff.[44]

Migrant newspapers in New York and Buenos Aires used women's fund-raising activities, especially the pro-wool drive, to fuse migrant communities together on the basis of a united Italian identity. In both countries female migrants took on new public roles as patriots in migrant marketplaces, consuming, spending, and producing as a transnational duty to the homeland, roles that would help feminize migrant food consumption during the interwar years. In most other facets of wartime mobilization, however, national differences in the economies, societies, and geopolitical priorities of the United States and Argentina produced distinctive experiences for migrants in New York and Buenos Aires.

In the more import-dependent Argentina, migrants rallied more often and with more intensity around items from Italy than did Italians in the United States, who by World War I had a wider array of locally produced industrialized foods and other items upon which to conceive of themselves as Italians and as consumers. In Argentina fund-raising events called specific attention to migrants as donors, buyers, and consumers of Italian foodstuff. Italian merchants and companies donated Italian imports such as olive oil, wine and spirits, and textiles to be included in gift bags for families of the recalled and to be auctioned off at fund-raising galas. In December 1915, for example, *La Patria* reported on a benefit held by the Rosario branch of the Italian War Committee, for which a number of Italian merchants and companies supplied imported items, including the Society for the Introduction of Cinzano Products, which contributed one case of Cinzano Italian vermouth, and the import firm Boero, Napoli, and Company, which supplied over a dozen bottles of Italian sparkling water.[45] These numerous fund-raising festivities depicted migrant consumption of Italian foodstuff as an expression of support for the Italian war effort and for Italian-Argentine trade, which was suffering as a consequence of the war.

The very prominent place of Italian imports in wartime mobilization in Argentina explains the chamber's official propaganda campaign that encouraged migrants to "buy Italian" exclusively during the war. "The home country does not only need soldiers," the chamber announced. Migrants must "remain on guard against those who mask products from other provenance with Italian brands."[46] Since the late nineteenth century, merchants had attempted to discourage migrant consumption of tipo italiano Greek olive oil, French wine, Spanish fish, and Swiss cigars. More than any other previous event, the "buy Italian" campaign provided merchants the opportunity to draw distinctions between foodstuff coming from Italy and those arriving from other European countries. The chamber also utilized the global conflict to try to usurp its European competitors. As part of a larger campaign to keep Italian-Argentine commercial relations active, the chamber undertook an extensive drive to identify goods that before the war had been imported from European nations but could now be substituted with Italian goods. *La Patria* applauded the chamber, urging them to use the war to "purge" the Italian market of the many inefficiencies that made imports susceptible to fraud. Such initiatives brought together Italy's heterogeneous people as Italians around the defense of imported foods. "We love, protect and defend unanimously the 'Italian brand,'" proclaimed *La Patria*. "And we respect it, so that it is respected."[47]

Italian merchants exploited the chamber's "buy Italian" campaign by making patriotic appeals to consumers in publicity. They equated the purchase of Italian products to expressions of Italian nationalism and used the war to depict themselves as heroic supporters both of their home country and their loyal consumer base in Buenos Aires. In an ad in *La Patria* for San Pellegrino sparkling water, importer and retailer Giuseppe Ferro promised to absorb the extra costs and risks associated with the transport of Italian goods during the war to assure consumers access to the popular beverage.[48] A 1916 ad for Florio wines sold by Francesco Jannello featured an illustration of a uniform-clad solider writing home to his sweetheart from the trenches of Thessaloniki, the Greek capital of Macedonia. Referencing Italian soldiers stationed in Greece, the soldier's letter read, "The Italians, my Nell, carried with them their enthusiasm, born of the sun, but also enthusiasm pressed from golden grapes, a sweet liqueur that warms cold veins: Florio Marsala and Malvasia wine."[49] Unlike earlier late nineteenth-century trademarks that portrayed men as detached from faraway families, the Florio publicity commodified these affective ties for migrant consumers. Migrant merchants like Jannello and Ferro employed World War I to bring a new and heightened awareness of migrants' national and consumer identities in ways that politicized buying as a patriotic obligation.

While migrants honored their homeland by buying Italian foods, they also redirected Argentine agricultural staples back to Italy. In a country where domestic food manufacturing continued to remain relatively small-scale, migrants capitalized on their host country's profitable agro-export sector to send not only wool but also wheat and frozen meat to connazionali in Italy. In 1915 Italians in Buenos Aires established the Italian Agrarian Committee to help Italy acquire grain, meat, wool, and leather from Argentina.[50] By organizing to send home such materials, Italian farmers would be undertaking "a work of great moral and economic importance for the homeland and obtain the approval of the authority and of the Argentine people," *La Patria* explained.[51] The plan benefited both Argentina's place in Italy's agricultural import market and the postwar Italian export market, as Argentina would continue to exchange their agricultural goods for Italian merchandise. In 1916 migrants helped the Italian War Committee send seventy-five hundred pesos' worth of bulk agricultural goods to Italy.[52] At the end of the year *La Patria* documented with great fanfare the departure of the Italian ship *Cavour*, which left Buenos Aires for Italy stocked with both Italian reservists and Argentine wheat.[53]

In Argentina, Latinità, cultivated over decades of transnational migration, and in part through consumption, produced expressions of solidarity in migrant marketplaces between Argentines and Italians, expressions largely absent between migrants and U.S. Americans. Women's war work made them central to reinforc-ing ties of Latinity. The migrant press pointed to wartime collaborations between Argentine and Italian women as proof of Argentina's support of the Italian cause, despite Argentina's official commitment to neutrality. In 1915 *La Patria* reported on the private Italian-Argentine Sewing Club, formed by a group of Argentine and Italian women in Rosario, "where Italians and Argentines live in fraternal agree-ment." The initiative, through which women raised money and made clothing for Italian soldiers, demonstrated that Argentine women were "sisters in the pious labor and bequeath to their work all the noble spirit of the people of the Latin race, understanding the grand cause of Italy and following with hope the fate of our army."[54] Collaboration between Italian and Argentine women proved that "our saintly war does not only move and excite us Italians, but it makes everyone's heart beat who knows Italy and knows the spirit of justice that guides the conquest of the irredentist land."[55] The war encouraged Argentine women, migrant women, and their Argentine-born daughters to join in support of their "brothers in Latin-ity" across the Atlantic.[56] Leading department stores in Buenos Aires and Rosario, including Gath & Chaves, La Favorita, and Harrods, donated wool and clothing for such drives and hosted lavish fund-raisers in their stores.[57] In September 1915 a group of migrant women carried out a benefit for the Red Cross at Harrods in Buenos Aires. The event, asserted *La Patria*, "was an extraordinary festival, for the remarkable occasion decorated with the flags of the Allied Nations outside, while inside it took on the appearance of an immense luxurious evening in which the elegant flowers and feminine beauty produced a mark of indescribable charm." Argentine department stores, which had already appealed specifically to female migrant consumers through their ads in *La Patria*, connected migrant and Argen-tine women together through fund-raising and female consumption during the war.[58] Although the war stitched together a triangular web of women as Italians in Buenos Aires, New York, and Italy, especially as "wool workers," nation-specific differences generated connections between migrant and Argentine women but not between migrants and U.S. Americans. New York's *Il Progresso* did not report on nor praise similar coordinated efforts, even after the United States entered the war on the side of the Allies in April 1917.

Italian merchants were not the only ones using nationalist sentiment to sell their wares; domestic manufacturers of foodstuff established or run by Argentines also employed the Italian-language press to publicize their support for Italy and mar-ket their goods among the largest foreign-born consumer block. In 1915 Argentine cigarette manufacture Piccardo and Company ran an ad in Italian for their popular 43-brand cigarettes depicting the crowned figure Italia standing on a map of Italy with a sword in her hand and the shield of the House of Savoy at her feet (see fig. 14).[59] The ad read simply, "Our emblem is freedom," while reminding readers that the

Figure 14. Ad for Piccardo & Co. 43-brand cigarettes, *La Patria degli Italiani* (Buenos Aires), September 20, 1915. Courtesy of the Hemeroteca, Biblioteca Nacional de la República Argentina.

cigarettes were "genuinely Argentine made." Notions of Latinità and the dominant presence of Italian migrants allowed Argentine manufacturers to assert the Italianità of their products by capitalizing on consumers' emerging national identities during the war. While U.S. food industrialists and department stores placed ads in Italian in New York's *Il Progresso* starting in the early twentieth century, it would not be until the mid-1920s that U.S. food conglomerates began targeting migrants specifically as Italians by referring to migrants' transnational social relations.

As in Argentina, the war brought newfound attention to migrant consumers in New York; however, the United States' more developed industrial and consumer society created a different setting in which migrants formed wartime identities through trade and consumption. New York's Italian Chamber of Commerce was no less concerned than their counterpart in Buenos Aires about sustaining Italian commerce. Just after war broke out in 1914, but before Italy's entrance, the New York chamber anticipated an opportunity for Italian imports, especially wine, to replace products from France and Germany, which were already embroiled in the conflict. Given that the war had "wreaked havoc on the vineyards of Eastern France," migrant merchants hoped that U.S. consumers would turn to Italian wine instead of domestically produced California substitutes. Just as merchants in Buenos Aires expected the war to benefit Italian imports over tipo italiano foods from Europe, importers in New York believed it would accomplish their long-standing goal of moving beyond migrants to reach the larger U.S. consumer market. "Before Americans allow patriotism to have the best of their taste," the *Rivista* hoped that U.S. Americans would challenge sellers of California wines who claimed their wines were "'just as good' as the products of Europe." Italian wines, already in demand at cafes, restaurants, and hotels catering to the "finest tastes of this cosmopolitan population," would please consumers "not only provisionally, but permanently, as American connoisseurs will relish them certainly as much as, and perhaps more than, the wines that have held sway on this market until the present."[60] Wishing in particular that Italian bubbly would came to replace French Champagne, the *Rivista* celebrated the characteristics of Italy's various regional sparkling wines, such as Moscato and Asti Spumante, "a lady's wine par excellence, owing to its harmless effects, luscious taste and nourishing qualities." As before the war, merchants continued to target Anglo-American women specifically as consumers of Italian imports.[61]

While hoping that the war would give Italian imports an advantage over those from other European nations, unlike in Buenos Aires, migrant merchants in New York focused the bulk of their efforts on stemming the production and sale of tipo italiano goods. Already before the war, the cheaper prices of tipo italiano foods had made them powerful competitors to imports in migrant marketplaces dominated by mainly male laborers who were intent on saving. The attenuation of Italian trade to both the United States and Argentina provided a powerful boost to the tipo italiano business in both countries, but especially in the United States, where an already robust industry existed. From 1915 to 1919 Italy's export of edible olive oil to Argentina dropped from 34,868 to 195 quintals; tomato preserves went from 35,516 to 11,151 quintals; and hard cheeses went from 27,212 to 485 quintals.[62] During those same years, the United States saw similar vertiginous drops: Italy's export of edible olive oil went from 89,830 to 406 quintals; tomato preserves decreased from 154,260 to 11,587 quintals; and hard cheeses plummeted from 84,672 to 1,456 quintals.[63] After 1916 Italy stopped sending cured meats and rice to the United States and Argentina for the duration of the war, opting to preserve those foods for Italian soldiers and civilians.[64] In 1916 the New York, Chicago, and San Francisco

Italian chambers of commerce prepared a report urging the Italian government to lift the ban on certain Italian exports, arguing that Italian companies would lose their hold over migrant marketplaces as tipo italiano entrepreneurs stepped in to fill consumer demand.[65] Their predictions proved correct; the number and variety of medium- to large-scale food companies expanded during and immediately after the war, and businesses were formed by migrants such as Vincenzo La Rosa, a Sicilian olive oil merchant who opened his pasta business during the war; by the mid-1930s, under the leadership of his children, the firm became the largest pasta supplier in the Northeast.[66] During the interwar years the number of migrant entrepreneurs and their children producing pasta, cheese, wine, cured meats, and other Italian-style foods throughout the United States mushroomed.[67]

Ironically, despite New York merchants' fears about tipo italiano foods, and unlike their counterparts in Buenos Aires, they did not carry out a massive campaign to "buy Italian" in order to discredit local products. Rather, they chose to focus their efforts on lobbying the Italian government to reduce restrictions on Italian exports.[68] By World War I tipo italiano foods and the migrants who consumed them were already too powerful a presence in migrant marketplaces, and already associated increasingly with a unified "Italian" national identity, to be depicted as unpatriotic by merchants. Instead, goods made in the United States by migrant entrepreneurs, and increasingly non-Italian businesses as well, were incorporated into transnational wartime campaigns.

Wartime initiatives in New York embraced the consumption of all goods circulating in its migrant marketplace—regardless of their origin—to build Italian identities around buying and consuming. Starting in November 1915, *Il Progresso* began a drive to raise money for Italian aid organizations for which the paper printed coupons for readers to cut out and present to business owners in New York. Italian migrants owned the majority of such establishments—grocery and food stores, bakeries, pharmacies, restaurants, butcher shops, and barbershops—including Giuseppe Gandolfo's grocery store, Francesco Altadonna's Italian butcher shop, and Polito Vincenzo's pasta factory and retail outlet.[69] In return for the coupon, buyers spending more than ten cents received a receipt that could be used as a raffle ticket for various prizes donated by *Il Progresso*. For every coupon distributed, participating retail establishments sent a one-penny donation to *Il Progresso*. The paper contributed the accumulated money to help the families of recalled soldiers.[70] This "pro patria initiative" that called out to "store owners and consumers" nationalized everyday exchanges between male store owners and mainly female migrant consumers, who used cash to buy groceries for their families in the United States and to assist Italy. Such campaigns encouraged migrants to "buy from Italians" rather than to "buy Italian" exclusively, reflecting the variety of consumer products—both domestic and imported—in New York's migrant marketplace.

Unlike in Buenos Aires, where migrants dispatched Argentine agricultural goods to Italy, migrants in New York organized to send mass-produced U.S. consumer items, including tipo italiano goods. One of the most popular transnational campaigns organized by *Il Progresso* consisted of collecting and sending U.S.-produced cigars and cigarettes to Italian troops. A cigar, explained the paper, is a soldier's "best companion." "In the wet trenches, in the camps, in the tiring marches, under the tents, in the barracks, in the hospital beds, during the rest hour or in the anxious vigils, nothing is dearer to soldiers than to smoke a good cigar or puff on a cigarette."[71] *Il Progresso* suggested that the purchase and consumption of tobacco products made in the United States linked migrant consumers transatlantically to Italian soldiers in the Alps. When soldiers smoked tobacco products from their connazionali in the United States, "they will think of their brothers beyond the Ocean and realize that even from a far distance Italians are united in one thought of love, in one wish of victory and in one heartbeat of hope for the triumph of the beloved homeland."[72] Cigarettes produced by tipo italiano tobacco companies, especially De Nobili Cigar Company in Long Island, were the most common brands donated. As with other campaigns, *Il Progresso* published a daily tabulated list of donors' names and quantities offered.[73] Migrant women featured centrally in the tobacco purchases and donations. In September 1917, for example, Miss L. Michelangelo of Hoboken, New Jersey, donated one hundred Tuscan-style cigars, and Mrs. A. Zeppini of Brooklyn, New York, gave twenty packets of tobacco. When the campaign ended on October 14, migrants had sent $4,027.09 worth of tobacco back to Italy.[74]

As the tobacco drive suggests, migrants in New York introduced their co-nationals in Italy to U.S. industrialized foodstuff during World War I. Another *Il Progresso* campaign rallied migrants to construct and send Christmas gift boxes to Italian troops in Italy. The paper called "all *Italians*, of whatever class, whatever social rank, both men and women, workers, industrialists, importers, merchants, artists, professionals, ladies and gentlemen," to assemble Christmas boxes for Italians back home. Each box contained a can of pears or peaches, a half-pound can of "chop meat" to *fare sandwiches* [make sandwiches], a box of "sweet crackers filled with fruit conserves (Fig Newton, fig bars, etc.)," a can of sardines, two Italian cigars, and a package of De Nobili tobacco.[75] As with the tobacco drive, Italian merchants and importers contributed large quantities of boxes. For example, *Il Progresso* thanked Antonio Zucca, former president of the Italian Chamber of Commerce in New York, for his 20-box donation, and Pasquale Pantano, New York importer, for his 10-box contribution.[76] When the drive closed on November 15, 1917, *Il Progresso* had collected 26,093 Christmas boxes.[77] In Buenos Aires migrants focused their efforts on sending unfinished agricultural staples such as meat, wheat, and wool to Italy, reserving gift boxes of foodstuff for the families left behind by their reservist husbands and fathers. In the United States, however, the war convinced increasing numbers of migrant consumers, who were especially skeptical of industrialized foods produced by non-Italians, to buy them not only for themselves but for con-nazionali in Italy as well.

Il Progresso portrayed campaigns involving U.S.-made foods as equally patriotic and "Italian" as *La Patria*'s coverage of fund-raisers and initiatives centered on Italian imports. "All Italians must enthusiastically support our initiative," *Il Progresso* wrote, urging readers to buy goods and services from Italian-owned stores participating in the coupon drive.[78] "How can we prove that our hearts beat in unison with our brave soldiers if we do not satisfy their every desire?" asked *Il Progresso*, goading migrants to buy U.S.-made cigars as evidence they had "done their duty" as Italians.[79] The United States' more industrialized, consumer-oriented economy allowed migrants in New York to experiment with a wider variety of available goods in the construction of wartime identities than in Buenos Aires. Buying and consuming U.S. products may have taken on increasing symbolic weight once the United States entered the war on the side of the Allies in April 1917, after which the consumption, and later rationing, of U.S. goods, especially food, became a way for migrants to display their national identity as Italians and their dual loyalty to both Italy and the United States. This was especially crucial within the context of rising xenophobic and nationalist sentiments during and immediately after the war, when calls for immigration restriction intensified and migrants experienced pressure to prove their "100 percent Americanism" by renouncing homeland ties.[80] Starting in 1917 *Il Progresso* and the Italian Chamber of Commerce in New York exhorted Italians to buy U.S. as well as Italian war bonds, and the press reported with pride the number of first- and second-generation Italians fighting in the U.S. Army.[81] Migrants in the United States expressed their identity as Italians and their allegiance to both their home and host nations by spending money on industrialized foodstuff made in the United States by Italian entrepreneurs and, increasingly, by large U.S. food corporations such as the National Biscuit Company, producers of the Fig Newton cookie.

While national contexts differentiated migrants' wartime experiences in Buenos Aires and New York, migrant print culture in both countries portrayed spending, consuming, and redirecting goods to Italy as part of migrants' duty to their homeland. The Italian-language press described migrants not only as transnational laborers, fixated on saving for families back home, but also as transnational consumers, united around a common national identity and financially capable of caring for Italians in Italy and in their communities abroad. In ways that revealed their middle-class status in migrant marketplaces, prominenti journalists hesitated to label migrants' donations to the wool drive a sacrifice: "In America, for even the most modest of workers, ten cents does not damage his savings even when he wants to throw it to the wind; therefore, we cannot call it a sacrifice."[82] In their coverage of wartime campaigns, the press often made comparisons between the cost of living and the quality of life for migrants abroad and Italians in Italy. *Il Progresso* explained that Italians in New York were better off than Italians in Italy, where "there is no work, no money, no resources; there in this moment there is only misery and grief, lessened only to the hope of better fortunes." Sending food and clothing to Italians offered migrants a vehicle for expressing their Italianness as well as their ability to consume in ways their co-nationals back home could not. The paper asked, "What will all those poor women, children and elderly in Italy do?"[83] While heightened sentiments of Italianità connected migrants to Italians across the ocean, the war called attention to differences in lifestyles, and especially consumer opportunities, between migrants and co-nationals in Italy. And while campaigns may have portrayed women in Italy as impoverished, migrant women abroad, as major protagonists in wartime saving, spending, and consuming initiatives, feminized migrant marketplaces in the immediate postwar years.

The Feminization of Migrant Marketplaces in the Interwar Period

A series of ads for Cinzano, the famed Italian vermouth, shows how gendered links between migration and trade crafted during the war strengthened postwar connections between women and migrant marketplaces. The ads in Buenos Aires' *La Patria* in 1915 featured silhouettes of people relaxing and eating at home, chatting at parties, and participating in sporting and recreational activities.[84] One of the many ads depicts a woman in elegant horse-riding attire galloping on her steed; another ad shows a playful scene between a man and a woman with a cocktail glass in her hand (see figs. 15 and 16).

These wartime Cinzano ads differed greatly from their prewar predecessors. Prior to World War I, Cinzano publicity called attention to the enormous quantities of vermouth entering the country. A 1905 ad bragged that during the previous year, 161,000 cases of the vermouth were sold in Argentina; five years later another emphasized that in 1909 Cinzano had imported to Argentina 517,000 cases.[85] This prominence given to international trade faded during and after the war in favor of publicity that focused on consumers and consumption. The 1915 Cinzano ads also contrasted sharply with the focus on Italy's industrial capabilities and male migrant heroics that were so prevalent in trademarks produced by Italian businesses in Italy in the late nineteenth and early twentieth centuries. While these trademarks had featured factories and military figures, advertising iconography in migrant newspapers after World War I was more likely to display people eating food and imbibing alcohol in festive, often opulent settings.

The female migrant consumer, emerging from women's war-related activities in migrant marketplaces during the war, led to a feminization of migrant food

Figure 15. Ad for Cinzano vermouth, *La Patria degli Italiani* (Buenos Aires), December 19, 1915. Courtesy of the Hemeroteca, Biblioteca Nacional de la República Argentina.

Figure 16. Ad for Cinzano vermouth, *La Patria degli Italiani* (Buenos Aires), November 23, 1915, 6. Courtesy of the Hemeroteca, Biblioteca Nacional de la República Argentina.

consumption in the postwar years. In publicity and discussion of Italian products, the post–World War I era witnessed a shift toward an emphasis on consumption that showcased ordinary people interacting with commodities. This new attention to consumption brought more women into ads as consumers of Italian products, making them, rather than male migrants abroad, the main receivers of Italian foods traveling along transatlantic commodity chains. By the late 1910s and 1920s, ads for imports targeted migrant women specifically and used them to generate transnational consumer identities.

This shift in gendered messages in the Italian-language press coincided with demographic trends that created more gender-balanced Italian communities in the United States and Argentina. The war had produced a temporary reversal in the gender ratios of Italian international migration. While male-predominant migrations resumed after the war, provisions for family unification in immigration legislation, increasingly restrictive immigration policy, and the global depression slowly increased the proportion of women to men in Italian migrant communities.[86] Immigration law in both countries conceived of women almost exclusively as dependents, as trailing mothers, wives, and daughters of male-headed households. These gendered conceptualizations of the family and women's and men's places within it made it extremely difficult for single females to migrate independently; women were considered morally and sexually suspect by Argentine and U.S. immigration officials and anti–white slavery societies.[87] By the 1920s migrants from Italy in the United States and Argentina increasingly took advantage of preferences for family reunification in immigration legislation to bring over female relatives, especially as more employment opportunities opened for migrant women and their daughters in light manufacturing, such as the textile and artificial flower industries, but also in tipo italiano sectors such as candy, tobacco, and vegetable and fruit canning. Indeed, relocation abroad became economically viable for Italian families as more wage-earning opportunities for women became available.[88] Furthermore, during the interwar years the now sizable second generation of Italians increased the number of women in migrant marketplaces. The consumer practices of second-generation Italians altered the global and gendered dynamics that defined migrant marketplaces as these new consumers brought their experiences growing up abroad to bear on their interactions with the foods and culinary traditions from their parents' homeland.

The war, combined with these demographic changes, proved instrumental in turning migrants toward consumption during the 1920s and 1930s. The transnational subsistence economies in which migrants were enmeshed had made Italians unreliable consumers of imports. Even among migrant families reunited abroad, low-wage jobs and suspicion of unfamiliar industrialized food products made most migrants conservative consumers. Already by the early twentieth century, Italian government representatives and migrant *prominenti* in the diasporas observed migrants' reluctance to spend money. One 1906 *Bollettino* report discussed the "sad and distressing fact" that every dollar remitted home "almost always represents the most humiliating subtraction of the most essential life needs" of migrants overseas. Suggestive of the way food was linked to notions of the body and racial fitness, the report highlighted how remittances affected the racial stock of Italians abroad: "Every dollar sent corresponds to a privation of food," leading to "a weakening of the race and a deterioration of an entire population."[89] Ironically, as Italian trade and migration promoters touted male migrant consumption as being key to Italian commercial success, they also sometimes joined U.S. American and Argentine elites in chastising migrants for their "stingy and self-abnegating" ways. The *Bollettino* reported on an unnamed U.S. congressman who described Italians as "birds of passage" who "scarcely spend one cent in our market, and hate all that is America except for the gold they take with them."[90] Similarly, José María Ramos Mejía, a leading Argentine physician and politician, denounced male migrants' reluctance to spend; after a full day of hard labor, "he returns straightaway to the stand or dwelling where he has wasted no time in hiding away his savings in some safe spot."[91]

Officials reporting from abroad saw in consumption a salvo for ameliorating anti-migrant prejudice, urging migrants to live among native-born workers "without stinginess and deprivation."[92] Gerolamo Naselli, from the Italian consulate in Philadelphia, contended in 1903, "The excessive love of savings is the only reason, I believe, that has blocked Italians from becoming popular in this country." He continued: "It is not, as some believe, job competition or other reasons that cause Americans to resist opening the doors to our emigration, but the fear of parsimony and the spirit of economy that distinguishes our emigrants." Unlike German and Irish immigrants who spent money in the United States, Italians sent it home to their families. "The American wants money earned in this country to be consumed here." He concluded, "There is not a doubt that if the emigrant, once here, settled for good and consumes all his earnings here, all the restrictive emigration legislation will lessen."[93]

While Naselli's optimistic predictions about immigration policy proved wrong, he anticipated correctly that if male migrants "settled for good" abroad—with their wives and children beside them—it would change Italians' spending and saving practices. Women's increased visibility during World War I made them a lucrative target market for the Italian-language press and for Italian companies and merchants, who began featuring female consumers regularly in their ads for imports and in discussion about spending and shopping more generally. The interwar feminization of consumption was not unique to Italy or Italian migrants, for it reflected a more global process during which women became gradually associated with consumerism.[94] Nineteenth-century industrialization had gendered consequences for women and men; production moved outside of the home and became linked to mainly male wage earners, whereas women became increasingly moored to the domestic sphere, where they engaged in unpaid reproductive labor and consumption for the household.[95] In the late nineteenth and early twentieth centuries, middle-class white women expressed their class and racial status through engaging in the public marketplace as buyers of industrialized goods.[96] While more a model than a reality for minorities and working-class households, by the 1920s the white, middle-class female consumer had become a generalized ideal in popular culture throughout cities in Europe and in industrialized and industrializing nations, including the United States and Argentina, where her presence simultaneously ignited debates about gender roles and the construction of national identities.[97] Women's magazines, in-house corporate advertising departments, and gradually more independent advertising agencies assisted in the construction of the female consumer, whose main contributions to the household economy were portrayed as rational spending on behalf of the family.[98] Also important in wedding women to consumption was the burgeoning field of home economics, which professionalized women's special role as efficient household managers whose purview included nutrition and cooking, child rearing, and cleanliness.[99]

Middle-class and affluent white women in industrialized countries remain the focus of research on the history of modern consumerism, a history that has linked women's buying and spending activities to the formation and consolidation of national identities.[100] And yet migrant women were also affected by the broader gendering of consumption; their marketplace interactions, however, reflected their experiences as transnational people whose consumer desires and identities transcended national borders while being affected by them.[101] Focusing on migrant marketplaces as global arenas complicates historical research that portrays consumption as a generally straightforward path toward nation building and migrant assimilation.[102] Unlike her Anglo-American and Argentine counterparts, however, the female Italian migrant consumer emerged out of transnational linkages to the Italian state during World War I. Supporting the homeland valorized overseas consumption of Italian products in ways that strengthened migrants' sense of Italianità during the interwar years while fostering a more distinct ethnic identity in their host countries.

During the interwar years, Italian companies and merchants capitalized on the increasingly powerful connection between women and consumption and on migrants' transnational ties. After World War I, ads in the Italian-language press portrayed women as the principal buyers of Italian imports and, through their consumer preferences, the main generators and sustainers of Italian identities among migrants in New York and Buenos Aires. Ads like the Cinzano publicity moved from the public, male-oriented working world of factories and exhibitions to more private scenes centered on the home. In 1927 an ad for Banfi Products Corporation, an Italian food-importing firm in New York, suggested that the consumption of Italian spirits preserved the Italianità of special holidays. "How difficult to pass another Christmas far from the land that nourished us as infants!" the ad began. "But—thank goodness—this year, at my house, there will be a purely Italian atmosphere . . . One hundred percent. Italian friends, Italian food, and, on the table, for the benefit of the stomach and for my delight, will shine five magnificent products from our Italy: Marsala Florio, Ramazzotti, Strega, Vermouth Cora, Fernet de Vecchi." Such publicity brought products into domestic arenas associated with family, merriment, and leisure in order to reaffirm the Italianness of consumption in migrant marketplaces. While by the early twentieth century tipo italiano manufacturers in the United States and Argentina had led the way in shedding regional distinctions in product labeling, after World War I growing numbers of merchants followed their lead by downplaying regional distinctions in the descriptions of imported foodstuff. As the Banfi ad illustrated, consuming imported liqueurs over the Christmas holiday kept both eaters ("*Italian* friends") and their meals ("*Italian* food") linked to an image of a unified homeland.[103]

Rather than highlighting Italy's production capabilities, many ads after World War I emphasized the importance of Italian consumers abroad in driving Italian trade. A 1925 Spanish-language ad for Cirio-brand canned tomatoes in *La Patria* shuffled viewers along the entire transatlantic commodity circuit, beginning with tomatoes from the "best lands in Italy" and ending around a dinner table where families "enjoy perfect health because they always use in their kitchens Cirio-brand tomato extract" (see fig. 17).[104] Publicity like Cirio's redirected attention from the starting point of the commodity chain, Italian factories and laborers in Italy, to the receiving end, migrant consumers abroad, while showing how diasporic eating bolstered the Italian food industry back home.[105] Similarly, an *Il Progresso* ad for Florio Marsala included photographs of male Italian workers attaching labels to bottles in the Florio plant in Sicily next to images of buxom peasant women harvesting grapes. The ad instructed readers that buying Florio Marsala not only conserved the Italianness of migrant's dining practices abroad but also sustained commodity chains that supported Italian workers in Italy.[106]

By the 1920s female consumers in migrant marketplaces—rather than the exclusively male jury of political and business leaders at international exhibitions suggested in turn-of-the-century trademarks—judged the quality of Italian imports while serving as Italian cultural envoys abroad. As gatekeepers of links between migration and trade, ads depicted women as possessing the instinct and authority to choose authentic, high-quality Italian imports over inferior substitute brands. A 1932 ad for Bertolli olive oil in Chicago's Italian-language *L'Italia* used a discussion

Figure 17. Ad for Cirio tomato extract, *La Patria degli Italiani* (Buenos Aires), September 3, 1925, 3. Courtesy of the Hemeroteca, Biblioteca Nacional de la República Argentina.

between two housewives, one knowledgeable and one inexperienced, to portray women as judicious custodians of Italian commerce while warning them against deception by dishonest retailers. The expert housewife scolds her friend, pointing to the English writing on the can indicating only a minimal amount of real olive oil. She asks, "You didn't realize that the grocer played a joke on you in that he sold you sesame seed oil flavored with the addition of the smallest amount of olive oil?"[107] She reminds her friend to demand only pure Italian olive oil from Italy. In contrast to early twentieth-century trademarks, which presented "Italia"—the embodiment of the Italian nation—guarding and facilitating Italian commerce, often from a celestial space above the Atlantic, these post–World War I ads depicted women protecting Italian trade and migration routes through their individual consumer preferences. As shown in an ad for Sasso-brand olive oil, publicity domesticated Italia, and the transnational economic spaces she oversaw, by reinventing her as a shopper, mother, and wife in stores and kitchens abroad. While the original Sasso trademark depicting the olive-harvesting woman in classical robes appears in small print on the can (see ch. 1, fig. 11), this postwar iteration presents a mob of female shoppers examining an oversized canister of Sasso oil (see fig. 18). Commercial iconography connected migrant women's everyday purchases—often represented as integral to the maintenance of Italian family traditions in the private domestic realm—to the very public sphere of transnational commerce and trade.[108] Women often symbolized the key and final point on an Italian commodity chain in which Italian laborers in Italy produced goods purchased by consumption-savvy migrant mothers and wives abroad.

While pre–World War I trademarks had created manly markets of emigration and trade by using male politicians, revolutionary heroes, and explorers in Italy and the Americas to condense time and space, interwar ads in the migrant press collapsed temporal and geographical divides through women consumers in homes and grocery stores. They showed migrant consumers actively engaged in

Figure 18. Ad for Sasso olive oil, *La Patria degli Italiani* (Buenos Aires), April 27, 1925, 4. Courtesy of the Hemeroteca, Biblioteca Nacional de la República Argentina.

transnational kin work: their purchases not only represented connections across Italian households in New York and Buenos Aires but also revealed generational links to families in Italy, where women used the same products.[109] Publicity for Bertolli olive oil, for example, implied that while women situated abroad could not physically participate in kin work across the Atlantic, they made imagined exchanges with extended family in Italy by consuming Italian brands. In a 1932 ad in Chicago's *L'Italia*, a woman buying olive oil from a grocer professes to always ask for Bertolli, "the purest, unsurpassed, made-in-Italy olive oil that my grandmother and mother used before me."[110] A 1916 ad for Sasso olive oil in *La Patria* called out to women, "Ladies! In Italy no food goods store lacks Sasso oil."[111] In ways that echoed the transatlantic solidarities between Italian women during wartime fund-raisers, postwar ads for imported Italian foods connected migrant households in the Americas to the kitchens and markets of their female relatives in Italy. Moreover, by identifying imports with Italian mothers, grandmothers, and sisters back home, ads constructed Italy as a consuming rather than exclusively producing economy. Migrant consumers saw in these ads not only male laborers working in Italy's industrial food sectors but also an increasingly consumer-oriented Italy of female buyers and food preparers who used the same ingredients and brands to build Italianità. This reflected actual changes in the eating habits of an expanding group of middle- and working-class consumers, mainly in northern cities, where industrialization, rising employment rates, and emigrant remittances boosted the standard of living, giving Italian consumers the means to spend more money on larger varieties and quantities of food.[112] Publicity for imports that portrayed Italian women in Italy with products and consumer expertise suggested that for both migrant women and women left behind, links between migration and commerce transformed women's position in a growing global consumer economy.[113]

Promotional material portrayed women using consumption not only to solidify transnational familial and commercial bonds but also to perpetuate distinctly Italian family structures revolving around women's domestic roles as mothers and wives. These ads resembled those for food products in English- and Spanish-language papers and magazines in the United States and Argentina, which frequently presented women using their consumerist skills to keep their families happy and healthy.[114] Like this mainstream literature, ads for Italian imports sometimes prescribed wives' agency in migrant marketplaces within the confines of husbands' palates, as in a 1915 Sasso ad that called out, "Mrs! Buy! Convince yourself that Sasso oil owes its taste to the best quality. . . . Prepare a meal with another oil that you have used and one with Sasso's. Your husband will tell you which is better."[115] Many ads targeted migrant mothers specifically, declaring that the purchase and use of Italian imported edibles allowed women to pass on generationally acquired nutritional and culinary knowledge originating in Italy. Brioschi Effervescent, a powdered Italian digestive aid, stated in its ad, "We believe that it is convenient to remain faithful to the discovery made by your grandmother and approved by your mother: when you need an antacid, or a refreshing thirst quencher, remember Brioschi Effervescent."[116] Italian companies capitalized on networks of food knowledge passed between generations of women to sell their products.[117]

Female migrant consumers began to replace the merchant explorers, military heroes, and notable men in turn-of-the-century export iconography. But they also began to take the place of migrant merchants and retailers themselves as custodians of migrant marketplaces. Earlier publicity in *La Patria* and *Il Progresso* spotlighted the names of merchants just as prominently, if not more so, than the brands of the Italian foods they sold. Italian businesses dealt with the complexities and financial risks involved in transatlantic commerce by giving exclusive import and distribution rights to individual merchants whose names became associated with the brand and its reputation abroad. This name recognition assisted companies to control and protect their merchandise as it passed through multiple hands and locations along international trade routes while helping guard against imitation, adulteration, and fraud. But the pairing of merchant and brand names also served an important cultural and economic function that facilitated migrant consumption in that it provided cautious migrant consumers a sense—however fictitious—of being shielded from the unfamiliar and sometimes hostile world of capitalist exchange and mass-produced foodstuff.[118] Migrant merchants as well as *tipo italiano* entrepreneurs employed their ethnic backgrounds as cultural and social capital to capture and sustain loyal consumers and labor markets, and migrants trusted them to help navigate novelties in migrant marketplaces. However, after World War I, well-known Italian imported brands became progressively unhinged from specific migrant sellers, as suggested by the Sasso, Cirio, and Cinzano ads. Merchants took a backseat while manufacturers made the brand name itself the publicity's main feature, tying these brands to images of consumers who were increasingly depicted as female. The advertising industry, still in its infancy in the early twentieth century, helped businesses speak directly to consumers by creating brand names accompanied by flashy imagery and persuasive messages aimed to both shape and reflect buyers' desires and feelings.[119]

While in both countries ads described women using foods from Italy to perpetuate transnational consumer patterns and identities, they also demonstrated differences in the way migrants used food consumption to express their incorporation into U.S. and Argentine societies. During the interwar years, publicity in *Il Progresso* began regularly referring to migrants' hyphenated identities as ethnic Americans. An ad directed at migrant mothers for Locatelli-brand cheeses in *Il Progresso* urged consumers to buy imported pecorino romano "so that a new Italian-American generation is able to perpetuate the tradition of taste, health, and nutrients."[120]

Similarly, a Caffè Pastene ad for imported Italian coffee, featuring a smiling young girl, read, "The good taste of eating, in dressing, and even in thinking is handed down from mother to daughter. The young girls of today—who will be the mothers of tomorrow—know that Caffè Pastene is good coffee—she acquires the taste of the mother, who after having tried many brands, comes to the conclusion that the most fresh and aromatic, the most exquisite is of course Caffè Pastene—an Italian type that is American."[121] Reflecting in part the growing second generation of Italian women who themselves were responding to Americanizing influences, merchants in the United States increasingly employed hyphenated monikers to describe Italians and the imported products they ate. Conversely, this hyphenization of Italian people and products in Buenos Aires was conspicuously absent in ads for Italian imports during the interwar years, notwithstanding the large second generation of Italians in Argentina by the 1920s. There, consumption kept female consumers, their families, and the cuisines they cooked Italian but rarely, if ever, "Italian-Argentine," demonstrating how differences in migrant incorporation and ethno-racial landscapes manifested in representations of identity making in the context of food consumption. Over the course of the twentieth century, both countries witnessed the slow evolution from regional to national designations of Italian people and products in migrant marketplaces; in both, sellers of imports continued to employ both national and regional labels to appeal to lingering local identities and migrants' rising awareness of a shared nationality to market their wares.[122] However, in the United States, but not in Argentina, the consumption of imports turned migrants into hyphenated ethnics during the interwar years.

*　*　*

World War I was a global event with long-lasting consequences for trade and migration patterns and for the consumer practices and identities of migrants abroad. Since the late nineteenth century, Italian leaders in Italy and migrant prominenti strove to wrench migrants from their village and regional ties and from the transnational family economies in which these ties were embedded in order to transform migrants into patriotic "Italian" consumers. However, it took Italy's entrance into the war to turn increasing numbers of migrants toward their homeland, culturally and commercially. Italian migrant communities organized and executed numerous campaigns to maintain international trade, collect and export monetary and material resources back to soldiers in Italy, and provide for their precariously positioned connazionali in the United States, Argentina, and Italy left behind by reservist husbands, brothers, and fathers.

During campaigns to help their homeland in need, migrants in Buenos Aires rallied more often and with more intensity around Italian imports than did migrants in New York. The wartime consumer identities they generated remained deeply linked to Italian-Argentine trade in Italian foods and other exports, and bulk agricultural goods such as wheat and wool that migrants sent home to Italy. In New York migrants used the more diverse resources at their disposal, which included some Italian imports but more frequently consumer items produced in the United States—including tipo italiano goods—to influence trade and consumption patterns. Migrants' wartime experiences did not completely unmoor migrants from their regional identities and transnational family economies, but neither did they foster complete incorporation into U.S. and Argentine society. Rather, consumption abroad—even of non-Italian products—in some cases strengthened migrants' commercial and cultural ties to their homeland while turning them into consumers of U.S. and Argentine goods.

These various transnational campaigns helped legitimize migrant consumption in the postwar years. By organizing, carrying out, and publicizing such campaigns, *La Patria* and *Il Progresso* defined Italians in Buenos Aires and New York as consumers whose monetary donations and purchases contained the power to influence international events. Women featured centrally in such initiatives, setting the stage for the feminization of consumption in migrant marketplaces after the war. World War I changed women's relationship with the migrant press when wartime campaigns launched women onto its pages as readers, workers, and consumers. Ads for Italian goods in newspapers during the interwar years connected migrant women's ordinary consumer choices to the preservation of Italian trade networks and to the maintenance of Italianità in grocery stores, around dinner tables, and at festive parties. Publicity for Italian products, as well as the identities they generated, became increasingly anchored simultaneously to consumption and to women.

The weakening of east-west transatlantic flows between Europe and the Americas during the war allowed for the strengthening not only of tipo italiano industries abroad but also of north-south hemispheric commercial ties between North and South America as U.S. politicians and industrialists hoped to fill in a gap left by Europeans in Latin America. During the interwar years, these ties between North and South America became more visible, especially in the Italian-language press in Buenos Aires, where savvy U.S. companies, having finally discovered migrant consumers in New York, began reaching out to Italians in Argentina.

CHAPTER 5

Reorienting Migrant Marketplaces in *le due Americhe* during the Interwar Years

During World War I, as geopolitical alliances shifted global associations among nation-states, Buenos Aires' *La Patria* debated whether Argentina should pursue *panispanesimo* by aligning with its former colonizer, Spain, or instead accept *panamericanismo*, led by its northern neighbor, the United States. Argentina, the paper argued in July 1917, should choose Spain as its ally. Not only had Spain remained neutral during the war—whereas the United States had entered on the side of the Allies by April—but U.S.-controlled Pan-Americanism increasingly threatened Argentine economic, political, and cultural independence. Comparing the Monroe Doctrine to German imperialist aims in Europe that had instigated the global war, the article concluded, "Panispanesimo means frankly neutrality, in opposition to panamericanismo that rings out—and how powerful a sound—intervention."[1] The Italian Chamber of Commerce in Buenos Aires agreed. While their counterpart in New York had worried that Italy's wartime moratorium on exports would advantage migrant manufacturers of tipo italiano foodstuff to the detriment of Italian trade, in Buenos Aires merchants feared that such prohibitions fostered panamericanismo. In the absence of Italian imports, U.S. companies would step in to conquer South American markets.[2] Simply put, panamericanismo threatened the consumption of Italian foods in Buenos Aires' migrant marketplace.

World War I, especially women's war-related work, helped normalize consumption and cultivate a more united national identity among migrants in New York and Buenos Aires as a transnational duty to Italy. While ushering in the Italian female consumer in migrant marketplaces overseas, the war also intensified a global reorientation in trade and migration patterns that affected migrants' consumer options and solidarities. As the *La Patria* editorial suggests, the war escalated long-standing discussions among migrants in Buenos Aires about U.S. expansion in Latin America and its influence on transatlantic connections between Italian people and products. For decades, migrants in the United States and Argentina had formed connections to each other and to Italy through Italian foods. But they also generated alternative subjectivities based on experiences and items particular to the western hemisphere. This included not only tipo italiano foodstuff but, increasingly, foods and other consumer goods produced by multinational U.S. firms. During the interwar years, U.S. food companies—which had previously disregarded migrants as a specific market within the larger U.S. consumer society—discovered Italians in New York's migrant marketplace as potential buyers of factory foods. After the war these mass-produced goods—processed meat, canned vegetables and fruits, tobacco products, pasta, liqueurs, sweets, and other industrialized edibles—appeared more frequently in New York's *Il Progresso* and other Italian-language papers in the United States. But they also emerged in Buenos Aires as U.S. industrialists, trade promoters, and investors turned with unprecedented enthusiasm to Latin American countries as importers and consumers of U.S. merchandise. As these U.S. enterprises moved their markets and factories south to Argentina, they too targeted Italians as consumers.

During the interwar years, the United States' increasing interest in Latin America as a receiver of U.S. exports and capital investment, a worldwide depression, intensifying restrictions against mobile people and goods, and rising nationalism changed the global geography of marketplace connections. To the consternation of Italian prominenti elites in Buenos Aires, continued Italian migration to Argentina collided with escalating north-south incursions of U.S. consumer products.[3] This collision, they feared, compromised the historic ties between Italian people and products that constituted migrant marketplaces and that had kept them critical sites for Italian nation and empire building. As they suspected, panamericanismo after the war signified U.S. unilateral intervention rather than equitable economic and political cooperation among countries of the western hemisphere. Pan-Americanism arrived in the form of U.S. imported foods and other consumer items in Argentina, which, while deepening hemispheric ties between migrant marketplaces of le due Americhe, simultaneously jeopardized the bonds between Italian migrants and trade products that had been fortified during the war.

Combating Yankee Imperialism: Migrants' Reactions to U.S. Expansion in Latin America during the Interwar Years

Well before World War I, when consumer items produced in the United States increasingly joined together Italians in New York and Buenos Aires, migrants fash-

Figure 19. Ad for Pineral *aperitivo, La Patria degli Italiani* (Buenos Aires), September 20, 1905. Courtesy of the Hemeroteca, Biblioteca Nacional de la República Argentina.

ioned diasporic connections linking consumers in the United States, Argentina, and Italy. A 1905 advertisement in *La Patria* for Pineral-brand aperitivo made by the Pini brothers, migrants from Lombardy, commodified hemispheric ties produced by the past migrations of famous Italian men (see fig. 19). The ad depicted a toga-clad Italia against the backdrop of a city skyline with her arms around an oversize bottle of Pineral liqueur. The right-hand corner features a scene at the candle factory in Staten Island in New York City where Italian nationalist leader and political exile Giuseppe Garibaldi worked for the Italian inventor Antonio Meucci in the early 1850s. The image of the candle factory was one of sixty-four illustrations in an album given to the Pini brothers' customers in honor of Italy's Venti Settembre holiday, commemorating Italian unification.

The Pineral ad sheds light on the global context within which prominenti journalists, advertisers, and businesses in Buenos Aires portrayed migrant consumers
in the early twentieth century. Garibaldi's own worldwide peregrinations made
him an especially powerful symbolic vehicle for connecting Italians in Argentina
both to their homeland and to Italians in the United States. Garibaldi, "the hero of
two worlds," led the Italian legion during post-independence civil wars in South
American countries during the early nineteenth century. After returning to Italy
and leading an unsuccessful republican defense of the Papal States against French
forces, Garibaldi went abroad as a political exile in New York, where he worked
with Meucci. From New York he traveled again to Latin America, eventually ending up in Peru, where he took command of a merchant ship that sailed from South
America across the Pacific and Indian oceans to the Philippines, southeastern
China, Australia, and then back to the U.S. East Coast, after which he returned
home to lead the Second Italian War of Independence.[4] The ad featuring Garibaldi
and Meucci, another Italian nationalist sympathizer settled abroad, used migrant
consumption of a tipo italiano beverage—made in Argentina but with ingredients
imported from Italy—to link migrant marketplaces in Buenos Aires to both Italy
and New York.

While the Pini brothers celebrated connections between the United States and
Argentina produced by migration, as the twentieth century progressed, Italians
in Argentina increasingly condemned relations between the United States and
Argentina brought on by U.S. economic expansion. Until World War I, migrant
prominenti in Buenos Aires had directed most of their concern toward European
countries as probable obstructions to Italian migration and trade. In 1901 the Buenos Aires–based Italian Chamber of Commerce expressed angst that Spain, having lost its colonies in the Caribbean and Pacific to the United States after 1898,
would refocus its efforts on strengthening commercial bonds to Argentina.[5] As
seen in debates over tipo italiano foods arriving from Europe in the early twentieth
century, migrants viewed Spain as a particularly menacing competitor; Spain's
long-standing imperial connections to Latin America, the large number of Spanish migrants that continued to enter Argentina, and Spain's export market, which
sent abroad foods similar to those produced in Italy, made Spain's rival migrant
marketplaces especially alarming. However, by the eve of World War I Italians in
Argentina had gradually turned away from Spain and other European countries
and toward the United States in their efforts to protect linkages between Italian
trade and migration.

Already by the turn of the century, journalists writing for La Patria, as well as
migrant merchants and businessmen associated with the Italian Chamber of Commerce, looked with skepticism at the expanding presence of U.S. investment, consumer goods, and other cultural products in Argentina. The Monroe Doctrine,

declared by President James Monroe in 1823 to protect both the newly independent countries of Latin America from European meddling and U.S. political and economic interests in the western hemisphere, received special criticism by Italians in Argentina. This was especially the case as Theodore Roosevelt and subsequent U.S. presidents employed the doctrine to justify increased U.S. activity south of its borders.[6] *La Patria* reminded its readers that the Monroe Doctrine encouraged Argentina to cut cultural and economic ties with Europe, essentially characterizing bonds between Italian migration and commerce as a danger rather than an aid to Argentine nation building.[7] In its coverage of Pan-American conferences held in the early twentieth century, *La Patria* criticized what it viewed as the United States' veiled acquisitive aims. At the fourth Pan-American Conference in Buenos Aires in 1910, *La Patria* argued that the "American spirit, American ideal" that the United States professed to cultivate through these meetings was in reality an "insidious movement" to discipline Central and South American countries into resisting European influence and therefore eliminating commercial competition in Latin America. The Monroe Doctrine, "waved around at every occasion," lined up the Americas against the "old world" and at the sole command of the United States, a reality that European countries, too distracted over fighting among themselves, did not understand.[8] The paper described U.S. goals in Latin America as "essentially imperialistic, occupying, assimilating, aggressive," citing Cuba and the Panama Canal as telling examples. Pan-American conferences had "become for the United States a means of propaganda and predominance," and other Latin American nations needed "to keep their eyes open," cautioned the paper.[9] *La Patria* praised articles in *La Nación* and *La Prensa* that confronted claims by the United States that Europe and Europeans, including Italy and Italians, symbolized threats to the integrity of Latin America.[10]

Swelling xenophobic sentiment in the United States gave Italians in Buenos Aires a platform for highlighting differences between the United States and Argentina that challenged notions of solidarity upon which panamericanismo rested. An editorial titled "The Lacuna of the Pan-American Conference" identified migration as conspicuously absent from the themes discussed at the 1910 meeting. The United States, believing their country did not need anymore migrants, increasingly looked to prohibit them. Echoing claims made by many Argentine statesmen, *La Patria* asserted that Latin America continued to consider migration from Europe "the natural foundation of prosperity and economic, political, and moral progress of America in general." In a manner that completely excised non-European migrants from discussions of nation building in either North or South America, the paper concluded that Argentina, unlike the United States, understood that regular European migration safeguarded the "American spirit" by neutralizing "every sentiment, be it Anglo-Saxon, Germanic, or Latin."[11] Nativist attitudes toward

European migrants allowed *La Patria* to differentiate rather than unite the United States and Argentina, calling into question whether U.S.-style panamericanismo advanced Argentine interests.

Visits by Italian and U.S. leaders provided *La Patria* occasions to rebuke U.S. commercial designs in Argentina and reminded readers of the demographic, commercial, and moral clout Italians held in their adopted country. In 1910 *La Patria* praised a speech given by the Italian criminologist Enrico Ferri at the Teatro Odeón in Buenos Aires in which he asserted that panamericanismo was inseparable from "North-American hegemony" and "North-American imperialism."[12] That same year, a trip by U.S. politician William Jennings Bryan elicited criticism from the paper, which claimed that while Bryan spoke publicly about cooperation, "in reality his was a trip of commercial exploration" for the exclusive benefit of the United States: "Until now the United States has neglected the markets of Latin America. Now it has decided to conquer, with its industrial production, this southern part of the continent." *La Patria* interpreted Bryan's visit as the beginning of a dangerous episode in Argentine history in which commercial relations with the United States would take precedence over those with Europe. The editorial concluded, "If European industrialists—and signally our Italians—will not know with all the means possible to beat in quality and price the terrible North American competition, they will be subjected to many bitter delusions in America."[13] The United States, *La Patria* asserted, used the "imaginary danger" of European intervention in South America to strengthen its hold on the continent and to "avert attention from that which is the true danger, the invasion of North America that threatens these Republics."[14]

While denouncing U.S. economic presence in Argentina, *La Patria* confidently pointed to Italian dominance in Argentine trade and industry. The paper reminded readers of Italians' great contribution to Argentina's national progress, a fact mentioned by many foreign visitors to Argentina, including a Russian who asserted, "If we could calculate all that Italian hands have turned out, regarding products, to the markets of the world, we would understand what enormous debt of gratitude European civilization owes to these audacious and valorous pioneers."[15] On the occasion of Argentina's centennial celebrations, *La Patria* proudly insisted that recent attempts by U.S. Americans, French, English, and Germans to monopolize Argentine commerce amounted to imitations of what Italy had already accomplished in its commercial relations with Argentina.[16] The Italian-language press used Pan-American conferences and the Argentine centennial to draw attention to the "intimate relations of interests and sentiments that unite Italy to the great country where its children find, as is said, a second homeland, and carry in exchange a continuous, precious contribution of the vigorous and refined Latin blood, of commercial and industrial genius and of honest work."[17]

"Latin blood," shored up by decades of recurring Italian migration, functioned as a stopgap measure against U.S. imperialism. A 1906 article in *La Patria* lauded a publication by Enrico Piccione, editor of *El Pensamiento Latino* in Santiago, Chile, in which he argued for a "Latin alliance" based on "blood ties" to stop U.S. encroachment. The strengthening of emigration and commercial bonds between Italy and other "Latin countries of Europe" would fend off German, but especially U.S., imperialism in Latin America, considered uniquely nefarious for its less overt qualities.[18] Latinità served as a means through which migrants manufactured ties between Argentines and Italians around food consumption, but it also offered a conduit for binding the United States and Germany together as a racialized Other—Anglo-Saxon and "Yankee"—who endangered Argentina racially and culturally. In 1925 *La Patria* covered the travels of Dr. Guglielmo di Palma Castiglione, Italian emigration administrator and senior officer of the International Labor Organization, who described the important protective "mission" Italians undertook in "young Argentina" besieged by "the invading Britannic, Teutonic and Slavic races." Employing patronizing and patriarchal language to describe Argentina as requiring safekeeping and tutelage, he added, "It is impossible that Argentines, intelligent and nobly ambitious as they are, would not appreciate the beauty of a mission so civil, a mission to help them realize their potential as youngsters, guided by us elders, to carry them toward peace, toward wealth, toward happiness." He assured readers that "Italians are embraced like siblings by the Argentines," which is why Italians preferred Argentina to the United States, even if they made less money in South America. Unlike the United States, Argentina "speaks to his [the Italian emigrant's] heart and adjusts to his spirit."[19] Notions of race, family, and patriarchy blended in discussions about the role Italian migrants played in containing U.S. commercial power in the western hemisphere. Italians' continued celebration of Latinity appeared as more Argentine thinkers began amending and even repudiating Europe as a racial and cultural yardstick for Argentine nation building; they championed Argentina's "Latin" heritage of racial hybridity as a contrast to North American racism, without, however, fully embracing *mestizaje* as official state doctrine.[20]

While migrants condemned the United States for jeopardizing Italy's stronghold in Argentina, they simultaneously and paradoxically expressed a sense of curiosity and even desire for U.S. consumer products and the modernity and affluence they symbolized. *La Patria* characterized consumerism in the United States—"the land of the dollar," where "the greed for money affects everyone"—as exerting a corrupting influence on Latin America, but it also included articles that praised the variety and quality of items available for purchase.[21] For example, after a visit to the United States, a *La Patria* journalist wrote glowingly about the wide array of goods—from ice cream sodas to sweets to medicines—that were obtainable at U.S. pharmacies, goods representing abundance, choice, and prosperity.[22]

These mixed sentiments paralleled debates among Latin American statesmen who, in their struggles to define themselves individually as nations and collectively as a continent, produced varied and shifting reactions toward enlarging U.S. investment and political sway in the western hemisphere.[23] By the turn of the century, a number of Argentines—particularly the growing middle classes at the helm of commerce and industry—sought to associate themselves not only with Europe but also with the increasingly powerful United States. In 1888 a group of industrialists in Argentina described Argentines as the "Yankees of South America" and U.S. Americans as Argentines' "big brothers," indexing how the country followed the United States in its "tireless work of progress."[24] Furthermore, to some nationalists hoping to diversify the Argentine economy and make it more self-sufficient, U.S. investment strategies and tariff policy offered productive industrial models.[25] At the same time, however, rising numbers of Argentine politicians expressed concern about U.S. hegemony in discussions over Argentine political and economic autonomy. In 1902 Argentine jurists passed the Drago Doctrine, which declared that foreign nations could not employ military force for debt collection against Latin America countries. Written mainly in response to the 1902 British, German, and Italian blockade and seizure of Venezuelan ports, the proclamation also functioned as a response to the Monroe Doctrine and as a stance against U.S. hemispheric dominance.[26] Incongruous images of the United States as a paragon of industrial progress and as a nefarious hemispheric bully circulated in Argentina among Argentines and Italians.

These condemnatory but often ambivalent attitudes increased as shifts in global trade made Argentina and other countries of the western hemisphere increasingly lucrative trade partners of the United States. In 1900 the United States had done most of its trade with Europe. On the eve of World War I, however, the United States had shifted its commercial gaze from Europe to the Americas and to Asia, receiving half of its imports from the western hemisphere and Asia and exporting many of its manufactured products and agricultural staples to these same areas.[27] Wartime slowdowns in investment and commodity flows between Europe and Latin America also intensified this shift. Before the war, Argentina had received most of its manufactured goods from Britain and other European countries in exchange for Argentine beef, wheat, wool, and other agricultural products.[28] World War I, however, dealt a heavy blow to European trade and investment capacities. This included the transatlantic trade in Italian foods and beverages, which slowed during the war, in some cases quite dramatically, as the Italian government redirected foods previously produced for export toward the home front. For example, Italy sent one-seventh the amount of wine to Argentina in 1919 as it did in 1915, and by the end of the war Italian wine imports to the United States had dropped by almost half.[29] This attenuation in transatlantic commerce fostered stronger economic partnerships between the United States and Latin America countries. During the war the value of Argentina's total exports to the United States quadrupled while the value of U.S. exports to Argentina experienced an almost sixfold increase.[30]

During the interwar years the increasing north-south orientation of trade patterns continued, with North and South America, as well as Asia, gaining ground on Europe as the United States' most important trade partner. Increasing numbers of U.S. companies looked for alternatives to the already saturated European consumer market, turning toward the less industrialized countries of Latin America and Asia as sites for consumption of U.S. products and capital investment. Latin American nations slowly incorporated their economies into North American commodity circuits as nations south of the U.S. border swapped raw materials for U.S. manufactured goods. By the end of the 1920s the United States received more unprocessed agricultural staples from South America than it did from anywhere else in the world.[31] In return, Latin America countries purchased U.S. machinery, industrial products, and foodstuff; by 1925 U.S. finished manufactured goods made up 76.8 percent of all U.S. exports to the region.[32] Argentina received the largest share of the United States' total exports to the continent, at 37 percent by the end of the 1920s, with Brazil next in line at just 20 percent.[33]

U.S. manufacturers, in their yearning for new markets, facilitated these transformations in geographies of trade. Leading U.S. industrial and business promoters championed Latin America as the United States' new commercial frontier. In 1938 James Farrell, past president of U.S. Steel and chairman of the National Foreign Trade Council, an organization formed in 1914 to promote U.S. foreign trade, advocated for a stronger commercial relationship among countries of the western hemisphere. Encouraging U.S. business leaders to turn their attention away from Europe, Farrell told U.S. industrialists, "We must pause in our gazing from East to West and follow instead the magnetic course of North and South."[34] While many U.S. companies exported their products to Latin America, other U.S. businesses established manufacturing branches there to take advantage of cheaper labor, to gain access to raw materials, and to avoid high tariffs against imported items. Professor Dudley Maynard Phelps's 1936 study on the "industrial migration" of North American manufacturing subsidiaries to Latin America shows that Argentina received by the far the majority of capital by U.S. firms after World War I. By the mid-1930s there were thirty-one branch plants of North American companies in Argentina as compared to nineteen in Brazil, the second most popular Latin American location for U.S. investment. Phelps estimated that by 1936 the total capital invested in U.S. manufacturing subsidiaries totaled $164,721,000 in Argentina and $47,165,000 in Brazil.[35] These companies represented a variety of industries that produced goods for domestic consumption within Argentina and for exportation abroad, such as automobiles and tires, construction materials and equipment, food and meat products, petroleum and pharmaceuticals, and phono-graph and radio equipment. Ford Motor Company, for example, opened a subsid-iary in Argentina in 1913, and in 1915 Goodyear Tire Company founded a branch in Argentina and established a manufacturing plant there in 1930.[36] The United Shoe Machinery Corporation installed a factory in Argentina in 1903, and three years later the Singer Sewing Machine Company also opened a branch.[37]

U.S. food, beverage, and tobacco corporations counted among some of the larg-est retail outlets and subsidiaries in Argentina. Chief among them were two leading U.S. meatpacking houses: Chicago-based Swift and Company and Armour and Company. Swift purchased the largest meatpacking plant in Argentina in 1907, and a couple years later Amour, Swift, and Morris and Company combined to buy out another major Argentine processing plant.[38] By the interwar years, U.S. and British meatpacking conglomerates battled against each other for control of the Argentine beef industry and against a number of Argentine interests, especially ranchers, who pressed the Argentine government to regulate the foreign-owned packers.[39] Meatpacking firms in and around Buenos Aires employed a large num-ber of Italian workers and administrators in their slaughterhouses and processing plants, such as Swift as well as Armour's large firm in the La Plata neighborhood of Berisso.[40]

Against the background of such changes in global trade and investment pat-terns, migrants' earlier twentieth-century concerns about panamericanismo took on new urgency during the interwar years. Italians in Buenos Aires discussed the rising presence of U.S. consumer goods, factories, retail outlets, and banks in Ar-gentina as threats to migrant marketplaces. And yet if reorientations in commerce worried migrants, reorientations in transatlantic migration patterns, which after more than thirty years made Argentina again the number one receiver of Italian migrants in the western hemisphere, gave trade promoters hope for strengthening links between migrants and food exports. While sustained migration excited Italian trade promoters in Buenos Aires, it also gained the attention of U.S. companies interested in enlarging their markets. After discovering eager migrant consumers in New York, they too would look southward and see Italians, as consumers and producers, in Buenos Aires' migrant marketplace.

Shifting Migration Flows to *le due Americhe* during the Interwar Years

During the interwar years the strengthening of trade and consumption patterns between the United States and Argentina combined with changes in transatlan-tic migration flows to transform migrants' options for identity making relative to consumption in migrant marketplaces. U.S. "gatekeeping" policy, a process that had begun in the late nineteenth century with the passage of the Chinese Exclusion

Act, intensified after and as a result of World War I, when nationalist and xeno-phobic sentiments reached new heights. Nativist legislators passed a number of restrictive immigration laws during the war that culminated in the Immigration Act of 1924, which perfected Asian exclusion and radically reduced the number of Southern and Eastern Europeans through a racially discriminatory quota system.[41] As a result of the act, in 1925 only 29,723 Italians entered the United States, down from 349,042 in 1910.[42] Simultaneously, however, the act instigated a move toward more gender-balanced migrations to the United States by allowing women and children, as dependents of male migrants, entry outside of the quota system.[43]

As had earlier restrictive immigration legislation, the passage of the 1924 act reverberated globally to affect other migrant-receiving nations in the western hemi-sphere.[44] Already in the late nineteenth century the United States had pressured Canada and Mexico to enact racist immigration laws modeled after U.S. policy in an effort to prevent Chinese migrants from entering the country through its north-ern and southern borders. During the 1920s and 1930s South American countries such as Peru and Brazil passed laws against Asian migrants while reaffirming the whiteness of Europeans by continuing to keep its doors relatively open to them.[45] Argentina begin enforcing restrictive administrative measures—but not formal exclusionary policy—directed at particular ethnic groups, first in the 1920s against *gitanos*, a group associated mainly with the itinerant Roma people from Central and Eastern Europe, and then in the late 1930s against Jewish refugees. However, un-like in the United States, organized labor in Argentina, made up mainly of Southern European migrants, remained relatively weak and collectively did not advocate for ethnic restriction against their continuously arriving co-nationals. Argentine politicians' insistence on its nation's racial whiteness, its long-standing history of and preference for European migrants, and the country's geographical distance from the United States made racial selection in immigrant policy less pressing, and the country kept its doors open to migrants during the interwar years.[46] Migration statistics suggest that Italians responded to the closing of U.S. borders by redirecting their paths southward. In 1922 Argentina overtook the United States as the most popular destination for western hemisphere–bound Italians. In 1923, two years after the passage of the Immigration Restriction Act of 1921, which first established the quota system as a temporary measure, more than twice as many Italians migrated to Argentina as to the United States. Argentina would continue to receive annually about double the number of Italians than the United States for almost a decade.[47]

This shift in Italian migration from the United States to Argentina in the post–World War I decades combined with two events to reduce Italian migration globally, especially after 1930. Starting in the mid-1920s, Italian prime minister Benito Mus-solini and his nationalist supporters discouraged Italian migration while express-ing interest in how Italians who were already abroad might assist in Italy's larger imperial plans.[48] Second, the 1930s depression curtailed international migration as employment opportunities dwindled abroad and as countries instituted more restrictive migration procedures, as well as higher tariffs, to protect domestic labor and industry against foreign threats in the form of cheap labor and commodities.[49] The early 1920s, therefore, witnessed an almost decade-long shift in transatlantic Italian migration from North to South America and then, especially after 1930, an attenuation in transatlantic Italian migration and trade to both countries.

Trade promoters in the United States after World War I saw connections between ongoing migration to the Americas and increasing U.S. investment abroad. Dudley Phelps, identifying conditions that either promoted or hindered U.S. economic growth in South America, argued that business achievements depended in part on each country's history of migration. He pointed to Argentina and Brazil—the two most important receivers of U.S. capital in South America—as foils to make his point. In Argentina, where by the mid-1930s U.S. firms had found the most success, racial uniformity, he mistakenly believed, had been achieved by generations of white European arrivals and mixing between migrant and indigenous groups. Disregarding Argentina's history of extermination and subjugation campaigns against native and mixed-race peoples, Phelps asserted that the country's racial homogeneity created a level of political stability that was necessary for U.S. investment. Brazil, on the other hand, was a problematic "racial melting pot" of Europeans, Asians, blacks, and indigenous people, a pot made particularly "interesting" due to its large Japanese population.[50] Similarly, William Ricketts, an executive with J. Walter Thompson Company—one of the first and largest advertising agencies in the United States—also looked with suspicion on Brazil, where the majority of the population consisted of "Portuguese, Negro, and Indian extraction" whom he described as "poor, ignorant, and lazy." Still, he depicted the cities of Rio de Janeiro and São Paulo as particularly promising locations for advertising work because of the large number of "Italian and German colonists."[51] Phelps and Ricketts implied that if Brazil had done a better job regulating its borders and the distribution of migrants within those borders, as had Argentina, the country would have made a better environment for U.S. capital and commodities.[52]

To Phelps and Ricketts, then, the performance of U.S. enterprises in South American nations largely depended on the racial "quality" of migrant and native populations. Echoing eugenicists in the early twentieth century, such business leaders viewed migrants in countries south of the U.S. border through a hierarchy of desirability that was based principally on race and through their assumptions about and experiences with migrants in the United States. Referring to South American countries, Phelps wrote, "Each country presents a composite of races and nationalities, much like in the United States, but is more truly a composite, for there has been less racial segregation and color discrimination, and the blending process has been carried much further."[53] He described the makeup of employees at U.S. manufacturing branches in Argentina, including one factory where 20 percent of laborers were Italian; another 20 percent German, Lithuanian, and Polish; and a smaller number Russian, Scandinavian, Spanish and Argentine, as proof that U.S. businesses could profit from a multinational workforce dominated by Europeans whom they considered white.[54] Countries like Argentina, he argued, which contained a higher percentage of Europeans in the nation's overall foreign-born population, increased the likelihood of industrial success. This was partly because U.S. firms preferred to employ European migrants as overseers, executive personnel, and salespeople over native-born South Americans, especially in countries with significant black and indigenous populations. Since the early twentieth century, as historians such as Julie Greene and John Soluri have shown, U.S. economic ventures in Latin America—especially banana production in Honduras and Ecuador and the building of the Panama Canal—depended on a large multinational labor force of Latin American, indigenous, West Indian, European, and U.S. American workers, whose jobs, pay scales, and lifestyles were determined through an inequitable racial system.[55]

Although Phelps considered Italians to be of "supervisory caliber" in Argentina, higher-paying management positions in urban factories of the United States were most often reserved for native-born white workers, including the descendants of Northern European migrants.[56] This incongruity was not lost on Italians in Buenos Aires, who during the interwar years discussed U.S. immigration legislation, racism, and job discrimination against "Latin emigrants" like themselves with disgust.[57] Even while companies touted European migrants—even Southern and Eastern European workers—as ideal laborers in South America, calls for restriction against these very groups resulted in discriminatory immigration legislation that limited their numbers in the United States and propelled migrants southward to countries like Argentina. In the mid-1920s restrictive laws in the United States and continued migration to South America became an opportunity for Italians in Argentina to reassert their commercial and cultural importance while reinforcing cultural and racial ties of Latinità between Italians and Argentines. *La Patria* and the Italian Chamber of Commerce employed Latinità mainly as an integrative force for producing similarities in identities and consumer practices between Italians and Argentines; conversely, in their attempts to discern what racial amalgamations translated best to U.S. commercial development in South America, U.S. business leaders dissociated Italians, as Europeans, from South Americans, Asian migrants, and indigenous workers, who were considered inferior. In fact, promoters of U.S. trade and business sometimes referred to Italians, but not Asians, in South America as "colonists" rather than "immigrants," perhaps because "immigrant" carried a negative connotation in the United States, employed by nativists to advocate for the exclusion of Asians and, eventually, Southern and Eastern Europeans.[58]

In their attempts to describe and evaluate the racial makeup of South American countries, U.S. business leaders exposed their confusion over how to make sense of their southern neighbors' complex demographic landscapes and their consequences for U.S. business endeavors in the region. They required a certain terminological flexibility as they tried to reconcile their own understandings and experiences of migration with a different pattern in countries like Argentina and Brazil. During the mid-1920s and 1930s, as U.S. companies turned their attention toward Argentina, migrants, products, and capital converged to connect migrant marketplaces in New York and Buenos Aires.

Targeting Migrant Consumers: U.S. Commodities in the Italian-Language Press

J. Walter Thompson's Russell Pierce, writing from Argentina in 1929, noted that despite Argentina's continued mix of "modern" and "primitive" peoples, the country was a "market that takes one's breath away." Pierce, director of the agency's Buenos Aires office, described that market: "Almost 11,000,000 souls now. A decade and a half ago there were 7,800,000. And each day a European boat brings in the hardy immigrants from Southeastern Europe who are now moving out into the farm lands to build homes, raise families, crops, and to *stay in the country*."[59] Pierce's special emphasis on the permanency of migrant communities in Argentina suggests an implicit critique of labor migrants in the United States whose high return rates and even higher remittance rates signaled to many an unwillingness to assimilate as consumers and future citizens. Pierce believed that European migrants in Argentina were more likely to remain in the country and to therefore serve as a trustier and more stable consumer base, even though return rates for Southern and Eastern European migrants in Argentina were about the same as those to the United States in the early twentieth century.[60]

It is not surprising that Pierce viewed "hardy migrants" in both Argentina and the United States as business assets during the interwar years. Sellers of Italian imports and tipo italiano foodstuff had been targeting Italian migrants in the Italian-language press since the late nineteenth century. And yet until the 1920s U.S. businesses largely ignored migrant consumers in New York; while they sometimes publicized in the Italian-language press, U.S. companies often advertised in English rather than Italian, and they tended to present images of white, middle-class consumers for all buyers—migrant and native-born. By the late 1920s, however, business and advertising firms like J. Walter Thompson had discovered Italians in U.S. cities such as New York, Chicago, and San Francisco. They began viewing such migrant marketplaces as lucrative consumer markets, different from Anglo-American markets yet equally accessible if the right techniques were employed.

Already by 1914 the advertising industry journal *Printers' Ink* reported that 20 percent of all advertising appeared in foreign-language newspapers.[61] In the early 1920s J. Walter Thompson began instructing its employees on strategies for reaching migrant buyers. A number of articles ran in the company's monthly newsletter under the umbrella title "Unrecognized Cities in the United States." The series began in 1924 when the agency reprinted a translated article from New York's *Il Progresso* titled "The Italian City Larger Than Rome." The article argued that even though U.S. exporters looked to Italy for consumers and business opportunities, "right here in America, before our very eyes, within easy reach of many metropolitan sales forces, there is an Italian city larger than Rome. It is the Italian section of New York City." Insisting on the slow, if gradual, upward mobility of Italian migrants and their children, which made them "a tremendous consuming market" by the 1920s, the article continued:

> It is more than an Italian City because it combines all the buying power and tastes of the Italian plus the greater average wealth and ability to spend money which favors the Italian in New York. . . . The Italian father is generally the father of a good sized family. He takes pride in his children. He and his family love music. They love nice things to wear. They take delight in a well set table. Their hospitality is proverbial. They not only live—they live well. This is a splendid market; a market in which the necessities of life find a ready sale and in which the luxuries, too, find quick and ready sale. While the frugality of the Italian shopper is too well known to require discussion, that frugality only tends to make it possible for the Italians to spread out his purchases and acquire much more than the bare needs.[62]

The characterization of Italians who "live well" was a far cry from earlier condemnations of excessively stingy, savings-oriented labor migrants by U.S. politicians and some Italian *prominenti*. By the interwar years, J. Walter Thompson had reevaluated migrants as consumers, consumers whose frugality was reimagined as giving U.S. industry an advantage in the long run. After the *Il Progresso* reprint, each month the newsletter ran an article on an "unrecognized city" in the United States such as "The Jewish city of New York" and the "Polish city of Chicago"— each a "worth-while market"—and provided its employees with cultural, religious, economic, and racial "facts" about the communities for businesses interested in attracting migrant consumers. As they had with Italians, J. Walter Thompson characterized migrants as passionate and well-equipped consumers, even insisting that in New York and other cities foreigners were better consumers than native-born U.S. Americans.[63] The newsletter noted that "Chicago Poles are heavy buyers" and that in Milwaukee and Detroit a person of "Polish extraction rapidly acquires substance and becomes a liberal consumer of everything the market offers"; the Jewish in New York represented a "rich market" with the "power of

tremendous sales"; the Swedes in Chicago "have the incomes to gratify their desires"; and the "Czechs of Chicago represent a real purchasing power."[64] This image
of migrants consuming with abandon, wholeheartedly embracing U.S. consumer
culture, represented an advertising ideal rather than the actual lifestyles of many
working-class migrants. While many Italians in New York found themselves in
less precarious financial positions, especially as their U.S.-born children found
jobs in low-skilled industrial work and, increasingly, in semi-skilled and white-
collar positions, their consumer habits remained constrained by the economic
needs of the working-class family, slowly rising wages, and high unemployment
rates.[65] For many migrants, economic goals continued to be oriented eastward,
toward family and relatives in Italy, even while the war, restrictive immigration
legislation, and the global depression curbed people and capital flows after 1925.
Italians certainly expressed their aspirations for social mobility in part through
participation in the 1920s consumer revolution, but their involvement did not
signify a wholehearted embrace of Anglo-American, middle-class culture or a
complete disappearance of ethnic traditions; instead they accepted industrialized, mass-produced items, including foodstuff, in ways that meshed with their
everyday material and symbolic needs as transnational working-class people.[66]
Furthermore, as Italians' experiences during World War I indicate, migrants expressed their duty to Italy by consuming foods and other products made and sold
by Italian companies, migrant entrepreneurs, and U.S. firms, all of which fostered
among migrants a more united identity as "Italians."

Following the publication of the *Il Progresso* article, J. Walter Thompson looked
specifically to migrant newspapers both for information on these "unrecognized
markets" and as the critical vehicle for reaching foreign-born consumers.[67] The
newsletter noted that in the German section of Philadelphia, stores selling goods
and services of all kinds advertised in the German press. "The German market
is clearly recognized locally as a rich field inaccessible except through the paper
which talks to these people in their native tongue."[68] The agency went so far as
to seek out ethnic journalists to provide input on these articles. For example, the
agency enlisted Israel Friedkin, publisher of the Yiddish-language *Jewish Morning
Journal*, to write on the "Jewish City of New York."[69] Overall, J. Walter Thompson
expressed a fervent, newfound interest in uncovering, describing, and exploiting migrant consumers and in employing foreign-language newspapers to target
ethnic buyers.

Advertisements for industrialized foodstuff and other consumer goods in *Il Progresso* shows that by the interwar years, U.S. businesses followed advertisers' lead
in targeting consumers specifically as migrants rather than as Anglo-Americans.
In November 1924 Helmar-brand cigarettes, manufactured by the American Tobacco Company, placed an ad in *Il Progresso* for their Turkish-style cigarettes. The

ad depicted an uncle gleefully embracing his newly arrived, luggage-laden nephew. The young man proclaimed, "Fellow citizens, listen. When I came to America for the first time and debarked in New York, my uncle took me to his house. There he offered me a Helmar cigarette. I will never forget the first impression I had from Helmar's."[70] American Tobacco commodified the migration narrative, associating U.S.-made cigarettes with positive portrayals of a successful migration story. The company had been advertising in *Il Progresso* starting in the early twentieth century, but it was not until after World War I that it began treating readers of *Il Progresso* as a distinct consumer market with cultural ties that extended across the Atlantic.[71] Ironically, the publicity ran the same year as the Immigration Act of 1924, which severely curbed the number of Italians who could enter the United States; fewer foreign-born consumers of Helmar cigarettes would experience the migration tale represented in the ad.

Ads for foodstuff often focused on consumers not only as migrants but as Italians by advertising in Italian and by tying products to Italians' transnational culinary cultures. In 1935 an ad for Heckers-brand flour presented a large picture of a smiling man leaning over a plate of cannoli made from Heckers "infallible" flour. The man declared, "My wife is a marvel! All her cannoli are exquisite."[72] Rather than reference white bread or other more typical Anglo-American baked goods, Heckers depicted Italian wives making a specifically Italian dessert while simultaneously affirming women's role as generators and guardians of their family's transnational identities. The ad also mimicked others in depicting the feminization of migrant marketplaces after World War I as part of a larger gendered process that linked women from different ethnic, class, and nationality backgrounds to consumption. This role afforded migrant women, like their native-born counterparts, a degree of authority and respect within the domestic realm while further marginalizing them from political and economic participation in the public sphere.[73]

Similarly, Armour and Company capitalized on connections between women, food, and ethnicity to emphasize the Italianness of both their meat products and their consumers. Armour's 1937 ad in *Il Progresso* depicted an older woman in an apron, happily displaying a platter of sliced La Stella–brand prosciutto. She testified in Italian, "When I serve Armour prosciutto, I know that dinner will be a great success!" (see fig. 20). The ad continued: "La Stella Armour has that delicious taste, that fragrance, that aroma, characteristic of the old country, that Italians love so much." In its use of signs and symbols comprehensible to consumers who were oriented gastronomically to tastes, ingredients, preparation styles, and models of commensality in the "old country," the ad echoed those for Italian imports and tipo italiano foods that had appeared in *Il Progresso* already. As had publicity for these products, the Armour ad linked production to consumption to assure readers that Armour, "the largest manufacturer in the world of Italian style dried

Figure 20. Ad for Armour La Stella–brand prosciutto, *Il Progresso Italo-Americano* (New York), December 6, 1937, 3.

sausage," employed expert Italian workers with "long-standing experience."[74] The female Italian consumer, the ad intimated, did not have to purchase imported dried sausage and cured prosciutto in order to guarantee the happiness and Italianità of her family; buying and consuming Armour-brand meat products—produced in the United States using Italian labor and preparation methods—reproduced Italian identities in New York's migrant marketplace.

Already by the late 1910s, Armour and Company produced meats prepared in the styles of various European national and regional specialties. A 1916 company catalog featured "Milan Salami," "Salami di Genoa," "Bari Salami," and "Caserta Pepperoni," all dried or smoked sausages seasoned with various "Italian spices," such as garlic and red pepper, and "popular with the Italian trade."[75] Paralleling the assertions of migrant entrepreneurs who had professed their products' superiority over imports by pointing to the lower costs of their tipo italiano foodstuff, Armour and Company suggested that migrant marketplaces democratized as consumers became less attached to costly imported foods from their homes. "The excessive cost of importation," the catalog declared, placed imported meats among "luxuries" until U.S. meatpacking plants like Armour and Company, "perceiving the great possibilities of the product," used "modern" methods of production and distribution and "brought it within the reach of all."[76] By the interwar years, firms like Armour and Company followed the lead of migrant entrepreneurs in producing tipo italiano foods using U.S. technology and distribution networks, and often low-wage migrant labor, in employing the foreign-language press to emphasize the Italianness of their products and, finally, in appealing specifically to migrant women as transnational consumers and keepers of Italian food traditions.[77]

U.S. business interest in foreign-born consumers extended to migrants outside U.S. borders after World War I. By the late 1920s, Armour and Company and other firms transferred their experience targeting migrants in the United States to South American countries with large migrant populations. The conviction that migrants were valuable not only as workers but also as consumers played a role in J. Walter Thompson's own expansionist strategies. Already by 1909 the company claimed that they placed 80 percent of all advertising for U.S. companies in South America.[78] The firm—whose accounts included such U.S. multinationals as Swift and Company, Ford Motor Company, General Motors, Frigidaire, Ponds, and National City Bank—opened branches in Buenos Aires, Argentina; São Paulo, Brazil; and Montevideo, Uruguay, in 1929.[79] Such companies, and the advertising companies that assisted them, were smart to take note of Italians in Argentina during the interwar years. Unlike in the United States, where Italians comprised a small proportion of the total foreign-born population and of the total U.S. population, Italians and their descendants in Argentina made up a significantly higher proportion of the total Argentine population and hence a formidable consumer market that U.S. companies could not afford to ignore. After the passage of U.S. immigration laws in the 1920s, the migrant marketplaces that some firms hoped to exploit in the United States literally migrated south as Italians during the 1920s redirected their travels to Argentina. Furthermore, the withdrawal of European capital from countries like Argentina during World War I presented new opportunities for U.S. firms, and European nations, slowly recovering from the war, could absorb only so many of the United States' manufactured products in the immediate postwar years. Argentina continued to remain a society dominated by an influential landed elite and a large, urban, mainly migrant working class, but by the 1920s the middle classes had expanded to support a rapidly maturing consumer society. By 1913 Fernando Rocchi estimated that Argentina's consumer market was nine times larger than that of 1873 (during these same years the U.S. consumer market, while overall much larger, grew by only five times).[80] Ongoing migration, rising incomes, urban growth, transformations in advertising and marketing strategies, shifting notions of femininity and masculinity, and, more generally, a reshaping of people's perspective toward consumption accompanied the growth of Argentina's consumer society.[81] More than any other Latin American country by the 1920s, Argentina offered a promising environment for U.S. firms hoping to reach consumers with money and a desire for imported products.

While guided by a basic assumption that consumer desire was universal, as U.S. businesses moved into Latin American markets, new strategies for reaching a diverse consumer community guided their marketing approaches. Who better prepared than U.S. business leaders—well versed by the 1920s in exploiting racially and ethnically distinct labor and consumer markets within its own borders—to begin doing the same in Latin America? "Markets are People—Not Places" became J. Walter Thompson's motto for work outside the United States. The export problem, agency executive Clement Watson declared in 1938, was "one fundamentally human."[82] Cultivating successful consumer markets abroad, he wrote, required "first-hand knowledge of the habits, customs, traditions, and living conditions of the people to whom the merchandise is to be sold." He urged U.S. businesses to understand the "factors of history, race, politics, geography, religion and economics" that determine whether a product would be accepted or rejected by consumers.[83] In Argentina this approach appears to have included an awareness of the large Italian-speaking population. U.S. companies' treatment of consumers as migrants in Argentina in the 1920s and 1930s paralleled strategies that had been relied upon by such businesses to appeal to migrants in the United States. Just as they had in New York's *Il Progresso*, ads in Buenos Aires' migrant newspapers for food and other commodities produced by U.S. firms often targeted its readers specifically as Italians by advertising in Italian, by stressing the Italianità of their products, and by alluding to migrants' ties to Italy.

U.S. companies began advertising in Buenos Aires' *La Patria* in the early twentieth century. At this time U.S. wares were largely limited to agricultural equipment, rubber products, and small domestic and electrical appliances, especially sewing machines, typewriters, and phonographs.[84] These early ads were in either Italian or Spanish and consisted of short informative phrases describing the function of the product. After the war, as U.S. firms increasingly shifted their commercial interests southward, a larger number and wider variety of U.S. consumer items began appearing in the paper alongside Italian imports and Argentine-produced goods. By 1920 *La Patria* ran advertisements for cars and motors made by Harley-Davidson, Oldsmobile, Saxon, Ford Motors, Overland, and Studebaker; tires manufactured by Goodyear, Michelin, and the United States Rubber Export Company; and tractors and milling machines produced by International Harvester Corporation; J. I. Case; and Fairbank, Morse, and Company.[85] Smaller consumer items produced by U.S. firms also appeared, such as Kodak cameras, Bayer aspirin, Keds shoes, B.V.D. men's underwear, and Westinghouse Electric International electrical appliances.[86]

Promotional material for these products developed stylistically, often including elaborate graphics of consumers interacting with them, as well as testimonials and text declaring a product's ability to satisfy a range of physical and psychological needs.[87] Comparing the Italian-language press in New York and Buenos Aires indicates that U.S. economic and cultural expansion into Latin America increasingly connected migrant marketplaces.

Armour and Company led the way in employing the Italian-language press to reach migrant consumers in Argentina. By the end of World War I, Armour had meatpacking plants in the United States, Argentina, Brazil, Canada, and New Zealand, and while most meat produced in Argentina during and immediately after the conflict fed hungry consumers in war-torn Europe, during the interwar years, as European demand slowed, Armour directed its focus on building domestic and regional consumer bases. By 1920 their largest plant globally was in Buenos Aires, a plant with the capacity to process 2,500 cattle, 1,500 hogs, and 5,000 sheep per day.[88] An ad for Armour's "Genoa-style tripe" in the Buenos Aires–based Italian-language *Il Mattino d'Italia* stressed the Italianness of its product by making favorable, romanticized references to Italian history and Italy's emigrants (see fig. 21). The 1935 ad featured an illustration of medieval Genoa and a text in Spanish that began,

Figure 21. Ad for Armour Genoa-style tripe, *Il Mattino d'Italia* (Buenos Aires), November 10, 1935, 5. Courtesy of the Hemeroteca, Biblioteca Nacional de la República Argentina.

"The Genoese, who today are many in this welcoming land, brought us something other than their proverbial industriousness and aptitude for work. They also brought their flavorful regional dishes and, of course, the most exquisite of all—Genoese-style tripe—that now Armour gives you ready to serve."[89] Armour disclosed their awareness of Argentina's own migration history, in which Italians loomed large, by praising Italian labor migrants and the regional foods they brought with them.

U.S. companies selling nonedible consumer items also frequently spoke to readers' specific experiences as migrants with continuing transoceanic ties to Italy by tying their products to Italian notables, and even Italian commodities. In 1925 the New York–based firm Tide Water Oil Company featured the testimonials of Italian race-car drivers in ads for their Veedol-brand car lubricant. Pietro Bordino, who competed in a number of U.S. car races and won the Italian Grand Prix in 1922, testified, "I learned about 'Veedol' in 1921 during my first trip to the Unites States of America, where this valuable lubricant produced by Tide Water Oil Company of New York is used by the majority of automobile drivers." Italian sports celebrities figured centrally in the Veedol advertisements, but so too did Italian Fiat cars, which competed with Ford and Oldsmobile for the attention of consumers in Argentina. In his testimonial, Bordino indicated that he had broken a record at the Italian Grand Prix in his Fiat car using Veedol-brand Gear Compound oil, while Felice Nazzaro, another Italian race-car driver, lauded both U.S.-owned Veedol and Italian-owned Fiat, a company that created "a perfect gem of a car, an envied global masterpiece of automobile construction."[90] By highlighting Fiat automobiles and the Italian sports stars who drove them, the U.S. Tide Water Oil Company Italianized their product for migrant consumers while simultaneously forging symbolic connections between Italians in the United States, Argentina, and Italy. Headline slogans such as "Another link between Italy and Argentina is Veedol oil" and "Soon Veedol will be as well known in Argentina as it is in Italy" played down the products' U.S. identity by giving readers the impression that the lubricant was a widely popular commodity in Italy.[91]

Nation-specific differences in the two countries' paths toward industrialization and in Italian migrants' occupational options allowed U.S. companies to employ not only food consumption but also food production to generate transatlantic cultural ties between Italians in Argentina and Italy. In both countries Italians overwhelmingly opted to work as manual laborers in urban areas rather than as farmers in the countryside. Still, in Argentina, whose economy continued to rely heavily on the lucrative agro-export sector, a significant number worked in agriculture on the vast Argentine pampas plains as farmers, sharecroppers, and rural wage workers.[92] This explains in part why ads for agricultural equipment proliferated in Buenos Aires' *La Patria* but not in New York's *Il Progresso*. In 1920 J. I. Case Threshing Machine Co., a Wisconsin firm specializing in agricultural machinery and construction equipment, ran an ad in *La Patria* that capitalized on migrants' drive to collect money and other resources to help Italy pay off its foreign debts and rebuild after the war. The ad featured an illustration of a man tilling a field in Argentina on a Case tractor adjacent to a contrasting poster for the Consolidated Italian Loan, which showed a similar scene but with a man in Europe laboring with difficulty behind a set of oxen. "Everyone wins with the work of repairing the ravages of war," the ad began. "It is the patriotic and humane work of every farmer of good will to produce as much as possible, because Europe needs cereals and will pay a good price." The Case ad played on the sympathies of Italians concerned with famished connazionali in Italy. And yet it also juxtaposed the two images to link Italians in Argentina with the "modern mechanical drive that had worked so effectively to help the Allies during the war," and Italians at home with the "old custom of using horses and oxen for plowing."[93] Purchasing a Case tractor allowed Italians in Argentina to produce wheat needed by relatives back home while reaffirming their own ability to buy "modern" food-producing equipment, which was unavailable in the war-torn "old world." Conversely, in the United States merchants encouraged consumers to help pay off Italy's war debts by buying Italian imports; such purchases would help resolve Italy's negative balance of payments while creating employment opportunities for Italian workers in Italy's food industries.[94] Joseph Personeni, seller of Italian foods in New York, instructed buyers in 1935 that it was their "duty" to help Italy pay its war debt by buying Italian specialties and promoting their use among both Italians and U.S. Americans.[95] In both New York and Buenos Aires, migrant newspapers encouraged Italians to aid Italy fiscally by keeping transnational food chains active. In Argentina, however, they were also invited to buy U.S. agriculture equipment to dispatch foodstuff in the other direction, across the Atlantic, to needy Italians back home as they had during the war.

Migrant consumption in New York and Buenos Aires constructed and reflected the different processes of migrant incorporation. Already at the turn of the century, Argentine companies had targeted Italian consumers specifically as Italians in ways that reflected Italians' powerful demographic presence as the largest foreign-born population. Argentine beverage, food, and cigarette companies, such as Quilmes beer and Piccardo and Company cigarettes, reached out to readers of *La Patria* as migrants by placing ads in Italian and by referencing Italian traditions, holidays, and notable persons. During this time, conversely, U.S. companies largely disregarded migrants as transnational consumers. In this much more populous and diverse country, where Italians accounted for only one of many foreign-born communities, U.S. companies considered Italians too small a market, and a weak, racially problematic market to boot. In short, Italians in Buenos Aires, but not in New York, were a "worthy market"—the term used to describe Italians in New York by J. Walter Thompson only in the mid-1920s.

By that time U.S. and Argentine companies had reversed their approach toward Italian consumers. Ads for Argentine consumer goods on the whole stopped speaking to readers and consumers as a distinct Italian market within the larger Argentine consumer economy.[96] So too did ads for Italian-style foods made by migrant entrepreneurs, which by the interwar years rarely made reference to Italy or homeland culinary ties and instead sometimes described their products and their prospective consumers simply as "Argentine."[97] By the late 1920s, while migrants continued to pour into Argentina, a substantial group of second-generation Italians had grown up in a country that had absorbed them as "Argentines" rather than as hyphenated people. Aiding in this process was Argentina's historic partiality to European migrants, including Italians, who were considered ultimately, if not without controversy, white, as well as almost four decades of interaction between Italians and Argentines surrounding Italian imports, which helped produce Latinità and justified linkages between Italian migration and exportation. As Argentine food producers and sellers deemphasized the Italianità of their products and consumers in the Italian-language press, U.S. companies looking for migrant consumers in both the United States and Argentina began conceiving of Italians as a distinct target market.

Therefore, it took until the mid-1920s for U.S. companies to pay as close attention to Italians (in both the United States and Argentina) as Argentine companies already had done in the early twentieth century. Only after World War I did U.S. businesses made a conscious effort to reach Italian consumers; they purchased a plethora of advertising space in the Italian-language press, where their publicity most often appeared in Italian, even in the 1930s, when in Argentina many if not most ads for Argentine goods appeared in Spanish. That ads for U.S. goods remained in Italian rather than Spanish suggests that U.S. firms thought of Italians in Argentina more as "foreigners" in Argentine society than did Argentine and even Italian companies, who had been marketing in Spanish in *La Patria* for over three decades. Only during the interwar years, when immigration policy decreased the number of Italians entering the United States by tens of thousands annually, did U.S. companies celebrate and commodify Italians' transnational links to tastes, experiences, and family back home, links perceived as less threatening and potentially profitable only after Italian migration slowed. Confident in their assumptions about migrant consumers in large metropolitan cities and about Latin America as a future site of U.S. economic and cultural domination, U.S. business promoters turned with ambition and excitement to migrants in Argentina.

*   *   *

The cities and neighborhoods described by J. Walter Thompson in the mid-1920s were "unrecognized" only to U.S. advertising executives and business leaders; to migrant merchants, entrepreneurs, and consumers these sites were visible and vibrant migrant marketplaces—locations where Italians' consumer patterns and identities shaped and were shaped by the economic and cultural ties migrants maintained to Italy. Rather than neighborhoods made meaningful only after U.S.-owned businesses discovered them, they had been furnishing sellers and buyers with foods and other consumer goods upon which Italians created meaning in their lives as transnational people.

Migrant marketplaces of mobile people and mobile goods played an important but overlooked role in the relocating of trade and migration paths that increasingly made north-south, rather than exclusively east-west, orientations central to the history of globalization as the twentieth century progressed. Starting in the mid-1920s, commercial flows between the United States and Argentina intensified while the migration patterns that so dominated the political economy of the late nineteenth and early twentieth centuries declined. U.S. business leaders labored hard, and with much success, to export consumer goods and capital outside U.S. borders toward Latin American countries; at the same time, U.S. legislators labored hard, and also with success, to close U.S. borders to Italians and other migrant groups. The confluence of changes in people and trade flows toward Latin America after World War I generated new opportunities for migrant consumers and U.S. exporters who were interested in expanding their markets abroad.

Cultural historians of U.S.–Latin American relations have argued that U.S. economic, cultural, and military expansion into Latin America in the first three decades of the twentieth century triggered, as historian Ricardo Salvatore wrote, "a new imagined scenario where the possibility of cultural assimilation of South Americans depended on the diffusion of U.S. products."[98] How transnational migrants in Latin America figured into the history of U.S. empire-building projects—as workers and consumers—however, is still relatively unknown. Even before World War I, Italians expressed varied but overwhelmingly skeptical sentiments toward the United States. *La Patria* and the Italian Chamber of Commerce in Buenos Aires saw in U.S. products and capital a major threat to Argentina's migrant marketplaces and claimed that the presence of Italian people and products protected Argentina from the United States' ever increasing hegemonic influence. By the interwar years, however, as suggested by the large number of ads for U.S. foodstuff, automobiles, agricultural machinery, beauty products, clothing items, and banking services in the Italian-language press, the U.S. industrial and consumer revolutions that had influenced New York's migrant marketplace were also transforming that of Buenos Aires. U.S. businesses sought to reach not only migrants in the United States but also migrants in other countries to which U.S. capital and commodities moved. By the 1920s and 1930s, migrants had a variety of "Italian" goods at their disposal—imports, tipo italiano, U.S., and Argentine—to invent and perpetuate transnational identities as people situated in the cultures and economies of both their home and host countries.

Over the course of the twentieth century, Italian migrants in le due Americhe reinvented their transatlantic subjectivities and buying patterns within the context of hemispheric flows of capital, consumer values, and imperial campaigns. They would continue to do so in the late 1920s and 1930s, when Mussolini's Fascist government endeavored to fortify Italy's relationship with migrants abroad. In Mussolini's paradoxical quest to both profit from migrant consumption and dissociate it from U.S.-style consumption—in which the female consumer commanded center stage—Fascist Italy faced a tough battle. The female Italian consumer, materializing during wartime fund-raising campaigns to spend money and send goods to Italy, had become stubbornly ubiquitous in ads for Italian, U.S., and Argentine consumer products. By the 1930s, Italy, the United States, and Argentina all vied for Italian consumers, especially women, who continued to serve as symbols through which migrants formed national and consumer identities.

CHAPTER 6

Fascism and the Competition for Migrant Consumers, 1922–1940

In the late 1930s Italian-language newspapers in New York and Buenos Aires solicited migrants to treat their relatives back home to a panettone, the traditional sweet bread enjoyed by Italians during the Christmas holiday. Advertisements for Motta-brand panettone in New York's *Il Progresso* publicized the company's special Christmas service that enabled Italians in New York to have panettoni sent to relatives and friends in Italy directly from the Motta factory in Milan.[1] In Argentina's *Il Mattino d'Italia* (hereafter *Il Mattino*), the Confetería del Molino, a Buenos Aires bakery run by Italian migrants, entreated readers to dispatch its panettone across the Atlantic to loved ones back home.[2] While the commodity chains along which these panettoni traveled originated in fields and factories separated by thousands of miles, both of them allowed migrants to forge a transnational consumer culture rooted in circulating networks of people and foods.

The two panettoni—one produced in Italy, the other in the diaspora—emerged out of enduring cultural and economic links between Italians abroad and their homeland established through decades of transatlantic migration. It built particularly on Italians' World War I experiences, during which migrants generated new identities as consumers and as Italians, in part by sending home agricultural staples like wheat, beef, and wool, and industrialized foodstuff like cigarettes, wine, and cookies. As they mobilized for their homeland in need, migrants coalesced around spending and saving in ways that regularized consumption as a duty to Italy while fostering a more intense sense of Italianità. Shifting gender ratios in international migration and in ethnic communities, combined with women's participation in fund-raising campaigns, gave women a new visibility in the Italian-language press as consumers of Italian foods and as guardians of Italian trade paths and culinary and familial traditions.

By the late 1930s Prime Minister Benito Mussolini's Partito Nazionale Fascista (National Fascist Party, or PNF), founded after his March on Rome in October 1922, endeavored to take advantage of Italian consumers in the transformed migrant marketplaces of New York and Buenos Aires. This proved especially the case in 1935 and 1936, when the League of Nations placed economic sanctions on Italy in response to that country's invasion of Ethiopia. However, by the mid-1930s Mussolini and his supporters faced a group of consuming connazionali abroad whose buying and eating practices—while still very much tied to products and traditions from back home—became less fixed to notions of duty to la patria. Migrants and their children after World War I had a surplus of foodstuff and other items upon which they developed consumer experiences as Italians but also as U.S. Americans and as Argentines. Much to Mussolini's chagrin, during the 1930s migrant marketplaces overwhelmingly featured female consumers making purchases in order to produce Italian identities even while in Italy he attempted to detach women from U.S.- and Western-style consumption.

The Global Depression, the Rise of Fascism, and Mussolini's Interest in Migrant Marketplaces

By the late 1930s the economic and geopolitical contexts in which transnational movements of people and products operated had drastically transformed. The Great Depression, and the nationalist economic policies it helped to produce, brought global trade to a standstill. In the United States panicked politicians passed the Smoot-Hawley Tariff Act, which increased trade barriers to new levels. Legislators had hoped that tariff hikes would protect U.S. jobs and manufacturing, but instead they largely exacerbated financial problems, as the legislation helped ignite a global trade war that stifled commerce worldwide.[3] The turn toward protectionism kept industrial and agricultural production levels low in the United States during the first five years of the Depression. Argentina too was hit hard, especially as the global prices for its agro-export commodities plummeted. During the 1930s Argentina experimented with an import substitution policy in order to advance domestic production, increase job opportunities, and diversify the economy to make it less dependent on fluctuating international markets, a policy that would intensify under nationalist Juan Domingo Perón in the 1940s and 1950s.[4] During and immediately after World War I, the Argentine legislature increased the tariff valuation upon which duties were based, giving manufacturers added protection from imports.[5] While commercial agriculture continued to serve as the primary engine of the nation's economic growth, manufacturing went from 15 percent of the gross national product on the eve of World War I to 21 percent by 1940.[6]

Coming to power amid the financial turmoil of the postwar years, Mussolini deepened his commitment to economic nationalism during the Depression. As part of his larger vision of an autarkic Italy, Mussolini limited trade, discouraged the consumption of foreign goods, and bolstered domestic manufacturing, including increased food production through agricultural land reclamation projects.[7] Italian exports to the western hemisphere began a slow decline after 1926 as Mussolini intensified trade relationships with fascist regimes in Europe and directed exports to Italy's colonies in North and East Africa. By 1934 the value of Italy's total exports to Argentina and the United States had decreased by about a fourth of their 1929 levels.[8] From the standpoint of the majority of Italians, Mussolini's austere trade and economic policies were largely unsuccessful; historians point to low consumption levels, especially among the working class, as evidence of autarky's inability to provide for its people and of the regime's willingness to prioritize state concerns over the basic needs of the Italian population.[9]

The rise and consolidation of the fascist state had major ramifications for Italy's migrants. Mussolini directed migrants to Italy's formal African colonies and re-conceptualized all Italian migration, even voluntary labor movements, as forms of Italian imperial expansion. While striving to curtail emigration and exportation to North and South America, he took an active interest in Italians abroad, especially in the United States and Argentina, where large populations of Italians and their descendants resided. By 1923 Mussolini had begun coordinating the activities of fascist supporters outside of Italy and limiting migrants' participation in the domestic politics of their host countries.[10] Fascists sought to tie Italian communities across the Atlantic to Italy through cultural-based propaganda campaigns organized by migrant organizations, including pro-Mussolini Italian-language newspapers and Italian chambers of commerce.[11] Mussolini hoped—as had his liberal predecessors—that rallying migrants in support of Italy would transfer into monetary support in the form of remittances and purchases of Italian goods abroad. Despite Mussolini's move to restrict Italy's trade relationship with the United States and Argentina at the diplomatic level, he encouraged migrant consumer demand for Italian commodities such as vegetable and fruit preserves because they were profitable and because most Italians in Italy could not afford such expensive, domestically produced items.[12]

While Mussolini looked to Italians in both the United States and Argentina as vital resources, fascists viewed Argentina as holding unique potential for amplifying Italian nation and empire building.[13] The United States' mature industrial sector, higher tariffs, and restrictive immigration legislation made it an increasingly inhospitable place for Italian people and products. Despite a shift toward economic self-sufficiency and import substitution, Argentina continued to require exports from abroad and the country kept its doors open to Italian migrants. Fascist supporters pointed to the powerful presence Italian migrants and their children continued to have in the Argentine economy. The 1935 Argentine industrial census established that of all foreign-born groups, Italians owned the largest number of factories, and their calculations did not include factories owned by second-generation Italians, who were counted as Argentine by census takers.[14] Italy also had a relatively large amount of fixed capital in Argentina by the 1930s. Like their U.S. counterparts, several major Italian companies, such as Cinzano, a vermouth producer; Fiat, a manufacturer of automobiles and agricultural machinery; Pirelli Tire and Company; and Gio. Ansaldo and Company, a producer of metallurgical goods, opened branches in Argentina during the interwar years.[15]

Furthermore, after 1930, when General José Félix Uriburu overthrew the democratically elected president Hipólito Yrigoyen of the Unión Cívica Radical (Radical Civic Union), Argentina appeared to be an even more favorable environment for the transfer of fascist political ideology. While Uriburu's dictatorship lasted only two years, historian Federico Finchelstein has suggested that Uriburu based his plans for Argentine national renewal on the Italian fascist model and that his military rule cultivated the presence of powerful right-wing nationalists, some of whom drew inspiration from Mussolini.[16] This ideological milieu blended with Mussolini's aggressive revitalization of Latinità to make Argentina a particularly fruitful site for expanding fascist economic and cultural influence. Already by the late nineteenth century, Italian leaders in both Argentina and Italy had celebrated Latinity to legitimatize Italian migration and trade. Mussolini built on these earlier racialized visions in the hopes that shared notions of Latinità between Italians and Argentines would result in support not only from Italians and their children but also from influential Argentine leaders.[17]

Mussolini found some level of backing from Italy's migrants in le due Americhe. Fascism drew support from many Italian nationalist prominenti in the United States and Argentina, especially members of Italian chambers of commerce but also elites in control of the Italian-language commercial press. Construction tycoon Generoso Pope, who assumed control of New York's Il Progresso in 1927, became outspokenly pro-Mussolini and reported favorably on fascist activities in Italy and abroad.[18] La Patria, which until 1930 had been the largest Italian-language newspaper in Argentina, shifted its politics toward a critical anti-Mussolini stance, reflecting many migrants' distrust or disinterest in fascism as an organized political party. By the end of the 1930s, the paper, which had reinvented itself as La Nuova Patria degli Italiani, had collapsed, dwindling from over fifteen pages of articles and ads in each edition to four or six pages by the end of the decade. The collapse was not accidental. After being named the Fascist Party delegate in Argentina, Vittorio

Valdani, an affluent industrialist who had migrated to Argentina in 1899, bought out *La Patria* only to kill it and divert remaining funds to the fascist daily *Il Mattino d'Italia*. Fascist supporter Piero Parini established the Buenos Aires *Il Mattino* in 1930, and under editor Michele Intaglietta's guidance, the paper's materials, aesthetics, and politics soon paralleled *Il Progresso*'s pro-Mussolini posturing.[19]

It would be shortsighted to characterize all Italians abroad who conveyed a sense of admiration for Mussolini during the 1930s as unequivocal fascists. Historians of fascism have acknowledged how little support the system actually enjoyed among Italians living in the United States and Argentina. The *Fasci all'estero* (Italian Fasci abroad), organizations established by the National Fascist Party to spread support for fascism and promote Italianità in migrant communities, were plagued with internal divisions and lacked widespread approval among migrants.[20] Fascists abroad also faced vocal anti-fascist organizations, mainly run by migrant socialist leaders, labor organizers, and anarchist groups.[21] Furthermore, scholars such as Matteo Pretelli and Stefano Luconi have asserted that prominenti at the helms of various migrant institutions tended to depoliticize fascist attitudes among migrants and their children by promoting an Italian identity rooted in love and respect for the customs, natural beauty, and rich history of their homeland.[22] Nevertheless, this culturally based understanding of Italianità promoted by fascist supporters came to influence migrants' consumer options and identities in migrant marketplaces. Despite pervasive indifference to fascism among migrants and their descendants, Italians abroad used and were used by fascism, and prominenti merchants, business leaders, and journalists achieved some success in cultivating an identity abroad anchored in a newfound Italian nationalism.[23] World War I first produced the formation of a more cohesive national identity among migrants who were drawn into transnational fund-raising drives to spend, save, and consume for the homeland. However, by the 1930s the end of mass migration, combined with an expanded second generation of "Italian Americans" and "Argentines" with more Americanized and Argentized palates, threatened Italian food trade, already weakened by the Depression. Against the background of global slowdowns in migration and trade and of rising nationalisms, fascists in Italy and abroad attempted to shore up their relationships with consumers in Buenos Aires and New York.

Debating Consumption in Migrant Marketplaces during the League of Nations' Boycott

Mussolini looked especially to migrant consumers in the wake of Italy's invasion and defeat of Ethiopia in 1935, an imperial campaign designed to expand Italy's colonial foothold in Africa. The League of Nations responded by instituting economic sanctions against Italy in November 1935. Lasting until July 1936, the sanctions included an arms and financial embargo and boycott against Italian trade goods. During the boycott Mussolini hoped that migrant demand for Italian goods would compensate for the decrease in Italian trade to the fifty-eight member countries in the League of Nations. Italians in the United States served as an especially critical outlet for Italian commodities; despite the U.S. government's increasing disapproval of Mussolini's imperial aims, the United States was not a member of the League of Nations and did not officially join the boycott.[24] Argentina, on the other hand, was a member of the league and formally instituted the ban, although historians have questioned the extent to which Argentina adhered to the sanctions, especially as the country grew irritated at the league's unwillingness to take the recommendations of Latin American countries seriously.[25]

Nation-based differences and changes in the ideological and political positioning of *Il Progresso* and *La Patria*—the two largest circulating Italian-language papers since the late nineteenth century—produced a diversity of migrant responses to Mussolini's overtures. In the early 1930s *La Patria*, now *La Nuova Patria*, became anti-Mussolini, expressing condemnatory attitudes toward the PNF but without embracing communist and socialist stances. Instead, both New York's *Il Progresso* and Buenos Aires' *Il Mattino* received direct financial assistance from the fascist regime in Italy through Italian consulates and embassies abroad. Coverage of the league's boycott therefore reflected the divergent ideological paths followed by the two migrant newspapers. Despite their contrasting political outlooks, both of these papers objected to the league's sanctions; their objections, however, were based on conflicting assumptions about Italy's commercial and export capacities, about Italian consumers in Italy, and about migrants' responsibilities as consumers to their homeland.

New York's pro-Mussolini *Il Progresso* positioned Italy as a powerful consumer and producer to argue that the boycott would severely disrupt international trade to all countries' disadvantage. By purchasing goods from abroad, Italy served as many nations' "best client," absorbing many international products. The paper even pointed to Italy's trade deficit, often viewed as an economic weakness, as evidence of Italy's ability as a powerful buyer to affect global trade.[26] Instead of harming Italy, *Il Progresso* predicted that the league's actions would prove disastrous to sanctioning countries like France and England, which would lose out on Italy's lucrative consumer market. *Il Progresso*'s Angelo Flavio Guidi wrote, "The day will come—not far from now—when the countries that wanted to apply the sanctions against Italy will understand that they have thrown a 'boomerang' that, without having touched its target, returned to hit he who threw it."[27] *Il Progresso* portrayed Italy as one of the United States' great trade partners and discouraged their host country from yielding to British pressure to apply the sanctions. The paper urged U.S. politicians to respect the Act of Neutrality, passed by Congress, under which the United States could legally prohibit the sale of arms and munitions to Italy but not the sale of primary materials such as the oil, cotton, steel, and iron that Italy greatly depended on. Articles often lauded the United States' right and necessity to remain independent of the league, especially of England, which the press consistently demonized.[28] Sanctions violated previous commercial pacts between Italy and the United States and denied U.S. businesses the Italian consumers in Italy on which their trade relied. *Il Progresso* calculated that, if applied, the sanctions would cost the United States $60 million per year in trade.[29] If the United States joined the sanctions, the paper threatened, Italy might retaliate in self-defense by refusing to buy U.S. goods in the future. The press argued that migrants had the support of the U.S. public, of English-language newspapers like the *Sun* and the *New York Herald*, and of U.S. industrialists and manufacturers, all of which backed Italy and migrants' calls for neutrality.[30] The paper's haughty depictions of the United States and Italy as commercially codependent masked the actual unequal commercial relationship between the two nations. While U.S. goods constituted 30 percent of all Italian imports, making the United States Italy's second most important trade partner behind France, Italian goods represented less than 2 percent of the total value of all U.S. imports and only 6.5 percent of the total European imports to the United States in 1935, the year the sanctions began.[31]

As it had during debates over tariffs in the early twentieth century, in objecting to the sanctions and encouraging U.S. neutrality in the 1930s, *Il Progresso* depicted migrants as an influential voice with the ability to affect international trade and diplomacy. *Il Progresso* editor Generoso Pope argued, "The Italians of America, because they are united, represent a large force, which the government and Congress cannot fail to take into account." Pope encouraged readers to send letters to their congressmen, suggesting that migrant protests kept the United States neutral.[32] The paper even insinuated that President Franklin D. Roosevelt would lose the Italian American vote in his 1936 bid for reelection if he supported the sanctions.[33] Naturalized Italians, but, more importantly, the large number of second-generation Italians born in the United States, would exercise their disapproval at the voting booth.

The anti-Mussolini *La Nuova Patria* in Buenos Aires also objected to the league's boycott, questioning the practicability and effectiveness of the sanctions. Like *Il Progresso*, the paper doubted that sanctioning countries would sacrifice the interests of big business to the league's authority and noted the boycott's potentially injurious consequences for smaller nations like Yugoslavia, whose economy depended on trade with Italy.[34] While condemning the sanctions, *La Nuova Patria* argued that migrants' delicate position in Argentina, a sanctioning country, precluded them from actively protesting the league's decision. The responsibility of Italians in Argentina regarding the sanctions, *La Nuova Patria* wrote, is, "in our humble opinion, nothing or very little."[35] When *Il Mattino* called on migrants to boycott products arriving in Argentina from sanctioning countries, *La Nuova Patria* pointed to the absurdity of doing so in such countries, where British-owned firms could "launch any day thousands of Italian workers on the streets."[36] The continued presence of British-owned businesses in Argentina turned battles between pro-Mussolini supporters and British capital in Argentina into one "between fleas and elephants." "Can you imagine an Italian who does not travel by train, tram, metro in order to damage British interests, who rejects the products of the British meat houses, who demands that his tailor is certain that the fabric he is using is not British, and who deserts the cinema?" asked the paper.[37] It was unreasonable for pro-Mussolini crusaders to ask migrants to sacrifice their consumer desires and needs, including their daily bread, in support of Italy's imperial ambitions.

Furthermore, Argentina's history of migration, *La Nuova Patria* implied, inhibited migrants as well as Argentines from applying the sanctions. A month before the boycott began, the paper praised an article by Rodolfo Rivarola, former president of the National University of La Plata, in the Argentine popular daily *La Nación*. Rivarola argued that Argentina's large Italian population made the application of sanctions impossible. He implied that breaking commercial ties with Italy would damage not only the relationship between Argentines and Italians within Argentina but also the relationship between Italians and the numerous other migrant groups in Argentina. Argentina's multinational society, composed of people with different economic priorities and national loyalties, excused the country from enacting the sanctions. The paper suggested that promoting harmonious relations among Argentina's diverse population took priority over the league's trade dictates, which, the paper predicted, would spark animosity among the country's many migrant and ethnic groups.[38]

Unlike discussion of economic protectionism at the turn of the century, which focused almost exclusively on migrant consumption, in their objections to the League of Nations both *Il Progresso* and *La Nuova Patria* concentrated on how the sanctions would affect Italian consumers in Italy. *La Nuova Patria* constructed Italian consumers as deprived and suffering, first, because of fascism's misguided economic policies, and, second, as guiltless victims of the league's actions. The paper pointed to the exorbitant cost of living in Italy, rising prices for food and other consumer items, reduced salaries, and the poverty under which most Italians lived; under fascism, the paper maintained, Italians "suffer and say nothing, fearing the worst."[39] The sanctions would not affect the Italian government, but rather the Italian people, especially women, children, and the elderly, "innocent victims" of hunger and malnutrition. The paper reminded readers, "Fascism is not Italy," observing that the Italian working class, not the fascist state, bore the material and spiritual effects of the sanctions.[40]

New York's pro-Mussolini *Il Progresso* painted an opposite picture of Italians' quality of life during the boycott. While representing the sanctions as "barbaric," the paper depicted Italians as prepared to face them "without a complaint, without a protest, without a defection," because Italians knew how to "live well on little."[41] Italians in Italy committed as consumers to fascism, willing to abandon "luxury and the whim of the palate, the flashy and unnecessary, without a drop of regret," and extolled Italian women specifically, who "protect the battle for the austere life."[42] Even after renouncing luxury goods, Italians actually saw their quality of life increase under fascism, the paper claimed. Italians profited from fascist economic policies; increases in the number of livestock per person and in the quantity of wheat per hectare boosted Italians' consumption levels of meat and bread. Any food restrictions on meat under the boycott, the paper added, would be compensated with increased cultivation and consumption of fish, game, fruits, and vegetables.[43] Even while the pro- and anti-Mussolini newspapers both disapproved of the sanctions, the league's boycott produced strikingly different opinions about Italians' ability to consume and live comfortably under fascism. And both made the Italian consumer in Italy a pivot around which diasporic debates concerning the sanctions turned.

As in migrant discussions about World War I, those about the league's sanctions underscored migrants' consuming and spending responsibilities vis-à-vis Italy. In response to the sanctions, Mussolini backers affiliated with the Italian-language press, and Italian chambers of commerce in both countries carried out campaigns to sustain Italian trade by encouraging migrants to buy goods from Italy.[44] While these drives were largely unsuccessful, they illustrate how the fascist state and its supporters in the United States and Argentina during the 1930s, like their liberal predecessors during World War I, sought to incorporate migrant marketplaces into transnational nation- and empire-building projects.

Unlike during World War I, when Buenos Aires' chamber of commerce ran an aggressive "buy Italian" campaign urging migrants to purchase Italian goods, during the sanctions New York's chamber instead joined its Argentine counterpart to rally migrants around Italian imports. *Il Progresso* covered with fanfare a series of radio addresses sponsored by the chamber, during which Italian *prominenti* in charge of key immigrant institutions pressured migrants to buy Italian goods.[45] The paper reported with praise a speech on the Italian radio station WOV New York by Antonio Corigliano, vice president of the Bank of Italy in New York, during which he pushed for the "defense and diffusion of Italian products in the United States."[46] Later that month, *Il Progresso* reported parts of another speech, by Vincenzo Vedovi, head of the Federation of Italian Veterans, in which Vedovi proclaimed, "Italians abroad have at their disposal a valid weapon for breaking the Geneva sanctions and for helping our Italy in this historic moment, which is the most difficult that has been seen since Italian unification was accomplished." That weapon, Vedovi insisted, was consumer choice. Italians should opt to buy all of Italy's exports, he instructed, not only food goods—the most popular Italian trade items—but other Italian products, such as hemp, raincoats, hats, and marble.[47]

Vedovi spoke specifically to women in migrant marketplaces. Echoing the paper's treatment of women during World War I campaigns, Vedovi suggested that women's purchases of Italian textiles and their domestic roles as housewives connected women of Italian descent across transatlantic spaces and generational lines:

> All of our housewives aspire to have a beautiful bed covering, whether of silk, cotton, or lace: Italy also makes these fabrics, and the difference from others is that they have great honor to have been made in a way that makes us remember a similar blanket seen one day, over a bed in a house far away, perhaps smaller, but so very special: the house where we spent the first years of our life, in the house of our beloved mother. Why shouldn't a similar blanket also adorn our house in this land where the events of life have carried us and give us a little Italian feel, the feel of our first years? With the Italian blanket that you bought, you will carry to your house an object that when you will see it, you will think of the faraway homeland, and you will have done something that gives back to the benefit to our country, during a moment in which it especially has needs.[48]

As buyers and preparers of food, migrant women also held the power to strengthen Italian commodity chains through the pressure they applied to sellers. Italians should always ask retailers for Italian goods, and "if he does not have them, when he is convinced that you want exclusively those goods, he will be obligated to procure them."[49] *Il Progresso* pressured its readers to purchase Italian imports in order to boost Italian commerce, but it also goaded readers to boycott goods from sanctioning countries like England and France. The paper guided migrants' purchases by publishing a list of countries, under the headline "Do not forget," divided into categories according to nations that adhered to the sanctions without reservations, those that adhered to the sanctions with reservations, and those that refused to adhere to the sanctions.[50] By channeling migrant consumption toward Italian merchandise, *Il Progresso* and the Italian Chamber of Commerce strove to keep migrants in line with fascist trade objectives.

Italian merchants used the sanctions, and growing nationalist sentiment more broadly, to reaffirm the Italianness of their products and to depict the purchase of food imports as acts of patriotism. An ad for Negroni-brand dried salami assured buyers that "the name NEGRONI is guaranteed to be an ITALIAN product," while food merchant Luigi Vitelli called out to buyers, "Italians! Prefer imported products from Italy."[51] Migrant entrepreneurs producing Italian-style foodstuff also saw the boycott as an opportunity to portray their products as genuinely Italian.

De Nobili and Company, the tobacco firm in Long Island City that had sent so many tipo italiano tobacco products to Italian troops during World War I, ran an ad reminding readers that for twenty-nine years the company had been "always entirely owned by Italians," even while De Nobili's tobacco products competed directly with those arriving from Italy.[52] Planter's Edible Oil Company, founded by Amedeo Obici from Venice, announced that the company's "Wings of Italy" peanut oil was the most popular cooking oil among Italian families in New York.[53] Migrants selling imported and tipo italiano foodstuff had been emphasizing the Italianità of themselves, as well as their products, employees, and consumers, since the early twentieth century; in most ways ads during the sanctions differed little from this earlier publicity in appealing to migrants as transnational consumers. However, the heightened sense of Italianità during the boycott gave makers and sellers of Italian foods an occasion to market their wares at a time when increasing numbers of migrants identified consumers in migrant marketplaces as "Italian," and when fewer goods and migrants were crossing the Atlantic.

Like their counterpart in New York, the Italian Chamber of Commerce in Buenos Aires ran a campaign in conjunction with the pro-Mussolini *Il Mattino* encouraging readers to consume Italian products and to avoid purchasing foods, beverages, and other items arriving from sanctioning countries, especially England.[54] The chamber formed a "blacklist" of British commercial houses and products and called for an "absolute boycott of all products originating from countries hostile to Italy in any way."[55] As in New York, fascist sympathizers in Buenos Aires politicized the consumption of imported foodstuff as a symbol of resistance to the league's "heinous" sanctions against Italy.[56] Supporting Italy meant not only preferring Italian imports but also rejecting and condemning imported English foods such as roast beef and whiskey. "All Italians worthy of their name," *Il Mattino* pronounced, "know very well that to drink a 'wisky' [sic] beyond ruining the stomach will give to England a bullet that tomorrow may strike an Italian chest." Genuine Italian patriots, the paper stated explicitly, abandoned whiskey and embraced *grappa*, a traditional alcoholic drink made from fermented grape pomace.[57] On the day the sanctions began, *Il Mattino* entreated Italians in Argentina to "consider all that emanates from Great Britain . . . whether products or traditions, merchandise or sports, poisonous alcohol or literature, the devil's excrement."[58] During the sanctions the criticism against U.S. economic and cultural hegemony in Argentina that was so prevalent in *La Patria* in the 1920s softened as migrant business authorities—particularly the pro-Mussolini supporters among them—branded England, rather than the United States, as the most malignant threat to migrant marketplaces.

Lest there be any confusion over where to shop, *Il Mattino* published lists of "Italian stores" where migrants were guaranteed to find imports from Italy. The paper assured readers that "the Italian trade in Argentina is perfectly equipped to respond to all the demands of the Italian consumer." The paper exhorted readers to "take up your role in the great Italian anti-sanctions battle by consuming exclusively Italian and Argentine products or products from non-sanctioning countries and renounce all others."[59] That the paper included Argentine as well as Italian products in their list of acceptable goods suggests that there were limits to the paper's nationalist stance; notwithstanding its patriotic appeals, the paper had to navigate its awkward position in a sanctioning country while acquiescing to the changing tastes and consumer options of Italians and their children.

As with New York's *Il Progresso*, *Il Mattino* characterized migrant consumption as a duty to the homeland, often singling out women specifically as it had during World War I. The paper described Italy's resistance to the sanctions as a transnational movement linking Italian women in the United States and Argentina. It ran articles about Italian women in Italy who were organizing to resist the boycott, and it printed Mussolini's special messages directed at Italian women during the sanctions. As migrants in various Argentine cities formed all-female "Pro Patria" groups to fund-raise for Italy, *Il Mattino* cheered them on, often sharing the amount of money or gold donated by women themselves or by the organizations and businesses they solicited for funds.[60]

Italian merchants, like their counterparts in the United States, exploited the national sentiment the boycott inspired to boost Italian trade and depict migrants' purchases as patriotic gestures. The Italian Tobacco Corporation in Buenos Aires prodded readers, "Italians! Prefer Italian products!"[61] Ads for Italian imports during the sanctions forged not only transnational links between Italians in Italy and those abroad but also imperial connections between Italians in the Americas and those in Italy's African empire. After Italy's declaration of victory over Ethiopia, an ad for Martini & Rossi announced, "Martini vermouth takes part in the jubilation of the Italian people for the victory achieved by its glorious army that, as a symbolic vessel, a messenger of strength, justice and civilization, reaches the port victorious with its sails fully unfurled."[62] Publicity for Italian imports associated Italian foodstuff and their consumers with Western imperialism and the racial assumptions it supported, inviting migrants to participate in and celebrate Italy's imperial triumphs.

In Argentina the "buy Italian" drive championed by *Il Mattino* and the Italian Chamber of Commerce inspired a vociferous defense in the anti-Mussolini *La Nuova Patria* of migrant entrepreneurs in Argentina, the tipo italiano foods they produced, and the consumers who bought them. In New York the pervasive, longstanding presence of tipo italiano foodstuff by the 1930s seems to have precluded criticism of migrant entrepreneurs in the United States by even by the most fervent pro-Mussolini apologists.[63] In Argentina, however, migrant entrepreneurs had been disadvantaged by a less industrially mature, import-dependent economy that delayed the development of the tipo italiano sector. And yet Argentine industrialization intensified after World War I, especially in the food sector, which by 1935 made up 47 percent of all domestic manufacturing.[64] The war had helped strengthen migrants' links to their homeland as Italians and as consumers of imports, but it also jump-started Argentine policies that supported a manufacturing sector producing a wider array of cheese, pasta, cigars, baked goods, and canned vegetables—many of them made by Italians.[65]

By the mid-1930s *La Patria* had moved far from the stance it had taken during World War I, when the paper commanded its readers to buy only Italian products. In response to the sanctions, *La Nuova Patria*, in its reincarnated anti-Mussolini form, printed a series of articles in 1936 under the headline "The Right to Make Italian Products Abroad" by Giuseppe Chiummiento, an anti-fascist Italian journalist and political exile.[66] In these articles Chiummiento objected to the Italian chamber's accusations that migrant makers and consumers of tipo italiano foods were unpatriotic and even traitorous. The paper framed its defense of migrant entrepreneurs within the context of Argentine economic development. In contrast to the predominantly agricultural nation of the early twentieth century, Argentina was as industrialized as the most advanced cities of Europe.[67] Chiummiento noted the "miracle of progress" in Buenos Aires over the previous twenty years, describing "the luxury of so many public places and of the accelerated pace of life."[68] Argentine industrial growth was "inevitable" and "natural" yet hastened by World War I, protectionist legislation, and Argentina's abundance of raw materials. In the early twentieth century *La Patria* had lobbied hard against high tariffs; by 1935 the paper claimed indifference to protectionism, a "product of an evolutionary process," and asserted that Argentina had the right to liberate itself from foreign imports, which it considered a form of "economic slavery."[69] Chiummiento's tone matched that of Argentine nationalists promoting industrialization and tariffs in the 1930s as a way to secure the nation's sovereignty.[70] Italians who made and consumed tipo italiano goods, the paper argued, were expressing their right to contribute to the industrial development and economic emancipation of Argentina.[71]

Just as migrants and their children had the right to employment in tipo italiano establishments, they were also entitled, as consumers, to reasonably priced food goods. "No one betrays one's country of origin: the immigrant producer, nor the one who consumes. As the descendants of the first, nor the second," Chiummiento wrote. The paper defended working-class migrant consumers against protectionist policies that put imported foods out of their reach. Who would buy imported Italian salami costing at least four times more than locally produced mortadella and prosciutto? the paper wondered. "Those who cannot buy a quarter pound of genuine parmesan cheese, because the family budget will not allow it, have at least the possibility . . . to eat spaghetti or ravioli with locally produced parmesan cheese," Chiummiento noted.[72] *La Nuova Patria* singled out for particular criticism affluent Mussolini supporters in Argentina, who could afford the more expensive Italian imports and who led the charge against poorer migrant manufacturers and consumers of Italian-style foods. These fascist supporters, while purporting to represent the Italian proletariat class, sought to prohibit working-class migrants' access to cheaper Argentine versions of Italian foodstuff and paper products.[73] While *Il Mattino* and the Italian Chamber of Commerce, like Mussolini's backers in New York, suggested that pro-Mussolini migrant supporters could commit equally to consumerism and to fascism, *La Nuova Patria* suggested that migrants' everyday consumer needs should trump fascist dictates.

Il Mattino's "buy Italian" campaign during the League of Nations' boycott pushed its readers to secure Italy's transnational trade routes by preferring Italian imports. However, in a somewhat paradoxical fashion, *Il Mattino* also advertised a number of Argentine commodities, including the tipo italiano foods that directly competed with Italian imports. During the sanctions, for example, *Il Mattino* advertised Mendoza wines as "The wines most similar to Italian wines," and "Italian-style *torroni*," a dense holiday cake made with honey, sugar, egg whites, and nuts.[74] As had migrant entrepreneurs in New York, those in Buenos Aires capitalized on the nationalism prompted by the boycott to sell their products and market them as "Italian." In 1935, Gaetano Graziosi, producer of Fox-brand prosciutto in Buenos Aires, ran an ad expressing New Year and Christmas wishes "to the Italian soldiers, to the Italian people, to its great conductor, the Duce, to the Italian colony in Argentina."[75] Despite admonishing readers to "Prefer Italian products," *Il Mattino* even used foods produced by U.S. corporations in campaigns to support Italy.[76] Starting in November 1935, the paper initiated a drive that prompted Italians in Argentina to "contribute with a material gesture of valor to the internal resistance of our people." As part of the drive, migrants purchased five- or ten-kilo packages filled with consumer goods of the "highest necessity" that had been reduced in Italy because of the sanctions and sent them to Italy. The chief items in the *pacchi tricolori* (three-color packages)—which suggested the red, white, and green Italian flag—included meat products made by the "renown North American Swift and Armour Meat Packing plants, products well known for their goodness and quality." *Il Mattino* reported that its readers had sent over six thousand packages on Italian steamships back to Italy only one month after the campaign began. After purchasing some ten tons of prosciutto for their homeland, *Il Mattino* bragged that Italian buyers had completely exhausted the annual stock of Armour and Swift meat products in Argentina.[77]

During the sanctions, *Il Mattino* characterized Armour and Swift not as U.S. American but as both Argentine and Italian, a strategy that had been employed by Armour itself starting in the 1920s, when it began targeting migrant consumers in Argentina. During the drive, *Il Mattino* pointed to the predominance of Italian migrant workers in Argentina's meat packaging plants while insisting that the packages were "composed solely of the products of the soil and of the Argentine industry, so closely connected to Italy by links of blood and labor."[78] By the late 1930s a convergence of global economic and geopolitical forces—protectionist economic policies, reorientations of trade and migration, and rising nationalisms—made possible the consumption of Swift and Armour meat products in the United States, Argentina, and Italy. If Armour joined other U.S. firms with an increasingly powerful presence in Argentina in deploying forms of U.S. imperialism in the 1920s and 1930s, the campaign also demonstrated how "on the ground" imperial exchanges could be absorbed and redeployed by locals. That Italian migrants incorporated the products of U.S. meatpacking plants into transnational campaigns to send food to Italy and to generate fascist support in migrant marketplaces suggests the "contact zone" quality of U.S. empire in Latin America defined not only by hegemony and unipolarity but also, as Gilbert Joseph writes, "of negotiation, borrowing, and exchange; and of redeployment and reversal."[79]

Furthermore, that *Il Mattino* employed Armour meat products to manufacture "links of blood" between Argentines and Italians signals not only how Italians co-opted the fruits of U.S. economic expansion to forge racial alliances in migrant marketplaces but also how the ethno-racial terrains of the United States and Argentina differentiated the way Italians generated identities based on buying and selling. Italian fascist supporters in Argentina, but not the United States, had at their disposal Latinità, which was cited during the sanctions to praise Italians' contributions to Argentine nation building, to spark pride in Italy and Italy's "civilizing mission" among Italians and their children in Argentina, and to once again justify Italian migration and trade.[80] During the League of Nations' boycott, pro-Mussolini trade promoters employed with fervor the discourse of Latinità to promote Italian-Argentine trade relations, the consumption of Italian products, and the boycott of British goods. Reminding readers of *Il Mattino* that Italy, "the great Latin Mother," "gave the world the highest and most enduring contribution to civilization throughout history," the paper regularly declared Italy's important role in civilization building worldwide but specifically in Argentina.[81] This heightened emphasis on race reflected Mussolini's preoccupation with protecting *la razza Italiana* (the Italian race) or *la stirpe* (race) at home and abroad through a variety of reproductive, imperial, and racial policies; in Latinità Mussolini saw a productive tool for building diplomatic and commercial bridges between Italy and Argentina that he hoped would serve as a transatlantic conduit for fascist propaganda.[82]

During the boycott, *Il Mattino* covered with great fanfare the arrival and activities of the Italian Commercial Mission to Argentina, sent by Mussolini to promote commercial relations between the two nations and to counteract Argentine enforcement of the sanctions. The mission was headed by fascist parliamentary member and commercial law expert S. E. Alberto Asquini and led by members of the Italian Chamber of Commerce in Buenos Aires. Asquini made much publicized visits not only to large, influential Italian firms and commercial establishments such as Cinzano, the Italian Tobacco Corporation, and Pirelli but also to major Argentine business, financial, and manufacturing institutions ranging from the Rotary Club to the Unión Industrial Argentina (UIA), Argentina's chief industrial association. In their coverage of these visits, *Il Mattino* called attention to the history of exchanges—cultural, commercial, and migratory—between Italy and Argentina that bound the two nations together. [83] The paper printed a speech by Davide Spinetto, president of the Rotary Club in Buenos Aires, who praised the Italian in Argentina as the "ideal migrant." He told members of the mission that the history of Italian migration in Argentina "created new bonds, and, with the related commercial trade that reached a remarkable figure, tightened always more the relations of cordial friendship between the two governments." The fact that Argentines and Italians were both "branches of the European civilization" and that Italians represented the most robust "graft" of that civilization in Argentine society explained the strong economic relationship between Italy and Argentina.[84] Ads for imported Italian foodstuff reinforced this message by suggesting that consumption united Italians and Argentines together in support of Italy during the sanctions. A Martini & Rossi ad in *Il Mattino* depicted a group of men and women raising glasses of vermouth with the headline "Greetings Argentina! With this exquisite vermouth let us commemorate magnificent Italy." While asking Argentines to honor Italy, the ad also toasted "the great Argentine nation." That the publicity invited Argentines and Italians to enjoy Martini & Rossi as a "symbol of brotherhood between the two powerful countries" shows that Latinità, forged over decades of migration, trade, and consumption, integrated both Argentines and Italians into campaigns to support Italy politically.[85]

The paper labored diligently to employ Latinità in various initiatives to garner the aid of Argentines, especially the sons and daughters of Italians. *Il Mattino* argued that the "legitimate relationship" between Italy and Argentina, both "Latin and Catholic" and evidenced by the some two million Italians and children of Italians in Argentina, made it "absurd" for Argentina to enforce the sanctions with vigor.[86] That Argentina joined the sanctions, even with reserve, was particularly appalling because Argentina was "a friendly nation, a sister nation, the only nation in the world through whose veins flow half-Italian blood."[87] The paper helped form the Comité Argentino Pro Italia, an Italian patriotic organization comprised of three hundred thousand second-generation Italians, and published the weekly column "La página de los hijos de italianos" (The children of Italy publication) in Spanish, aimed at cultivating their loyalty.[88] *Il Mattino* covered "Italy's Week" in mid-December 1935 to drum up wide-scale support for Italy during the sanctions, citing Latinità as justification. The paper called out in Spanish to "Italians! Argentine children of Italians! Latins!" to participate.[89] During "Italy's Week," *Il Mattino* asked all stores in downtown Buenos Aires to fly an Italian flag as proof of retailers' "faith in the fated destiny of Latin civilization, that is the Italian civilization, that is Argentine civilization." These bonds of amicable fraternity went only so far; the paper threatened violent reprisal against noncompliant Argentine-owned business owners who failed to fly the flag: "Our co-nationals take note of those stores and won't forget them. Remember the old Italian saying: 'eye for an eye, tooth for a tooth.'"[90]

The League of Nations' sanctions against Italy in the mid-1930s provoked discussion and debate in migrant marketplaces over Italy, the Italian people, and Italians abroad as global food consumers. While both pro- and anti-Mussolini leaders objected to the boycott, they presented contrasting images of Italians under fascism. New York's *Il Progresso* depicted Italy as able to meet the basic consumer and nutritional needs of its people and, paradoxically, represented Italians as prepared and committed to living an austere life. Buenos Aires' anti-Mussolini *La Nuova Patria* instead emphasized Fascist Italy's failure to feed its people while portraying Italians as deprived and hungry. Pro-Mussolini supporters in both countries pressured Italians to support Italy from afar during the sanctions by preferring Italian imports and by rejecting goods from sanctioning countries. In Argentina—a sanctioning country—*Il Mattino* softened its rhetoric against Argentine manufactures and even co-opted Argentine producers and consumers, as well as U.S. manufacturers, into their initiatives. As during World War I, migrants called on one another to consume and produce for their homeland in need, but by the late 1930s fascists faced two migrant marketplaces that had been much transformed by trade and migration and by the politics, economics, and cultures of their host countries. A close look at the Italian-language press suggests that despite intensified efforts by Fascist Italy and Italian trade promoters to nationalize migrant marketplaces and bring migrant consumption in line with fascist goals, the competition for migrant consumers was fierce; migrant entrepreneurs as well as U.S. and Argentine companies all competed for the dollars and pesos of Italians and their children in le due Americhe. In this battle over migrants' consumer loyalties, women, so visible in ads for Italian, Argentine, and U.S. foodstuff by the 1930s, exposed the limits of transnational nation and empire building through consumption.

Fascism, Gender, and Consumption: The Competition for Female Migrant Consumers

During the League of Nations' sanctions against Italy, San Pellegrino ran an ad in *Il Mattino* featuring a woman who, having partied too hard, required immediate digestive relief (see fig. 22). The ad depicted a scene from a festive costume ball; a

Figure 22. Ad for San Pellegrino sparkling water, *Il Mattino d'Italia* (Buenos Aires), February 26, 1936, 1. Courtesy of the Hemeroteca, Biblioteca Nacional de la República Argentina.

woman in the foreground adjusts her mask while behind her couples dance beneath balloons. "After the party if you ate or drank too much," the publicity instructs readers, "regulate your digestive system by drinking San Pellegrino."[91]

While the ad portrayed immoderate food and beverage consumption tying migrants to Italian imports, in Italy Mussolini condemned displays of decadence suggested in the ad and demanded that Italians, women in particular, subordinate their consumer desires to the needs of the autarkic state. Appearing in *Il Mattino*, which received financial support from Fascist Italy, the ad is particularly ironic because it shows Italian women abroad engaging in the type of overindulgent material and bodily consumption associated with the United States and Western Europe but reviled by fascists in Italy. By the 1930s the pro-Mussolini migrant press in both Argentina and the United States used female consumption to generate Italian identities; at home fascists instead attempted to separate Italian women from U.S.- and Western-style consumption. The disconnect between messages about femininity and consumption abroad and at home reveals the peculiar position female consumers held in migrant marketplaces as generators of national identities under fascism.[92]

As advertising became more omnipresent in Italian-language newspapers in New York and Buenos Aires in the late 1920s and 1930s, women took an increasingly prominent place in promotional material for a large variety of consumer items, especially industrialized foodstuff, but also nonedible products such as automobiles, clothing, and beauty products. The press's interest in mobilizing migrant women during World War I as transnational producers, savers, and spenders had launched women onto the pages of the migrant press; the war brought about a new set of regular columns, articles, and entire pages directed specifically at female readers. During the interwar years, this space devoted to women enlarged; Buenos Aires' *La Patria* ran a regular column called "Vita Femminile" (Feminine life), and by 1925 an entire women's page called "La Donna in Casa e Fuori" (The woman inside and outside the home) appeared, devoted to fashion, housekeeping, cooking, and motherhood.[93] In New York's *Il Progresso*, "Per Voi, Signore" (For you, ladies), which began as a column in the early 1920s, expanded to an entire women's page in the Sunday section by 1927, covering, like its Buenos Aires counterpart, topics related to food, fashion, and homemaking.[94] The paper also published a shortened version of "Per Voi, Signore" called "Vita Femminile"; this column, while running sporadically since the early 1920s, became a daily column by 1937.[95] This increased attention to women reflected growing literacy rates among women migrants and their daughters as well as changing gender ratios in migrant communities during the interwar years, when more women entered the country and as a more gender-balanced second generation of Italians came of age in the United States and Argentina.[96] By the mid-1920s women had gained a new visibility in Italian-language newspapers, as readers and consumers.

As discussed in chapter 4, merchants after World War I began featuring women in ads using imports to sustain and guard transnational trade routes and to reinforce the Italianità of migrant households. The migrant marketplaces of New York and Buenos Aires during the 1920s and 1930s became feminized as women, rather than men, became the main conduits for generating Italian national identities and consumer patterns around food. And yet by the mid-1920s, Argentine and U.S. companies too sought to capitalize on intensifying linkages between femininity and consumption to sell their products in the Italian-language press. Already in the early twentieth century, companies in Argentina, especially large department stores like British-owned Harrods and Gath y Chaves, ran ads in *La Patria* that marketed foodstuff, fabrics, clothing, and home furnishings specifically to women.[97] After World War I, as Argentina's manufacturing sector grew, greater numbers of Argentine food, beverage, and cigarette manufacturers, such as Piccardo company's 43-brand cigarettes and Diadema cooking oil, also gave prominence to women in ads appearing in *La Patria* and, later, *La Nuova Patria*.[98] Similarly, New York department stores, such as Alexander's, Wanamaker's, and May's, advertised their clothing and furniture to female readers of *Il Progresso*.[99] Furthermore, by the interwar years a wide variety of U.S. companies, such as Heckers flour, Armour meat products, and American Tobacco, began speaking specifically to women, with some passing their products off as "Italian."[100]

It is not surprising that for a regime that was preoccupied with masculinity, male-centered ads for Italian imports emerged in the early 1930s. In many ways these ads resembled trademarks from the late nineteenth and early twentieth centuries that portrayed manly markets of merchant explorers, warriors, and prized symbols of Italian industrial, technological, and cultural genius.[101] Fascists in the 1920s and 1930s, like nationalists in the late nineteenth century, would again resurrect the image of imperial Rome as the central vision for a "New Italy." Italy's defeat of Ethiopia actualized Francesco Crispi's late nineteenth-century failed colonial attempts in East Africa. Mussolini crafted and represented a new fascist model of masculinity grounded in aggressive virility and a strict division of gender roles.[102] However, by the 1920s, as a powerful symbol of transnational relationships between migration and trade, Italian women had indelibly shaped migrant food consumption in ways that U.S. and Argentine companies, and even pro-Mussolini businesses and migrants abroad, could not ignore. In New York's *Il Progresso* and Chicago's *L'Italia*, for example, ads for Bertolli olive oil, Buitoni pasta, Pastene coffee, and Florio Marsala all gave special prominence to women.[103] Similarly, in Argentina's *La Patria*, ads featuring women ran for Cirio canned tomatoes and sauces, Cinzano vermouth, and Sasso olive oil.[104]

The feminization of consumption became particularly visible and politicized with the rise of fascism in Italy when Mussolini attempted to strengthen Italy's relationship to migrant marketplaces in the United States and Argentina—in part by supporting the Italian-language press and Italian chambers of commerce abroad—while disassociating consumption from womanhood in Italy. Economically, the female consumer assisted Italy as the money from purchases of Italian canned tomatoes, boxed pastas, cheeses, and bottled wines fortified Italian companies and the Italian economy more generally. Politically and culturally, however, she represented a paradox. For while in New York and Buenos Aires the pro-Mussolini Italian-language press depicted the female consumer as producing Italian identities linked to Fascist Italy, Mussolini curbed female consumption at home while co-opting it to advance the interests and policies of the fascist state.

Ads for Italian foodstuff during the late 1920s and 1930s created transnational and generational ties by picturing Italian women in the United States and Argentina using the same wide variety of brand-name products that mothers and grandmothers cooked with back home. However, fascist food and trade policy demanded that women in Italy make do by feeding themselves and their families monotonous meals with little nutritional value. Carol Helstosky explains how fascists built a national cuisine based on austerity. As part of Mussolini's larger, inward-looking, autarkic aspirations, the fascist state managed the purchase, preparation, and consumption of food; limited imports; and increased wheat and cattle production. These policies, rather than providing adequately for Italy's people, led to major agricultural and industrial food shortfalls. Food shortages especially intensified during the League of Nations' boycott, forcing most Italians to do without flour, cornmeal, and other basic staples.[105] State-sponsored cookbooks, home economics literature, and magazines aimed at middle-class women promoted a diet that reflected the national imperative of feeding Italians with simple and cheap dishes; women were encouraged to prepare meals that relied heavily on carbohydrates, domestically grown produce, and very little meat and imported ingredients. This literature celebrated Italian women's thriftiness and sobriety while critiquing female consumers in the United States and Western Europe, women portrayed as immorally materialist, greedily individualistic, and insalubriously skinny.[106] Mussolini's food policies aimed at women buttressed his parallel battle against foreign beauty, body, and fashion models. As Victoria de Grazia has shown, fascists juxtaposed the Italian "authentic woman"—a wholesome, fecund, full-figured woman, eager to procreate for the state—against the image of the thin U.S. woman, the *donna crisi*, or crisis woman, who was genderless and sterile, a deprived boy-like waif subsisting mainly on cigarettes, cocktails, and canned goods.[107] Mussolini attempted to control women's identities, labor, spending habits, and reproductive choices as part of his larger aim to nationalize women as self-sacrificing mothers and wives rather than as consumers of U.S. industrialized foods, leisure, and cultural products.[108]

Ads for foodstuff in the pro-Mussolini Italian-language press abroad in some ways substantiated gendered imagery emanating from Italy. In these ads women supported transatlantic commodity chains by buying Italian products, which they used to maintain and police the Italianità of their families, themselves, and the dishes they served at the dinner table. Other times, however, publicity advanced messages about femininity and Italianità that directly contradicted fashion and food directives aimed at women in Italy. Like the woman soothing her dyspeptic stomach with San Pellegrino mineral water at a festive soirée, these ads associated Italian women with overconsumption and with bodily ideas directly at odds with the *Nuova Italiana*, the New Italian woman of Fascist Italy. In an ad for Florio Marsala in *Il Progresso* in the mid-1930s, a group of men stare open-mouthed at a modish, slender woman in lingerie. A man poses a question to the nearly nude svelte lady: "At this age how do you conserve the beautiful complexion of a young girl?" She maintains her diaphanous shape, she tells her admirers, by drinking eggnog made with Florio Marsala every morning, "which nourishes me, without making me gain weight, and keeps my body slim and my spirit awakened."[109] The ad encouraged women to purchase the Italian beverage as a device for achieving a particular form of fashion—and especially weight-conscious femininity rejected by fascists in Italy. Similarly, an ad in Buenos Aires' *Il Mattino* for Pirelli-brand shoes marketed a number of styles for women, including the "Alicia," the "Lavinia," and the "Anita," as well as others made especially for yachting. Pirelli's ads for shoes made with Pirelli rubber heels—"il taco degli Italiani" (the heels of Italians)— suggested that female identities, rather than being fixed within the confines of the Italian nation-state, or any nation-state for that matter, were forged by the changing preferences of the female migrant consumer.[110] Ads for Italian imported foods and other products depicted women buying Italian goods and using them to maintain a distinctly Italian household that revolved around women's roles as mothers, wives, and food preparers, images that complemented fascist ideals of domesticity. But they also showed women using imports to achieve identities and bodily standards in contrast to fascist dictates while drawing on ideas about consumption and femininity circulating in the United States, Argentina, and other countries with mature and burgeoning consumer societies.[111]

Furthermore, the larger culinary landscapes within which migrants and their children consumed imports differed from Italian food culture in Italy. For over three decades, migrants and their children had been experimenting with new ingredients in the United States and Argentina, integrating them into regional traditions from back home.[112] The integration of these novel foods into migrants' everyday eating patterns is visible in recipes on women's pages and columns in the Italian-language press during the 1920s and 1930s. Not only did newspapers include recipes for dishes that added new foods, especially meat and fish, into Italian regional specialties, but they also included recipes and ads for dishes and beverages associated with typical U.S. and Argentine food cultures, such as Thanksgiving Day turkeys and apple pies in the United States and empanadas and yerba mate in Argentina.[113] While ads for Italian imports portrayed Italian women in Italy and those in the United States and Argentina eating the same meals, the food practices of migrants and their children abroad—and into which Italian imports were incorporated— differed from those in Italy. Access to novel and cheap ingredients, new culinary experiences, and economic and familial imperatives differentiated migrant cuisines in the United States and Argentina from Italy and from each other. Moreover, by the late 1930s fascist food policies deprived Italians of the varied meals and of the expensive imported Italian coffee, pasta, wine, canned tomatoes, and cheeses portrayed in migrant newspapers and used to generate Italian identities.

That migrants associated women with the consumption of Italian and U.S. goods in order to generate Italian identities in Argentina must have been particularly frustrating for fascists who saw in Argentina the most promising location for strengthening economic, political, and cultural ties to Italians abroad. Mussolini and his fascist supporters both in Italy and Argentina stressed the shared racial and cultural bonds that united Argentines and Italians to encourage Argentine support for the "New Italy" and to compel migrants and their children to favor fascist policies.[114] The proliferation of U.S. consumer items in Argentina, as well as their accompanying messages about womanhood, however, challenged Mussolini's visions of Argentina as the most favorable overseas Italian "colony" for the transfer of fascist ideologies. By the 1930s fascists encountered an Italian-language press in Buenos Aires that was saturated with U.S. consumer goods. Already in the early twentieth century Argentina's increasingly consumer-oriented society began associating women, especially upper-class and middle-class women, with the nation's inchoate consumer society.[115] Once U.S. companies began marketing heavily in the Italian-language press in Argentina, women's presence in *La Patria*, and later *La Nuova Patria* and *Il Mattino*, rose exponentially. As more U.S. firms moved south to seek markets in South America, so too did the ties between femininity and consumer culture that were becoming so prevalent in U.S. print cultures. Because portrayals of U.S.-style feminization of leisure, consumption, and beauty so contrasted the ideals of the fascist *Nuova Italiana*, and because Argentina held a prominent place in the fascist imaginary, Argentina serves as a particularly unique location for exploring triangular interactions between consumption and gender amid the intense nationalism characterizing the interwar years.

By the early 1920s New York's *Il Progresso* discussed U.S. consumer goods in the context of institutions, styles of consumption, and female bodily standards that were popular in the early twentieth century United States. Articles and ads on "Per Voi, Signore" pages, especially those related to clothing and beauty products, were tied to images and discussions of college campuses, sporting events, and especially Hollywood stars. These articles connected migrant women's purchases to ideas about modernity and individuality circulating in larger U.S. society and, in particular, in women's magazines designed for middle-class white audiences.[116] By late 1920s and 1930s these links between femininity, consumption, and U.S. cultural products became increasingly prominent in Buenos Aires' *La Patria*. In the mid-1920s, for example, the newspaper began to regularly cover articles about Hollywood actresses and actors, U.S. fashion and beauty icons, and jazz music, often in a regularly published page titled "Tra cinematografi e 'films'" (Between cinema and "films").[117] Ads for U.S. goods in *La Patria* and *Il Mattino*, everything from foods to electrical appliances to beauty products, overwhelmingly portrayed women buying in ways that suggested affluent leisure patterns, progressive gender arrangements, and consumer abundance.[118] As explained in chapter 5, at times these companies treated readers as migrants by Italianizing their products. Other ads for U.S. goods in the Buenos Aires' Italian-language press—like their counterparts in New York—presented homogenous images of white middle- or upper-class lifestyles, especially in ads for U.S. automobiles, which depicted affluent women tooling around town in their cars, often unaccompanied by men.[119]

The omnipresence of U.S.-style models of womanhood and consumption in Buenos Aires' migrant marketplace made images of Italian women especially contradictory in the pro-Mussolini *Il Mattino*. While the paper ran a women's page expressly for female readers in the early 1930s, by 1935 it had removed almost all columns and material about fashion, consumption, and, in particular, Hollywood and other trappings of U.S. culture that had previously prevailed in *La Patria*. Instead, *Il Mattino* mentioned women in articles almost exclusively as they related to larger transnational fascist campaigns.[120] During the League of Nations' boycott, *Il Mattino* covered the fund-raising activities of female auxiliaries of various "Pro Patria" organizations. As with *Il Progresso* in New York, *Il Mattino* extolled migrant women during "Day of Faith" ceremonies, during which women donated wedding rings, gold jewelry, and other precious metals to the fascist cause.[121] In 1935 *Il Mattino* reprinted fascist and futurist leader Filippo Tommaso Marinetti's "ten commandments for Italian women during the Africa war." The commandments asked women "to practice a rigid discipline and severe economy" and "to live a simple and austere life."[122] A visit by Marinetti to Argentina during the sanctions served as a particularly opportune moment to urge women in Argentina toward fascist ideas of domesticity, cuisine, and consumption and away from the U.S.-inspired models. Marinetti hoped to use food to generate a more modern Italian identity under fascism, a goal concretized with the publication in 1932 of *La cucina futurista*, a culinary treatise that, quite controversially, denounced pasta for lighter, more creative dishes.[123] His travels through Argentina began with an enthusiastic reception and breakfast hosted by *El Hogar*, Argentina's most popular women's magazine.[124] In *El Hogar*'s salon, Marinetti prepared futurist dishes for the female director of the magazine along with a group of Argentine and Italian women and notable Italian businessmen. Marinetti even invented a dish in honor of the occasion: a meat roll stuffed with oranges called "Da mangiare a cavallo" (food for horse riding) and dedicated especially to gauchos, the cowboy figure once spurned by elites as barbaric and provincial but by the 1930s resurrected as a proud symbol of Argentine national identity.[125] Marinetti's visit to *El Hogar* and his fascisization of an Argentine culture figure promoted pride in fascist-inspired foodways among Italian and Argentine women while forging bonds between them.

Even while the paper encouraged values of patriotism and frugality among Italian migrant women, it continued to run ads for numerous U.S. products that linked migrant consumption to messages that were at odds with *Il Mattino*'s official treatment of women. Publicity in *Il Mattino* for Kent cigarettes, owned by the American Tobacco Company, featured four provocatively dressed, slender blonde women, one for each of the four letters making up the "Kent" brand name. The ad played on the word *bionda* (light) to describe both the hair color and complexion of the women and the bionda cigarettes sold by Kent. In what amounted to a confusing sensory and sensual overload, the ad invited readers to "conquer," "possess," and "please" themselves, the cigarettes, and the four women the cigarettes embodied by purchasing the "American cigarettes of superlative quality."[126] Similarly, an ad for Armour pasta sauce depicted a happy housewife preparing to serve her husband an enormous dish of steaming pasta. The ad tells readers that the canned sauce "requires no work, because it comes properly prepared."[127] This modern, fashionably dressed, slender housewife, released from the time-consuming burden of preparing foods from scratch by using a costly, canned meaty tomato sauce—"hot and ready to serve in only two minutes"—suggested a model of domesticity, consumption, and womanhood that fascists at home found threatening. Ads such as these for U.S. items, running in the pro-Mussolini migrant press, appeared incongruous next to articles touting the official party line. Argentina proved not to be the ideal location for building the *Nuova Italiana* of Fascist Italy. The extension of U.S. commercial influence into Argentina, and the overwhelming visibility of women in ads for all consumer goods, created expressions of womanhood and Italianità in New York and Buenos Aires that resembled each other much more than those of Fascist Italy.

In the 1930s, Italy, the United States, and Argentina all battled for the attention of the migrant consumer, especially the female migrant consumer, who after World War I became ubiquitous in ads for Italian imports and domestically produced foodstuff and other goods in Argentina and the United States. Women's overwhelming presence in ads made female migrants important vehicles for mobilizing Italians abroad in relation to imports. And yet ties between consumption and femininity in migrant marketplaces of the Americas posed a problem for Mussolini, as he was attempting to break associations between consumption and womanhood at home. Fascist Italy restricted the consumption of consumer goods—especially liqueurs, coffees, fashionable clothing, and other lavish items—and denounced U.S. forms of entertainment and leisure pursuits among women in Italy. Migrant newspapers, even those receiving financial backing from the fascist state, however, encouraged the consumption of these items and experiences among migrants. In presenting women buying imported and domestically made consumer luxuries rejected by fascists and unavailable in Italy, the female consumer in the pro-Mussolini Italian-language press was indeed a paradoxical creature.

*   *   *

Some expressions of fascism kept their ideological and organizational form as they moved from Italy to the Americas and emerged in pro-Mussolini migrant newspapers. During the League of Nations' boycott, *Il Progresso* and *Il Mattino* portrayed migrants answering fascist calls to oppose the sanctions by buying Italian imports, by spurning and even demonizing products from sanctioning countries, and by putting pressure on U.S. and Argentine politicians to maintain favorable trade policies with Italy. However, Mussolini struggled to transplant intact across the Atlantic fascist discourses about women and food consumption. Women's increasingly prominent role in ads for and discussion about imports during the 1920s and 1930s discloses the importance of female consumers in sustaining a transatlantic network of Italian trade and consumer behavior, even after the rise of fascism, when economic and political ties between Italy and the Americas became increasingly restricted and when fascists attempted to disrupt ties between women and consumption both in Italy and abroad. While rejecting images of U.S. and Western European–style consumption and beauty ideals among Italians, Mussolini helped expand these very images by supporting migrant newspapers like *Il Progresso* and *Il Mattino*. These papers employed the consumption of both Italian and non-Italian goods by migrant women to generate Italian identities. This transatlantic disconnect reveals mixed messages about womanhood, consumption, and nationhood in migrant newspapers.

During the 1930s, migrant newspapers depicted Italians in the United States, Argentina, and Italy all buying and eating Armour meat products. That ads for Armour prosciutto appeared in New York's *Il Progresso* and in Buenos Aires' *La Patria* and *Il Mattino* points to how U.S. goods increasingly linked Italian diasporas in North and South America to each other, as well as to Italy. Furthermore, pro-Mussolini Italians in Buenos Aires sent more than ten tons of Armour-brand meat on steamships to Italy at the exact period when Mussolini himself strove to make the country economically independent by limiting the importation of foreign goods and Italians' consumption of meat. Despite fascists' endeavors to monopolize migrants' interactions in migrant marketplaces, by the 1930s Italian, U.S., and Argentine companies all jockeyed, with growing success, for the attention of migrant consumers. Fascist Italy strove to turn migrants' heads with pride toward a new imperialist and economically formidable homeland; however, campaigns such as the "*pacchi tricolori*" drive reveal that Italy remained a diasporic nation as it had at the turn of the twentieth century, a nation that was dependent on support from its peoples abroad.

Epilogue

Sugary, slightly effervescent, licorice-flavored mouthwash. Both indelicate and desirable, Fernet con Coca has a taste that is difficult to describe. The cheerful sweetness of the carbonated Coca-Cola cuts the sharp bitterness of the seriously herbal fernet, producing a surprisingly balanced libation. A sip of the frothy Fernet con Coca, also known as a "Fernandito," conjures up images of Julie Andrews as Mary Poppins crooning to the impressionable Jane and Michael Banks, "Just a spoonful of sugar makes the medicine go down . . . in the most delightful way."[1] By the early 1990s Fernet con Coca had taken on the status of Argentina's national beverage and an emblem of Argentine pop culture. Ubiquitous in bars, clubs, and restaurants in cities throughout the nation, Fernet con Coca even served as the name of a 1994 music hit by the neo-folk Argentine band Vilma Palma e Vampiros.[2]

The Fernandito's standing as the national drink of Argentina, however, belies its more global origins in the country's migrant marketplaces of traveling people and trade goods in the early twentieth century. As its name implies, the drink is traditionally made from two products that were initially imported into Argentina: fernet, introduced to Argentina by migrants, and Coca-Cola, the bubbly soft drink from the United States. The beverage—with its bizarre hybrid flavor—divulges much about how the historic movements of products and people have shaped the eating and drinking habits, food cultures, and identities of migrants and nonmigrants alike.

Fernet's widespread popularity in Argentina originated in the demand for such homeland tastes by Italian migrants in Argentina starting in the early twentieth century. By this time advertisements for Fernet-Branca—which today remains the most popular brand of imported fernet in Argentina—appeared on the pages of both the Italian-language *La Patria* and the Spanish-language dailies. Ads lauded Fernet-Branca as an excellent appetite stimulant and digestive aid, perfect for calming nerves and dyspeptic stomachs. Publicity for the product resembled advertisements for other imported Italian foodstuff, such as pasta, olive oil, and canned tomatoes, in presenting the *digestivo* as a "grand Italian specialty" and celebrating migrant consumers' transatlantic cultural ties to the homeland they left behind.[3] Yet the fact that publicity for Fernet-Branca materialized in the mainstream Spanish-language press also shows the permeability of Buenos Aires' migrant marketplace as a space that brought Italian and Argentine consumers into liquor shops run by migrant merchants.[4] This same permeability was less prevalent in New York, where racial and cultural differences thwarted migrants' attempts to build shared consumer experiences with Anglo-Americans around imports like fernet.

Throughout most of the twentieth century, consumers in Argentina, like those in Italy, enjoyed fernet as a postprandial digestivo. It would not be until the 1980s, however, that fernet transformed into a symbol of *argentinidad* when it was combined with another imported beverage—Coca-Cola. Unlike fernet, Coca-Cola traveled to Argentina independent of labor migrants and their consumer preferences. During World War II the Coca-Cola Company, looking to expand its markets globally, turned to Argentina and other Latin American countries.[5] They followed the well-trodden footsteps of Armour, American Tobacco, and other U.S. companies who, during the interwar years, targeted migrant and native-born consumers in Latin America and opened manufacturing plants there. Indeed, Fernet con Coca aptly embodies the historic reorientations of trade and migration after World War I that were brought on by responses to progressively protectionist policies against mobile people, capital, and goods. Restrictive immigration legislation in the United States redirected western hemisphere–bound Italians and other migrant groups to Argentina at the same time as increasing numbers of multinational companies saw in Latin America a source of cheap labor and a burgeoning consumer market. In the 1940s the Coca-Cola Company established its first bottling plants in Buenos Aires and Córdoba, and by the 1970s the soft drink had gained widespread popularity across Argentina.[6]

Italian companies also continued their expansionist efforts in Latin America during the postwar years. Capitalizing on the robust consumer demand for the herbal digestivo fostered through decades of transnational migration, the Milan-based Fernet-Branca, founded in the mid-nineteenth century, established a distillery near Buenos Aires in 1941, making it the first branch established by Branca Brothers Distillery.[7] Branca trailed Italian companies like vermouth maker Cinzano and the Pirelli Tire Company, both of which had established manufacturing subsidiaries in the early twentieth century. Both Fernet-Branca and Coca-Cola arrived as Argentina embraced state-directed import substitution industrialization, a nationalist economic policy designed to decrease foreign dependency through domestic production and high tariffs. As part of this plan, Argentina encouraged foreign investment by companies like Fernet-Branca and Coca-Cola, whose manufacturing outlets helped industrialize the country while bolstering nationalist movements under populist president Juan Domingo Perón.[8] These historic collisions of mobile consumers, products, and capital marking twentieth-century globalization merged in the 1980s, when, as the legend goes, bartenders in Córdoba first began mixing fernet and Coca-Cola, thus igniting a national fad.[9]

Despite their humble beginnings in the migrant marketplaces of New York and Buenos Aires, Fernet con Coca, pasta, pizza, and other foodstuff have been mainstreamed as typical Argentine and U.S. family foods and identified with high-end gustatory Italianità worldwide. That they did so only decades after mass Italian migration ended suggests that migrant marketplaces of mobile Italians and foods today exist more in the imaginary and in commodified form than in the actual, embodied movements of people and foods from Italy. While Italian migration to Argentina saw a return to prewar levels immediately after the war, by the mid-1950s the age of Italian proletariat migration to the Americas had come to an end.[10] Two decades of fascist rule and the ravages of World War II produced new economic migrants from Italy, but they traveled to different places. In cities like Toronto, Berlin, and Melbourne, Italians forged new migrant marketplaces. As they had almost a century before in New York and Buenos Aires, Italian labor migrants in these cities opened up trade routes in wines, pastas, cheeses, canned tomatoes, and other homeland foods. Gender ratios in these postwar Italian diasporas gradually began to equalize, especially among transoceanic migrants to Canada and Australia. Like their pre–World War II predecessors in New York and Buenos Aires, the presence of wives and families transformed the way migrants and their children shopped, ate, and identified.[11]

In the two decades following the end of the war, Italy experienced an economic boom characterized by industrialization, job growth, and expanded welfare services and access to education, consumerism, and rising standards of living.[12] By the late 1950s increased job opportunities in northern and central Italy during Italy's "economic miracle" transformed international emigration into internal, rural-to-urban and south-to-north migrations.[13] In cities such as Turin, Milan, and Genoa, Italians from different regions met, as they had in Buenos Aires and New York years before; they visited each other's market stalls, peered into each other's cooking pots, and sampled each other's street foods, often in mistrust. Over time, these sometimes fractious culinary interactions aided in nationalizing Italy's people and a distinctive "Italian" cuisine for domestic consumption and export while solemnizing localism as the hallmark of "authentic" Italian cookery.[14]

As more and more Italians found *pane e lavoro* (bread and work) at home rather than in le due Americhe, the migrant marketplaces forged decades ago by mobile, hungry laborers became unmoored from their original geographies and social relations. In 1905 Italian American journalist and lawyer Gino C. Speranza, writing for New York's *Rivista*, had proclaimed, "For Italy, at the present moment, the most important export to the U.S. is the emigrant, infinitely more important than oil, wine, and macaroni."[15] Over a century later, as emigration slowed, and as second- and third-generation Italians moved solidly into the middle classes and out of their grandparents' tenement buildings and *conventillos*, that order of importance has been reversed.[16] Today it is generally middle-class and affluent consumers in the Americas and globally who drive the transnational trade in Italian foods associated with high culture and refined consumption: award-winning *piemontese* wines, herb-infused olive oils, San Marzano canned tomatoes, syrupy balsamic vinegars from Modena, and pungent slabs of aged *parmigiano* cheese.[17] These culinary and cultural goods are a postwar product of Italian economic success, the "discovery" of the healthful Mediterranean diet, and Italian food revolutionaries like Carlo Petrini. Petrini's slow food movement has championed, with much success, locally sourced ingredients and home-cooked meals as the antithesis of corporate factory foods and the loss of commensality, tradition, and the food diversity they represent.[18]

The mass popularity of Italian foodstuff today suggests, perhaps, that only once foods are divorced from the mass labor migrations they originally trailed, and the diasporic nations from which they left, do migrant-sending countries' foods gain elite status.[19] Slow food advocate Oscar Farinetti established the first Eataly store in Turin in 2007. This mammoth Italian food emporium, which sells Italian food specialties made by small- and medium-size artisanal producers—producers often protected by hard-to-get Italian and European Union certifications guaranteeing their products' place of origin and quality—now has locations in New York, Chicago, São Paulo, Munich, Istanbul, Dubai, Seoul, and Tokyo. It is easy to forget that in some of the cities decades ago, Italians, their foods, and the culinary traditions they brought with them were stigmatized as undesirable symbols of working-class culture.[20]

Nevertheless, imagined migrant marketplaces continue to play a critical role in the performance of ethnicity for the descendants of Italians and in the consumption of Italianità for Argentines and U.S. Americans. While some items like Fernet con Coca and pizza have been naturalized into Argentine and U.S. culinary landscapes, third- and fourth-generation Italians have sought to rediscover their roots in part by buying and cooking with imported ingredients from their great-grandparents' birthplaces. Ironically, those kinfolk may have been too poor to eat such specialties both before and after migration, relying instead on the cheaper tipo italiano substitutes.[21] Since the 1980s these now middle-class consumers of Italian heritage, as well as consumers with no family connection to Italy, have celebrated the hybrid migrant creations that emerged out of New York's and Buenos Aires' migrant marketplaces earlier in the century—classic meat, tomato, and cheese-based dishes like veal parmigiana in the United States and *milanesa napolitana* in Argentina.[22] For others, however, consuming Italian imports offers an alternative to those down-market culinary amalgamations invented by migrants, which they perceive as "not real Italian." Their cultural and economic capital lies in the knowledge they possess about imported wines, cheeses, and pastas from Italy's various regions; their ability to purchase them; and their patronage of restaurants in Buenos Aires and New York that have disassociated their fare from its modest migrant beginnings in favor of the "authentic" dishes of Italian regional cuisines.[23]

In their pursuit of true Italian eating experiences, consumers have the help of a host of first-, second-, and third-generation Italian celebrity chefs, restaurateurs, cookbook authors, and owners of upscale grocery stores, many of whom, including Mario Batali, have procured their cultural capital through tourism back to Italy. And today, as in the past, Italian food production and consumption in imagined migrant marketplaces of New York and Buenos Aires are not wholly gender-neutral. Smaller mom-and-pop Italian eateries and corporate giants like Olive Garden deliberately craft an atmosphere of ethnic domesticity, a largely feminized space where the food, décor, and music conjure up warm images of informal Sunday dinners prepared by a skillful *mamma* and where family, tradition, and nostalgia for the "old country" endure.[24] Since the 1980s, however, notwithstanding the presence of Italian female culinary personalities such as Marcella Hazan and Lidia Bastianich, the transplantation and commercialization of "authentic" regional ingredients and dishes, and their modernized cutting-edge reinterpretations, are dominated by professional male chefs, restaurateurs, and cookbook authors like Batali in New York and Donato De Santis in Buenos Aires.

Because the current passion for Italian food in the United States and Argentina emerged out of past migrant marketplaces, the Italian nation continues to capitalize on global connections between Italian people and goods, although they are detached from Italy's imperial ambitions of the early twentieth century. Back then, Italian labor migrants in le due Americhe helped create a *più grande Italia*, a nation whose emigrants pushed the geographical and cultural borders of the nation-state, building it from without by buying Italian products, sending home remittances, and returning to their home villages.[25] Italy's foods have now achieved the prestige status so desired by Italian economic and political elites on both sides of the Atlantic who desperately sought to portray Italian foodstuff as emblems of European civilization.

Today Italy and Italian food is being remade again from without, and not only by its citizens and their descendants dispersed throughout the world. Since the 1980s Italy has transformed from a migrant-sending to a migrant-receiving country.[26] Labor migrants from Eastern Europe, North Africa, and Asia, and, most recently, refugees from the Middle East, have created vibrant migrant marketplaces in small and large cities throughout Italy. Among these new arrivals are the ancestors of Italians born in Argentina, who have acquired Italian citizenship by descent and a European Union passport in the hopes of better prospects in Europe after Argentina's economic depression and currency collapse in 1998.[27] Just as homeland foods helped Italian construction workers and factory hands in New York and Buenos Aires create value in their lives as transnational people, so too have such foods for Filipino nurses in Rome, Chinese textile workers in Prato, and Tunisian street traders in Palermo today. These newcomers bring with them foods and traditions, they experiment with Italian foodways, and they introduce Italians to multinational eating experiences. Chinese restaurants run by migrants, with their Italianized versions of traditional fare and modest prices, are now commonplace in large cities and small towns throughout Italy.[28]

Migrants' visibility (and often invisibility) in the nation's food industries as agricultural workers, grocery store and restaurant owners, chefs and kitchen staff has provoked both curiosity and discomfort among Italians. Italy's multicultural eaters do not always support multicultural policies related to migration and naturalization. Migrants have been met with xenophobic and racist sentiments and calls for the restriction of immigration, residency permits, and work visas.[29] Unsurprisingly, fear of the perceived economic, political, and social menace posed by migrants finds expression in food politics. In 2004, for example, the Lega Nord, the right-wing, anti-immigrant regional party, staged a polenta protest in Como, Lombardy, during which some eighty pounds of polenta were handed out to bystanders as a defensive stance against exogenous threats to traditional regional foodways. Similarly, in 2009 the Tuscan town of Lucca banned kebabs and other foods considered non-Italian from the historic city center.[30] A century ago foods like polenta marked Italians as impoverished "others" abroad; today such foods are used in the othering of foreigners and their foods in Italy. Like they were in New York and Buenos Aires years ago, migrant marketplaces in Italy today are sites where racial differences are articulated to justify exclusion and celebrate pluralism.

Today's food fights point to the danger that lurks behind imagined migrant marketplaces detached from their embodied, material, and territorialized past. The risk lies in a historical amnesia that uncouples food commodities from the actual cross-border movements of the people who facilitate them, movements achieved against the backdrop of global, gendered, and often inequitable processes. The popularity of Fernet con Coca in Buenos Aires and pizza globally is not the simple outcome of Italian food culture's "expansion" or "spread" throughout the world. Histories of culinary globalization that rely on abstract diffusionist theories tend to divorce cuisines from the political, legal, and economic policies that allow for or discourage the mobility of food producers and consumers and to ignore the food hybridity and experimentation such mobility creates.[31] As this book argues, Italian migrants' choices and strategies drove the history of Italian food in the Americas, and they did so within large structures in Italy, Argentina, and the United States— labor demands, gendered assumptions about production and consumption, trade and migration policies, and racial hierarchies—that constrained and guided what, where, and how people ate. As in the past, today's mobile people and products, and impediments to such mobility, change the way migrants and nonmigrants eat, the identities forged by migrant eaters and inscribed upon them, and, ultimately, the larger culinary infrastructures in which they are enmeshed.

The history of Italian migration, trade, and consumption in North and South America suggests that there is much to learn about globalization and its consequences for food cultures by focusing on migrant marketplaces. The interconnections between people and products on the move that so shaped global integration in the late nineteenth and early twentieth centuries are just as important to twenty-first-century globalization, despite the reluctance of policy makers and multicultural eaters to see these connections. Several binational or multinational accords passed over the last thirty years have secured rights for products to enter and exit borders. Lower trade barriers, the foremost goal of these trade agreements, and other neoliberal economic policies aim to simultaneously promote national development and global economic integration. And they have given products (commodities, capital, services, technologies, and information) more freedom to traverse national boundaries than the people, particularly the low-wage workers, whose labor remains instrumental to the global economy. Furthermore, the planners of such accords have largely failed to anticipate or have remained indifferent to the consequences these trade partnerships have had on migration patterns as roadblocks to products, services, and capital lessen while barriers to mobile people remain unchanged.[32]

The North American Free Trade Agreement (NAFTA), passed in 1994 by the United States, Canada, and Mexico, serves as a telling example of this failure. While NAFTA generated hundreds of thousands of mainly low-paying jobs in *maquiladora* factories along the U.S.-Mexican border and in big retail stores like Wal-Mart in cities across Mexico, migration to the United States since 1994 has expanded. The opening of borders to commodities and investment made it increasingly impossible for small and medium-size Mexican farms and factories to compete, displacing large numbers of laborers hungry for work. Many Mexicans responded to economic changes brought on by NAFTA by migrating to the United States, where their labor continues to be in high demand, particularly in low-paying, seasonal agricultural work—including work in Midwestern corn fields that make possible the mass production of industrialized corn flour used in Mexican-style fast-food eateries like Taco Bell—even while obstacles to migration and U.S. citizenship have become more arduous. International Studies scholar Peter Andreas points to the challenges posed to countries like the United States and Mexico, which promote neoliberal initiatives to advance economic flows while intensifying law-enforcement policies to protect their nations from illicit movements of people and goods, especially drugs. Because neoliberal globalization itself promotes undocumented migration and illegal drugs, the state's border policing is more performative and symbolic, designed to assuage anxious citizens by projecting images of sovereignty and control, than it is an effective strategy for upholding the state's territorial rule.[33] Unable to see international trade and migration patterns as connected, or simply unmoved by NAFTA's effects on low-wage migrant laborers, its architects have all but guaranteed that migration will remain an unresolved, contentious topic in United States, Mexico, and Canada for decades to come.

European regional accords on trade and migration have raised similar debates over globalization and national sovereignty. The legal building blocks upon which the EU was established, the 1985 Schengen Agreement and the 1986 Single European Act, eliminated many of the region's internal borders to people, capital, and goods. These treaties and their amended versions, while allowing for increased mobility of products, including food and agricultural exports, have intensified both the migration incentives for non-EU labor migrants and refugees and the surveillance of the bloc's external borders.[34] The historical formation of New York's and Buenos Aires' migrant marketplaces, emerging from connections between people and commerce and obstructions to these connections, suggests that solutions depend on studying these inextricable links between migration and trade, and by recognizing both, together, as crucial value- and wealth-generating facets of globalization and nation building, whether in the nineteenth or twenty-first century.

Notes

Introduction

1. Ad for Cella's, *Il Progresso Italo-Americano* (New York), November 29, 1925 (hereafter *Il Progresso*).

2. Ad for Cella's, *Il Progresso*, December 13, 1925.

3. On the global scattering of Italy's people, see especially Donna R. Gabaccia, *Italy's Many Diasporas* (Seattle: University of Washington Press, 2000).

4. Gianfausto Rosoli, ed., *Un secolo di emigrazione italiana, 1876–1976* (Rome: Centro Studi Emigrazione, 1978), 353–55.

5. This percentage is based on trade statistics from several Italian government sources. For the years 1880–1902, see *Movimento commerciale del Regno d'Italia*, issued by the Ministero delle Finanze, Direzione Generale delle Gabelle; for the years 1903–1933, see *Annuario statistico italiano*, issued by Direzione Generale della Statistica e del Lavoro (1911–1915), issued by Ufficio Centrale di Statistica (1916–1918), issued by Direzione Generale della Statistica (1919–1921), issued by Istituto Centrale di Statistica del Regno d'Italia (1927–1943); for the years 1934–1938, see Istituto Centrale di Statistica, *Commercio di importazione e di esportazione del Regno d'Italia* (Rome: Tipografia Failli, 1939), 638, 640.

6. While historians of migration have long been aware of such cross-border processes, the pioneering work of anthropologists during the early 1990s ignited a multidisciplinary torrent of studies on transnational migration. See Nina Glick Schiller, Linda Basch, and Christina Blanc-Szanton, "Towards a Definition of Transnationalism: Introductory Remarks and Research Questions," *Annals of the New York Academy of Sciences* 645 (July 1992): ix–xiv; Nina Glick Schiller, Linda Basch, and Christina Blanc-Szanton, "From Immigrant to Transmigrant: Theorizing Transnational Migration," *Anthropological Quarterly* 68, no. 1

(1995): 48–63. Earlier work on Italian and European migration from a transnational perspective include Robert F. Foerster, *The Italian Emigration of Our Times* (Cambridge: Harvard University Press, 1919); Frank Thistlethwaite, "Migration from Europe Overseas in the Nineteenth and Twentieth Centuries," in *A Century of European Migrations*, 1830–1930, ed. Rudolph J. Vecoli and Suzanne Sinke (Urbana: University of Illinois Press, 1991), 17–57; and Ernesto Ragionieri, "Italiani all'estero ed emigrazione di lavoratori italiani: Un tema di storia del movimento operaio," *Belfagor, Rassegna di varia umanità* 17, no. 6 (1962): 640–69.

7. A short list on Italian migration to the United States and Argentina includes Virginia Yans-McLaughlin, *Family and Community: Italian Immigrants in Buffalo, 1880–1930* (Ithaca, NY: Cornell University Press, 1977); John W. Briggs, *An Italian Passage: Immigrants to Three American Cities, 1890–1930* (New Haven, CT: Yale University Press, 1978); Dino Cinel, *From Italy to San Francisco: The Immigrant Experience* (Stanford, CA: Stanford University Press, 1982); Donna R. Gabaccia, *From Sicily to Elizabeth Street: Housing and Social Change among Italian Immigrants, 1880–1930* (Albany: State University of New York Press, 1984); Donna R. Gabaccia, *Militants and Migrants: Rural Sicilians Become American Workers* (New Brunswick, NJ: Rutgers University Press 1988); Carol Lynn McKibben, *Beyond Cannery Row: Sicilian Women, Immigration, and Community in Monterey, California, 1915–99* (Urbana: University of Illinois Press, 2006); Piero Bevilacqua, Andreina De Clementi, and Emilio Franzina, eds., *Storia dell'emigrazione italiana. Partenze* (Rome: Donzelli, 2001); Piero Bevilacqua, Andreina De Clementi, and Emilio Franzina, eds., *Storia dell'emigrazione italiana. Arrivi* (Rome: Donzelli, 2002); Donna R. Gabaccia and Fraser M. Ottanelli, eds., *Italian Workers of the World: Labor Migration and the Formation of Multiethnic States* (Urbana: University of Illinois Press, 2001); Donna R. Gabaccia and Franca Iacovetta, eds., *Women, Gender, and Transnational Lives: Italian Workers of the World* (Toronto: University of Toronto Press, 2002); Fernando J. Devoto, *Historia de los italianos en la Argentina* (Buenos Aires: Editorial Biblos, 2006); Gianfausto Rosoli, ed., *Identità degli italiani in Argentina: Reti sociali, famiglia, lavoro* (Rome: Edizioni Studium, 1993); Fernando J. Devoto and Gianfausto Rosoli, eds., *L'Italia nella società argentina: Contributi sull'emigrazione italiana in Argentina* (Rome: Centro Studi Emigrazione, 1988).

8. Christiane Harzig and Dirk Hoerder, with Donna Gabaccia, *What Is Migration History?* (Cambridge, MA: Polity Press, 2009), 3.

9. On regionalism in Italy, see Carl Levy, ed., *Italian Regionalism: History, Identity, and Politics* (Oxford: Berg, 1996); John E. Zucchi, "Paesani or Italiani? Local and National Loyalties in an Italian Immigrant Community," in *The Family and Community Life of Italian Americans*, ed. Richard Juliani (New York: American-Italian Historical Association, 1983), 147–60; *Estudios Migratorios Latinoamericanos*, Special Issue: *Las cadenas migratorias italianas a la Argentina* 3, no. 8 (1988).

10. By focusing on the role of gender in organizing links between trade, migration, and consumption in New York and Buenos Aires, this book builds off of but moves well beyond Mark Choate's *Emigrant Nation: The Making of Italy Abroad*. Choate's study examines Italy's attempts to nation- and empire-build through governmental and nongovernmental programs, including Italian chambers of commerce abroad, that linked emigration and colonialism. Choate, however, is largely unconcerned with gender and with the large literature on consumer culture and food studies. He also favors a global approach rather

than the in-depth place-based comparative perspective privileged in *Migrant Marketplaces*. See Mark I. Choate, *Emigrant Nation: The Making of Italy Abroad* (Cambridge: Harvard University Press, 2008).

11. This book was in part inspired by Samuel Baily's comparative work on Italians in New York and Buenos Aires. I depart from Baily by shifting attention to identity formation and consumer practices around food exports in migrant marketplaces. Samuel L. Baily, *Immigrants in the Lands of Promise: Italians in Buenos Aires and New York City, 1870–1914* (Ithaca, NY: Cornell University Press, 1999); Gabaccia, *Italy's Many Diasporas.*

12. I follow the lead of Lok Siu, who has studied Chinese migrants in Panama within the context of both migrants' ongoing ties to their homeland and to the history of U.S. economic expansion in Panama. Lok C. D. Siu, *Memories of a Future Home: Diasporic Citizenship of Chinese in Panama* (Stanford, CA: Stanford University Press, 2005). On the integration of hemispheric and transnational approaches, see Sandhya Shukla and Heidi Tinsman, eds., *Imagining Our Americas: Toward a Transnational Frame* (Durham, NC: Duke University Press, 2007); Erika Lee, "Orientalisms in the Americas: A Hemispheric Approach to Asian American History," *Journal of Asian American Studies* 8, no. 3 (2005): 235–56.

13. Very few scholars have made connections between the historic movements of labor migrants and trade goods their central concern. Important exceptions include Kevin H. O'Rourke and Jeffrey G. Williamson, *Globalization and History: The Evolution of a Nineteenth-Century Atlantic Economy* (Cambridge: MIT Press, 1999); Donna R. Gabaccia, *Foreign Relations: American Immigration in Global Perspective* (Princeton, NJ: Princeton University Press, 2012); Alejandro Fernández, *Un "mercado étnico" en la Plata: Emigración y exportaciones españolas a la Argentina, 1880–1935* (Madrid: Consejo Superior de Investigaciones Científicas, 2004); Hasia R. Diner, *Roads Taken: The Great Jewish Migrations to the New World and the Peddlers Who Forged the Way* (New Haven, CT: Yale University Press, 2015).

14. See, for example, Patrick Manning, *Migration in World History*, 2nd ed. (New York: Routledge, 2013); Dirk Hoerder, *Cultures in Contact: World Migrations in the Second Millennium* (Durham, NC: Duke University Press, 2002); Jeffry A. Frieden, *Global Capitalism: Its Fall and Rise in the Twentieth Century* (New York: Norton, 2006).

15. A good introduction to sociological perspectives on immigrant enclaves and entrepreneurship is Alejandro Portes, ed., *The Economic Sociology of Immigration: Essays on Networks, Ethnicity, and Entrepreneurship* (New York: Russell Sage Foundation, 1995); Roger Waldinger, Howard Aldrich, and Robin Ward, *Ethnic Entrepreneurs: Immigrant Businesses in Industrial Societies* (Newbury Park, CA: Sage, 1990). More recent transnational approaches have challenged previous depictions of ethnic enclaves as closed economies by showing how migrant businesses facilitate global flows of workers, capital, and products. See Min Zhou, "Revisiting Ethnic Entrepreneurship: Convergencies, Controversies, and Conceptual Advancements," *International Migration Review* 38, no. 3 (2004): 1040–74; Alejandro Portes, Luis Eduardo Guarnizo, and William J. Haller, "Transnational Entrepreneurs: An Alternative Form of Immigration Economic Adaption," *American Sociological Review* 67, no. 2 (2002): 278–98.

16. Mary Douglass and Baron Isherwood, *The World of Goods: Towards an Anthropology of Consumption* (London: A. Lane, 1979).

17. See, for example, Yong Chen, *Chop Suey USA: The Story of Chinese Food in America* (New York: Columbia University Press, 2014); Donna R. Gabaccia, *We Are What We Eat: Ethnic Food and the Making of Americans* (Cambridge: Harvard University Press, 1998); Hasia R. Diner, *Hungering for America: Italian, Irish, and Jewish Foodways in the Age of Migration* (Cambridge: Harvard University Press, 2001); Robert Ji-Song Ku, Martin F. Manalansan IV, and Anita Mannur, eds., *Eating Asian America: A Food Studies Reader* (New York: New York University Press, 2013). Some important exceptions include Richard Wilk, *Home Cooking in the Global Village: Caribbean Food from Buccaneers to Ecotourists* (New York: Berg, 2006); Jeffrey M. Pilcher, *Planet Taco: A Global History of Mexican Food* (New York: Oxford University Press, 2012).

18. For an introduction to this literature, see Katharine M. Donato et al., "A Glass Half Full? Gender in Migration Studies," *International Migration Review* 40, no. 1 (2006): 3–26; Sarah J. Mahler and Patricia R. Pessar, "Gendered Geographies of Power: Analyzing Gender across Transnational Spaces," *Identities* 7, no. 4 (2001): 441–59; Donna Gabaccia, *From the Other Side: Women, Gender, and Immigrant Life in the U.S., 1820–1990* (Bloomington: Indiana University Press, 1994).

19. An important exception is Diane C. Vecchio, *Merchants, Midwives, and Laboring Women: Italian Migrants in Urban America* (Urbana: University of Illinois Press, 2006). See the work of Saskia Sassen for an example of how social scientists have more recently used gender and feminist perspectives to explore migrants in urban informal economies and in the wider global economy. Saskia Sassen, *Globalization and Its Discontents: Essays on the New Mobility of People and Money* (New York: New Press, 1998).

20. Leonore Davidoff and Catherine Hall, *Family Fortunes: Men and Women of the English Middle Class, 1780–1850* (Chicago: University of Chicago Press, 1987); Erika Diane Rappaport, *Shopping for Pleasure: Women in the Making of London's West End* (Princeton, NJ: Princeton University Press, 2000); Lisa Tiersten, *Marianne in the Market: Envisioning Consumer Society in Fin-de-Siècle France* (Berkeley: University of California Press, 2001); Kristin L. Hoganson, *Consumers' Imperium: The Global Production of American Domesticity, 1865–1920* (Chapel Hill: University of North Carolina Press, 2007).

21. Simone Cinotto, ed., *Making Italian America: Consumer Culture and the Production of Ethnic Identities* (New York: Fordham University Press, 2014); Nan Enstad, *Ladies of Labor, Girls of Adventure: Working Women, Popular Culture, and Labor Politics at the Turn of the Twentieth Century* (New York: Columbia University Press, 1999); Kathy Peiss, *Cheap Amusements: Working Women and Leisure in Turn-of-the-Century New York* (Philadelphia: Temple University Press, 1986); Elizabeth Ewen, *Immigrant Women in the Land of Dollars: Life and Culture on the Lower East Side, 1890–1925* (New York: Monthly Review Press, 1985); Andrew R. Heinze, *Adapting to Abundance: Jewish Immigrants, Mass Consumption, and the Search for American Identity* (New York: Columbia University Press, 1990); George J. Sanchez, *Becoming Mexican American: Ethnicity, Culture, and Identity in Chicano Los Angeles, 1900–1945* (New York: Oxford University Press, 1993); Krishnendu Ray, *The Ethnic Restaurateur* (New York: Bloomsbury Academic, 2016).

Important new work on consumption in Argentina has focused almost exclusively on the post–World War II years and pays little attention to migration. See Eduardo Elena, *Dignifying Argentina: Peronism, Citizenship, and Mass Consumption* (Pittsburgh: University of

Pittsburgh Press, 2011); Natalia Milanesio, *Workers Go Shopping in Argentina: The Rise of Popular Consumer Culture* (Albuquerque: University of New Mexico Press, 2013); Rebekah E. Pite, *Creating a Common Table in Twentieth-Century Argentina: Doña Petrona, Women, and Food* (Chapel Hill: University of North Carolina Press, 2013).

22. Victoria de Grazia, with Ellen Furlough, eds., *The Sex of Things: Gender and Consumption in Historical Perspective* (Berkeley: University of California Press, 1996).

23. "L'emigrazione transoceanica, le correnti commerciali, e i servizi marittimi," *Bollettino dell'emigrazione* 11 (1904): 109.

24. On the heuristic quality of food, readers might begin with foundational texts in food studies by cultural theorists in anthropology, sociology, geography, and semiotics. See Roland Barthes, "Toward a Psychosociology of Contemporary Food Consumption," in *Food and Culture: A Reader*, 3rd ed., eds. Carole Counihan and Penny Van Esterik (New York: Routledge, 2013), 23–30; Pierre Bourdieu, *Distinction: A Social Critique of the Judgment of Taste*, trans. Richard Nice (Cambridge: Harvard University Press, 1984); Claude Lévi-Strauss, "The Culinary Triangle," in Counihan and Van Esterik, *Food and Culture*, 40–47; Jack Goody, *Cooking, Cuisine, and Class: A Study in Comparative Sociology* (Cambridge: Cambridge University Press, 1982); Mary Douglas, "Deciphering a Meal," *Daedalus* 101, no. 1 (1971): 61–81.

While not about food specifically, *The Social Life of Things* offers a productive reflection on the cultural and symbolic value of commodities, reminding us that the value of food extends beyond the sum of labor required to produce it, and that commodity exchange involving food is not neutral but rather embedded in particular social relations and cultural contexts. Arjun Appadurai, *The Social Life of Things: Commodities in Cultural Perspective* (Cambridge: Cambridge University Press, 1988).

25. Simone Cinotto, *The Italian American Table: Food, Family, and Community in New York City* (Urbana: University of Illinois Press, 2013); Simone Cinotto, *Soft Soil, Black Grapes: The Birth of Italian Winemaking in California*, trans. Michelle Tarnopoloski (New York: New York University Press, 2012); Diner, *Hungering for America*; Gabaccia, *We Are What We Eat*; Vito Teti, "Emigrazione, alimentazione, culture popolari," in *Storia dell'emigrazione Italiana. Partenze*, ed. Piero Bevilacqua, Andreina De Clementi, and Emilio Franzina (Rome: Donzelli Editore, 2001), 575–600. On Italians' changing food habits in the United States and Argentina, see also Tracy N. Poe, "The Labour and Leisure of Food Production as a Mode of Ethnic Identity Building among Italians in Chicago, 1890–1940," *Rethinking History* 5, no. 1 (2001): 131–48; Paola Corti, "Emigrazione e consuetudini alimentari. L'esperienza di una catena migratoria," in *Storia d'Italia*, Annali 13. *L'alimentazione nella storia dell'Italia contemporanea*, ed. Alberto Capatti, Albero De Bernardi, and Angelo Varni (Turin: Einaudi, 1998), 683–719.

26. Peter H. Smith, *Politics and Beef in Argentina: Patterns of Conflict and Change* (New York: Columbia University Press, 1969); Roger Horowitz, *Putting Meat on the American Table: Taste, Technology, Transformation* (Baltimore: Johns Hopkins University Press, 2006); Jimmy M. Skaggs, *Prime Cut: Livestock Raising and Meatpacking in the United States, 1607–1983* (College Station: Texas A&M University Press, 1986).

27. On gendered food practices among Italians and other migrant groups, see Cinotto, *Italian American Table*, 47–71; Suzanne M. Sinke, *Dutch Immigrant Women in the United States,*

1880–1920 (Urbana: University of Illinois Press, 2002), 64–95; Valerie J. Matsumoto, "Apple Pie and *Makizushi*: Japanese American Women Sustaining Family and Community," in *Eating Asian America: A Food Studies Reader*, ed. Robert Ji-Song Ku, Martin F. Manalansan IV, and Anita Mannur (New York: New York University Press, 2013), 255–73; Sonia Cancian, "'Tutti a Tavola!' Feeding the Family in Two Generations of Italian Immigrant Households in Montreal," in *Edible Histories, Cultural Politics: Towards a Canadian Food History*, ed. Franca Iacovetta, Valerie J. Korinek, and Marlene Epp (Toronto: University of Toronto Press, 2012), 209–21; Meredith E. Abarca, *Voices in the Kitchen: Views of Food and the World from Working-Class Mexican and Mexican American Women* (College Station: Texas A&M University Press, 2006).

28. Rudolph J. Vecoli, "The Immigrant Press and the Construction of Social Reality, 1850–1920," in *Print Culture in a Diverse America*, ed. James P. Danky and Wayne A. Wiegand (Urbana: University of Illinois Press, 1998), 17–33.

29. On the Italian-language press in the United States, see Peter G. Vellon, *A Great Conspiracy against Our Race: Italian Immigrant Newspapers and the Construction of Whiteness in the Early 20th Century* (New York: New York University Press, 2014); Nancy C. Carnevale, *A New Language, A New World: Italian Immigrants in the United States, 1890–1945* (Urbana: University of Illinois Press, 2009); George E. Pozzetta, "The Italian Immigrant Press of New York City: The Early Years, 1880–1915," *Journal of Ethnic Studies* 1 (Fall 1973): 32–46; Philip V. Cannistraro, "Generoso Pope and the Rise of Italian-American Politics, 1925–1936," in *Italian Americans: New Perspectives in Italian Immigration and Ethnicity*, ed. Lydio F. Tomasi (Staten Island: Center for Migration Studies of New York Inc., 1985), 265–88. On Argentina, see Ronald C. Newton, "Ducini, Prominenti, Antifascisti: Italian Fascism and the Italo-Argentine Collectivity, 1922–1945," *Americas* 51, no. 1 (1994): 41–66; Mirta Zaida Lobato, "*La Patria degli Italiani* and Social Conflict in Early Twentieth-Century Argentina," trans. Amy Ferlazzo, in *Italian Workers of the World: Labor Migration and the Formation of Multiethnic States*, ed. Donna R. Gabaccia and Fraser M. Ottanelli (Urbana: University of Illinois Press, 2001), 63–78. For this book, I look exclusively at the mainstream commercial press. Italian-language print culture was also a key voice for anarchists, socialists, syndicalists, and radicals. This study does not include a consideration of the *sovversivi*, or leftist, press because as anticapitalist outlets, these newspapers less frequently discussed the experiences of Italian migrants as consumers.

30. In 1908 *Bollettino Mensile* became *Bollettino Ufficiale Mensile*. For an overview of Italian chambers of commerce abroad, see Giovanni Luigi Fontana and Emilio Franzina, eds., *Profili di Camere di commercio italiane all'estero*, vol. 1 (Soveria Mannelli, Italy: Rubbettino Editore, 2001); and Giulio Sapelli, ed., *Tra identità culturale e sviluppo di reti. Storia delle Camere di commercio italiane all'estero* (Soveria Mannelli, Italy: Rubbettino Editore, 2000).

Chapter 1. Manly Markets in *le due Americhe*, 1880–1914

1. *Dall'Italia all'Argentina: Guida practica per gli italiani che si recano nell'Argentina* (Genoa: Libreria R. Istituto Sordo-Muti, 1888), 119.

2. Luigi Tonissi, "Progetto per un banco del commercio italo-americano," in *L'esplorazione commerciale e L'esploratore* (Milan: Premiato stabilimento tipografico P. B. Bellini, 1896),

386. For migration, see Dirección General de Inmigración, *Resumen estadistico del movimiento migratorio en la Republica Argentina, años 1857–1924* (Buenos Aires: Talleres Gráficos del Ministerio de Agricultura de la Nación, 1925), 5.

3. Luigi Einaudi, *Un principe mercante: Studio sulla espansione coloniale italiana* (Turin: Fratelli Bocca, 1900), 86–87.

4. Ibid., 127–38.

5. Mark Choate skillfully describes these debates between nationalists and liberals over Italy's "emigrant colonialism." Mark I. Choate, *Emigrant Nation: The Making of Italy Abroad* (Cambridge: Harvard University Press, 2008), 21–56.

6. Einaudi, *Un principe mercante,* 23.

7. On the 1901 emigration law, see Choate, *Emigrant Nation*, 59–62; Mark I. Choate, "Sending States' Transnational Interventions in Politics, Culture, and Economics: The Historical Example of Italy," *International Migration Review* 41, no. 3 (2007): 728–68; Caroline Douki, "The Liberal Italian State and Mass Emigration, 1860–1914," in *Citizenship and Those Who Leave: The Politics of Emigration and Expatriation*, ed. Nancy L. Green and François Weil (Urbana: University of Illinois Press, 2007), 91–113.

8. Frank J. Coppa, *Planning, Protectionism, and Politics in Liberal Italy: Economics and Politics in the Giolittian Age* (Washington, DC: Catholic University of America Press, 1971); Alexander De Grand, *The Hunchback's Tailor: Giovanni Giolitti and Liberal Italy from the Challenge of Mass Politics to the Rise of Fascism, 1882–1922* (Westport, CT: Praeger, 2001).

9. On the Emigration Commissariat and the *Bollettino dell'emigrazione,* see Choate, *Emigrant Nation*, 59–62; 75; Douki, "Liberal Italian State," 91–113.

10. "L'emigrazione transoceanica, le correnti commerciali, e i servizi marittimi," *Bollettino dell'emigrazione* (hereafter *Bollettino*) 11 (1904): 107, 109.

11. Giovanni Luigi Fontana and Emilio Franzina, eds., *Profili di Camere di commercio italiane all'estero*, vol. 1 (Soveria Mannelli, Italy: Rubbettino Editore, 2001), and Giulio Sapelli, *Tra identità culturale e sviluppo di reti. Storia delle Camere di commercio italiane all'estero* (Soveria Mannelli, Italy: Rubbettino Editore, 2000) are two recent edited collections arguing for the centrality of Italian chambers of commerce abroad to the history of Italian global expansion.

12. Choate, *Emigrant Nation*, 82–89.

13. Guglielmo Godio, *L'America ne' suoi primi fattori: La colonizzazione e l'emigrazione* (Florence: Tipografia di G. Barbèra, 1893), 11, 17–18.

14. Ibid., 140.

15. Aldo Visconti, *Emigrazione ed esportazione: Studio dei rapporti che intercedono fra l'emigrazione e le esportazioni italiane per gli Stati Uniti del Nord America e per la Repubblica Argentina* (Turin: Tipografia Baravalle e Falconieri, 1912), 25.

16. Luigi Fontana-Russo, "Emigrazione d'uomini ed esportazione di merci," *Rivista coloniale* (1906), 30.

17. Torsten Feys, *The Battle for the Migrants: The Introduction of Steamshipping on the North Atlantic and Its Impact on the European Exodus* (St. John's, Nfld: International Maritime Economic History Association, 2013); Drew Keeling, *The Business of Transatlantic Migration between Europe and the United States, 1900–1914* (Zurich: Chronos, 2012).

18. Kevin H. O'Rourke and Jeffrey G. Williamson, *Globalization and History: The Evolution of a Nineteenth-Century Atlantic Economy* (Cambridge: MIT Press, 1999), 145–66; Dirk Hoerder, ed., *Labor Migration in the Atlantic Economies: The European and North American Working Classes during the Period of Industrialization* (Westport, CT: Greenwood Press, 1985).

19. "L'emigrazione transoceanica," 109.

20. Ministero delle Finanze, *Movimento commerciale del Regno d'Italia nell'anno 1907*, vol. 2 (Rome: Stabilimento tipografico G. Civelli, 1909), xxxi.

21. Vera Zamagni, *The Economic History of Italy, 1860–1990* (Oxford: Oxford University Press, 1993), 35. On the history of industrial food production in Italy, see Silvano Serventi and Françoise Sabban, *Pasta: The Story of a Universal Food* (New York: Columbia University Press, 2002); Francesco Chiapparino, "Industrialization and Food Consumption in United Italy," in *Food Technology, Science and Marketing: European Diet in the Twentieth Century*, ed. Adel P. den Hartog (East Linton, Scotland: Tuckwell Press, 1995), 139–55; David Gentilcore, *Pomodoro!: A History of the Tomato in Italy* (New York: Columbia University Press, 2010), 109–115, 136–42; Giorgio Pedrocco, "La conservazione del cibo: Dal sale all'industria agro-alimentare," in *Storia d'Italia, Annali 13, L'alimentazione,* ed. Alberto Capatti, Albero De Bernardi, and Angelo Varni (Turin: Einaudi, 1998), 419–52.

22. On Italian chambers of commerce in Argentina and the United States, see Fernando J. Devoto, *Historia de los italianos en la Argentina* (Buenos Aires: Editorial Biblos, 2006), 218–31; Mil Vassanelli, "La Camera di commercio italiana di San Francisco," in *Profili di Camere di commercio italiane all'estero*, vol. 1, ed. Giovanni Luigi Fontana and Emilio Franzina (Soveria Mannelli, Italy: Rubbettino Editore, 2001), 123–48; Sergio Bugiardini, "La Camera di commercio italiani di New York," in *Profili di Camera di commercio italiane all'estero*, vol. 1, ed. Giovanni Luigi Fontana and Emilio Franzina (Soveria Manelli, Italy: Rubbettino Editore, 2001), 105–121.

23. Guido Rossati, "La vendemmia in Italia ed il commercio dei vini italiani cogli S.U. nel 1905," *Rivista Commerciale* (New York) (hereafter *Rivista*), December 1905, 43.

24. Fontana-Russo, "Emigrazione d'uomini ed esportazione di merci," 26.

25. Italian American Directory Co., *Gli italiani negli Stati Uniti d'America* (New York: Andrew H. Kellogg Co., 1906), 158.

26. See, for example, Visconti, *Emigrazione ed esportazione*, 32, 48, 52; and "L'emigrazione transoceanica," 107.

27. Amy A. Bernardy, "L'etnografia della 'piccole italie,'" in *Atti del primo congresso di etnografia italiana*, ed. Società di Etnografia Italiana (Perugia, Italy: Unione Tipografica Cooperativa, 1912), 175. On Bernardy, see Amy Allemand Bernardy and Maddalena Tirabassi, *Ripensare la patria grande: Gli scritti di Amy Allemand Bernardy sulle migrazioni italiane* (Isernia, Italy: C. Iannone, 2005).

28. Fontana-Russo, "Emigrazione d'uomini ed esportazione di merci," 30.

29. Visconti, *Emigrazione ed esportazione*, 5.

30. Godio, *L'America ne' suoi primi fattori*, 122.

31. "Statistica degli emigranti curati durante l'anno 1906 nelle inferie di bordo ed appunti sul servizio dell'emigrazione," *Bollettino* 2 (1908): 182.

32. *L'Italia nell' America Latina: Per l'incremento dei rapporti industriali e commerciali fra l'Italia e l'America del Sud* (Milan: Società Tipografica Editrice Popolare, 1906), viii. On the im-

proved diets of peasants in Calabria due to migration, see Piero Bevilacqua, "Emigrazione transoceanica e mutamenti dell'alimentazione contadina calabrese fra Otto e Novecento," *Quaderni storici* 47 (August 1981): 520–55.

33. Cesare Jarach, *Inchiesta parlamentare sulle condizioni dei contadini nelle provincie meridionali e nella Sicilia*, vol. 2, *Abruzzi e Molise*, Tomo I (Rome: Tipografia nazionale di Giovanni Bertero e C., 1909), 264; Oreste Bordiga, *Inchiesta parlamentare sulle condizioni dei contadini nelle provincie meridionali e nella Sicilia*, vol. 4, *Campania*, Tomo I (Rome: Tipografia nazionale di Giovanni Bertero e C., 1909), 612.

34. Adolfo Rossi, "Vantaggi e danni dell'emigrazione nel mezzogiorno d'Italia," *Bollettino* 13 (1908): 6.

35. Giovanni Lorenzoni, *Inchiesta parlamentare sulle condizione dei contadini nelle provincie meridionali e nella Sicilia*, vol. 6, *Sicilia*, Tomo I (Rome: Tipografia nazionale di Giovanni Bertero e C., 1910), 705–706.

36. Fontana-Russo, "Emigrazione d'uomini ed esportazione di merci," 40.

37. On the development of Italy's consumer society, see Emanuela Scarpellini, *Material Nation: A Consumer's History of Modern Italy*, trans. Daphne Hughes and Andrew Newton (New York: Oxford University Press, 2011).

38. Einaudi, *Un principe mercante*, 10.

39. Visconti, *Emigrazione ed esportazione*, 5.

40. "Relazione sulla convenienza morale finanziaria che le camere all'estero accentrino ogni manifestazione economica, sia ufficiale che non ufficiale," *Rivista*, November 1911, 6.

41. Camera di Commercio Italiana in New York, *Nel cinquantenario della Camera di Commercio Italiana in New York, 1887–1937* (New York, 1937), 65.

42. I deliberately use the terms "U.S. American" rather than "American," and "United States" rather than "America," throughout the book to acknowledge that "America" and "American" are contested designations among scholars of the western hemisphere. In this I follow Mary Renda's lead in her work on the U.S. occupation of Haiti to remind readers of the hegemonic and imperialist assumptions underpinning the term "America" and "American." In the sources explored for this study, Italians in Argentina sometimes described Argentines and themselves as "American" as do some Argentines and Latin Americans today. See Mary A. Renda, *Taking Haiti: Military Occupation and the Culture of U.S. Imperialism, 1915–1940* (Chapel Hill: University of North Carolina Press, 2001), xvii.

43. On Candiani, see *L'Italia nell' America Latina: Per l'incremento dei rapporti industriali e commerciali fra l'Italia e l'America del Sud* (Milan: Società Tipografica Editrice Popolare, 1906), 131–34.

44. On "Italia," see Stephen Gundle, *Bellissima: Feminine Beauty and the Idea of Italy* (New Haven, CT: Yale University Press, 2007), 19–20, 28–29.

45. Einaudi, *Un principe mercante,* 23.

46. The government passed the first law regulating and protecting trademarks in 1868 and revised them in 1913. Leggi 30 agosto, 1868, n. 4577; Leggi 20 marzo 1913, n. 526. Applications for trademark registration were required to include reproductions of the actual trademark, a detailed description of the trademark's distinctive qualities (colors, form, language, etc.), and an indication of the intended mode of application.

47. Gian Paolo Ceserani, *Storia della pubblicità in Italia* (Rome: Laterza, 1988), 28, 32. In the late nineteenth century, before advertising emerged as a specialized field, business owners themselves maintained a considerable amount of control over the creation and content of their publicity. See also Pamela Walker Laird, *Advertising Progress: American Business and the Rise of Consumer Marketing* (Baltimore: Johns Hopkins University Press, 1998), 38–44.

48. Serventi and Sabban, *Pasta*, 159.

49. Trademark for Giacomo Costa Dante-brand olive oil, Archivio Centrale dello Stato, Ministero dell'Industria, del Commercio e dell'Artigianato, Ufficio Italiano Brevetti e Marchi (hereafter ACS, UIBM), fasc. 8415, 1907. For another example, see trademark for G. B. Martino Dante-brand olive oil, ACS, UIBM, fasc. 8618, 1908.

50. For trademarks featuring Mercury, see Luigi Parpaglioni olive oil, ACS, UIBM, fasc. 4964, 1901; Amoruso & Co. olive oil, ACS, UIBM, fasc. 4906, 1901; Franco & Lamb, Carroni "Cotone Mercurio" fabrics, ACS, UIBM, fasc. 8487, 1907.

51. Alfonso Scirocco, *Garibaldi: Citizen of the World*, trans. Allan Cameron (Princeton, NJ: Princeton University Press, 2007), 99–103, 122–37; Lucy Riall, *Garibaldi: Invention of a Hero* (New Haven, CT: Yale University Press, 2007), 41–56.

52. See, for example, trademark for Fratelli Branca e C. Fernet-Branca, ACS, UIBM, fasc. 3946, 1898.

53. Edoardo Pantano, "Conclusione," *Bollettino* 11 (1904): 119.

54. "Parte prima. Condizioni presenti dell'emigrazione italiana," *Bollettino* 11 (1904): 13.

55. *L'Italia nell'America Latina*, 37–43.

56. On Luigi Amedeo and his expedition, see Luigi Amedeo and Umberto Cagni, *On the Polar Star in the Arctic Sea*, trans. William Le Queux (London: Hutchinson & Co., 1903).

57. Trademark for Fratelli Branca e C. Fernet-Branca, ACS, UIBM, fasc. 5636, 1902.

58. On Latin American independence movements, see Jeremy Adelman, *Sovereignty and Revolution in the Iberian Atlantic* (Princeton, NJ: Princeton University Press, 2006); Jay Kinsbruner, *Independence in Spanish America: Civil Wars, Revolutions, and Underdevelopment*, 2nd rev. (Albuquerque: University of New Mexico Press, 2000).

59. The *fasce*—a bundle of sticks joined with an axe blade—was a symbol of republicanism during the Roman Republic, and of peasant cooperatives and working-class solidarity during the nineteenth and twentieth centuries. In the 1920s the Italian Fascist Party exploited the fasce as its chief symbol. Simonetta Falasca-Zamponi, *Fascist Spectacle: The Aesthetics of Power in Mussolini's Italy* (Berkeley: University of California Press, 1997), 95–99.

60. New technologies in printing, transportation, and distribution methods, as well as changes in business organization and culture, helped universalize progress discourse in advertising material. Laird, *Advertising Progress*, 101–151; Jackson Lears, *Fables of Abundance: A Cultural History of Advertising in America* (New York: Basic Books, 1994), 102–133.

61. On world expositions, see, for example, Peter H. Hoffenberg, *An Empire on Display: English, Indian, and Australian Exhibitions from the Crystal Palace to the Great War* (Berkeley: University of California Press, 2001); Robert W. Rydell, *All the World's a Fair: Visions of Empire at American International Expositions, 1876–1916* (Chicago: University of Chicago Press, 1984); Robert W. Rydell, *World of Fairs: The Century-of-Progress Expositions* (Chicago: University of Chicago Press, 1993).

62. Choate, *Emigrant Nation*, 87, 104–105; Einaudi, *Un principe mercante*, 3, 10.

63. "Esposizione internazionale di Torino 1911," *Bollettino Ufficiale Mensile della Camera Italiana di Commercio in Buenos Aires* (Buenos Aires, Argentina) (hereafter *Bollettino Mensile*), December 1910, 11.

64. Donna Gabaccia, "In the Shadows of the Periphery: Italian Women in the Nineteenth Century," in *Connecting Spheres: Women in the Western World, 1500 to Present*, ed. Marilyn J. Boxer and Jean H. Quataert (New York: Oxford University Press, 1987), 166–76.

65. Lears, *Fables of Abundance*, 17–39; Gundle, *Bellissima*, 48–54.

66. For examples of women as agricultural workers, see trademarks for V. del Gaizo tomato extract, ACS, UIBM, fasc. 6982, 1905; Muratorio & Martino olive oil, ACS, UIBM, fasc. 2904, 1895; Modesto Gallone butter, ACS, UIBM, fasc. 2932, 1894. Trademark for Poirè & Balletto "la Perfetta" pasta, ACS, UIBM, fasc. 8437, 1907.

67. Choate, *Emigrant Nation,* 50–51.

68. *L'Italia nell'America Latina*, 235–38.

69. Trademark for Raffaello & Pietro Fortuna olive oil, ACS, UIBM, fasc. 8384, 1907. For an overview of the history and historiography of the Risorgimento, see Lucy Riall, *Risorgimento: The History of Italy from Napoleon to Nation-state* (New York: Palgrave Macmillan, 2009).

70. *L'Italia nell'America Latina*, xxxi.

71. Einaudi, *Un principe mercante*, 7, 14, 16–18.

72. Visconti, *Emigrazione ed esportazione*, 7, 40.

73. "Assemblea Ordinaria del 15 Luglio 1900," *Bollettino Mensile*, August 1900, 2.

74. Richard J. Bosworth, *Italy, the Least of the Great Powers: Italian Foreign Policy before the First World War* (New York: Cambridge University Press, 1979), 1–38.

75. Ibid.

76. As part of the 1901 emigration law, the Italian state contracted the Bank of Naples to establish branches abroad and to transmit economic remittances from the Americas back to Italy at low costs. On remittances and the Bank of Naples, see Francesco Balletta, *Il Banco di Napoli e le rimesse degli emigrati (1914–1925)* (Naples: ISTOB, 1972); Dino Cinel, *National Integration of Italian Return Migration, 1870–1929* (New York: Cambridge University Press, 1991), 122–49.

77. Zamagni, *Economic History of Italy*, 38.

78. Loretta Baldassar and Donna R. Gabaccia, "Home, Family, and the Italian Nation in a Mobile World: The Domestic and the National among Italy's Migrants," in *Intimacy and Italian Migration: Gender and Domestic Lives in a Mobile World*, ed. Loretta Baldassar and Donna R. Gabaccia (New York: Fordham University Press, 2011), 21; Bosworth, *Italy, the Least of the Great Powers*, 6–7.

79. Lorenzoni, *Inchiesta parlamentare*, 514.

80. Linda Reeder, *Widows in White: Migration and the Transformation of Rural Italian Women, Sicily, 1880–1920* (Toronto: University of Toronto Press, 2003), 56.

81. Victoria de Grazia, with Ellen Furlough, eds., *The Sex of Things: Gender and Consumption in Historical Perspective* (Berkeley: University of California Press, 1996); Leonore Davidoff and Catherine Hall, *Family Fortunes: Men and Women of the English Middle Class, 1780–1850* (Chicago: University of Chicago Press, 1987); Erika D. Rappaport, *Shopping for Pleasure:*

Women in the Making of London's West End (Princeton, NJ: Princeton University Press, 2000); Lisa Tiersten, *Marianne in the Market: Envisioning Consumer Society in Fin-de-Siècle France* (Berkeley: University of California Press, 2001).

82. Scarpellini, *Material Nation*, 67–80.

83. Walter F. Willcox and Imre Ferenczi, *International Migrations*, vol. 1 (New York: Gordon and Breach Science Publishers, 1969), 820–21, 835.

84. Samuel L. Baily, *Immigrants in the Lands of Promise: Italians in Buenos Aires and New York City, 1870–1914* (Ithaca, NY: Cornell University Press, 1999), 64.

85. On family economies in late-nineteenth and early-twentieth century Italy, see especially David I. Kertzer, *Family Life in Central Italy, 1880–1910: Sharecropping, Wage Labor, and Coresidence* (New Brunswick, NJ: Rutgers University Press, 1984); Jane Schneider and Peter Schneider, *Culture and Political Economy in Western Sicily* (New York: Academic Press, 1976). On late nineteenth-century living conditions in rural Italy, see the Italian government's voluminous parliamentary inquest *Atti della giunta per la inchiesta agraria e sulle condizioni della classe agricola*, 15 vols. (Rome: Forzani e C., 1881–1886).

86. Gabaccia, "In the Shadows of the Periphery," 166–76.

87. Lorenzoni, *Inchiesta parlamentare*, 469–70; Ernesto Marenghi, *Inchiesta parlamentare sulle condizioni dei contadini nelle provincie meridionali e nella Sicilia*, vol. 5. Basilicata e Calabrie, Tomo II Calabrie (Rome: Tipografia nazionale di Giovanni Bertero e C., 1909), 592–602.

88. On the use of remittances by Italian migrants and returnees, see Cinel, *National Integration of Italian Return Migration*, 122–176.

89. Hasia R. Diner, *Hungering for America: Italian, Irish, and Jewish Foodways in the Age of Migration* (Cambridge: Harvard University Press, 2001), 61–77; Simone Cinotto, *The Italian American Table: Food, Family, and Community in New York City* (Urbana: University of Illinois Press, 2013); Donna R. Gabaccia and Jeffrey M. Pilcher, "'Chili Queens' and Checkered Tablecloths: Public Dining Cultures of Italians in New York City and Mexicans in San Antonio, Texas, 1870s–1940s," *Radical History Review* 110 (Spring 2011): 109–126; Vito Teti, "Emigrazione, alimentazione, culture popolari," in *Storia dell'emigrazione Italiana. Partenze*, ed. Piero Bevilacqua, Andreina De Clementi, and Emilio Franzina (Rome: Donzelli Editore, 2001), 575–600.

90. Samuel L. Baily and Franco Ramella, eds., *One Family, Two Worlds: An Italian Family's Correspondence across the Atlantic, 1901–1922*, trans. John Lenaghan (New Brunswick, NJ: Rutgers University Press, 1988), 44–45.

91. Reeder, *Widows in White*, 142–67. On Italian women in transnational family economies, see especially Donna R. Gabaccia, "When the Migrants Are Men: Italy's Women and Transnationalism as a Working-Class Way of Life," in *American Dreaming, Global Realities: Rethinking U.S. Immigration History*, ed. Donna R. Gabaccia and Vicki L. Ruiz (Urbana: University of Illinois Press, 2006), 190–206.

92. Jarach, *Inchiesta parlamentare*, 258. Italian migrants in the Americas or those who returned from the Americas to Italy were often called *americani* by locals.

93. Jarach, *Inchiesta parlamentare*, 159; quoted in Gabaccia, "When the Migrants Are Men," 199.

94. Bordiga, *Inchiesta parlamentare*, 612–13.

95. On women's role in fostering consumption in Italy, see Reeder, *Widows in White*, 142–67.

96. Historians have rejected simplified and homogenized depictions of the Italian south as economically, politically, and culturally backward, instead arguing for the existence of an emerging commercial economy alongside the continuation of subsistence agriculture. While economically the south lagged behind the north, capitalist and commodity markets touched Southern Europe and the South of Italy. John Davis, "Changing Perspectives on Italy's 'Southern Problem,'" in *Italian Regionalism: History, Identity, and Politics*, ed. Carl Levy (Oxford: Berg, 1996), 53–68; Robert Lumley and Jonathan Morris, *The New History of the Italian South: The Mezzogiorno Revisited* (Exeter: University of Exeter Press, 1997). On Italy's incipient consumer society, see Scarpellini, *Material Nation,* 3–80; Reeder, *Widows in White,* 223–31.

97. Quoted in Choate, *Emigrant Nation*, 73; and Richard Bosworth, *Italy, the Least of the Great Powers*, 422.

Chapter 2. Race and Trade Policies in Migrant Marketplaces, 1880–1914

1. C. A. Mariani, "Tariff Revision and the Italians in America," *Rivista Commerciale* (hereafter *Rivista*), December 1911, 44–45.

2. Ad for Luigi Bosca & Figli, *La Patria degli Italiani* (Buenos Aires, Argentina) (hereafter *La Patria*), May 25, 1910, 12.

3. Historians of Italian migration to Argentina have recognized Argentina's heavy dependence on Italian laborers as compared to other migrant groups. See Fernando J. Devoto, *Historia de los italianos en la Argentina* (Buenos Aires: Editorial Biblos, 2006); Fernando J. Devoto and Gianfausto Rosoli, eds., *L'Italia nella società argentina: Contributi sull'emigrazione italiana in Argentina* (Rome: Centro Studi Emigrazione, 1988); Eugenia Scarzanella, *Italiani d'Argentina: Storie di contadini, industriali e missionari italiani in Argentina, 1850–1912* (Venice: Marsilio, 1983).

4. Samuel L. Baily, *Immigrants in the Lands of Promise: Italians in Buenos Aires and New York City, 1870–1914* (Ithaca, NY: Cornell University Press, 1999), 54, 58–59.

5. Gianfausto Rosoli, ed., *Un secolo di emigrazione italiana, 1876–1976* (Rome: Centro Studi Emigrazione, 1978), 353; Scarzanella, *Italiani d'Argentina*, 25–30; Fernando J. Devoto, "Programs and Politics of the First Italian Elite of Buenos Aires, 1852–80," in *Italian Workers of the World: Labor Migration and the Formation of Multiethnic States*, ed. Donna R. Gabaccia and Fraser M. Ottanelli (Urbana: University of Illinois Press, 2001), 41–59.

6. "Atti della camera," *Bollettino Mensile della Camera di Commercio in Buenos Aires* (hereafter *Bollettino Mensile*), July (1901): 2.

7. Comitato della Camera Italiana di Commercio ed Arti, *Gli italiani nella Repubblica Argentina* (Buenos Aires, 1898), 17.

8. República Argentina, *Tercer Censo Nacional,* Tomo 8, *Censo del Comercio* (Buenos Aires: Talleres Gráficos de L. J. Rosso y Cía., 1917), 132, 135–36.

9. On Italians in Argentina's class system and the growing middle class, see Baily, *Immigrants in the Lands of Promise*, 73–75; Scarzanella, *Italiani d'Argentina*, 25–69; Eugenia Scarzanella, "L'industria argentina e gli immigrati italiani: Nascita della borghesia industriale

bonaerense," in *Gli italiani fuori d'Italia: Gli emigrati italiani nei movimenti operai dei paesi d'adozione, 1880–1940*, ed. B. Bezza (Milan: F. Angeli, 1983), 583–633; Devoto, *Historia de los italianos en la Argentina*, 204–230, 283–291.

10. Ad for Florio, *La Patria*, April 5, 1895, 3; ad for Florio, *La Patria*, January 21, 1905, 2; ad for Florio, *La Patria*, January 1910, 17.

11. On Jannello, see Comitato della Camera Italiana di Commercio ed Arti, *Gli italiani nella Repubblica Argentina* (1898), 303. See also *L'Italia nell'America Latina: Per l'incremento dei rapporti industriali e commerciali fra l'Italia e l'America del Sud* (Milan: Società Tipografica Editrice Popolare, 1906), 57; Comitato della Camera Italiana di Commercio ed Arti, "La Casa di Rappresentanze Francesco Jannello," *Gli italiani nella Repubblica Argentina all'Esposizione di Torino 1911* (Buenos Aires: Stabilimento Grafico della Compañia General de Fósforos, 1911).

12. On the working-class experiences of Italian urban laborers in the United States, see especially Jennifer Guglielmo, *Living the Revolution: Italian Women's Resistance and Radicalism in New York City, 1880–1945* (Chapel Hill: University of North Carolina Press, 2010); Donna R. Gabaccia, *Militants and Migrants: Rural Sicilians Become American Workers* (New Brunswick, NJ: Rutgers University Press 1988); Dino Cinel, *From Italy to San Francisco: The Immigrant Experience* (Stanford, CA: Stanford University Press, 1982); Virginia Yans-McLaughlin, *Family and Community: Italian Immigrants in Buffalo, 1880–1930* (Ithaca, NY: Cornell University Press, 1977); Robert F. Foerster, *The Italian Emigration of Our Times* (Cambridge: Harvard University Press, 1919), 342–62.

13. Baily, *Immigrants in the Lands of Promise*, 102.

14. For Italians in agricultural importing and wholesale businesses, see Donna R. Gabaccia, *We Are What We Eat: Ethnic Food and the Making of Americans* (Cambridge: Harvard University Press, 1998), 65–73. On Italian food businesses in New York, see especially Simone Cinotto, *The Italian American Table: Food, Family, and Community in New York City* (Urbana: University of Illinois Press, 2013).

15. Louise Odencrantz, *Italian Women in Industry* (New York: Russell Sage Foundation, 1919), 12.

16. On L. Gandolfi & Co., see Italian American Directory Co., *Gli italiani negli Stati Uniti d'America* (New York: Andrew H. Kellogg Co., 1906), 292–95. See also ad for L. Gandolfi & Co., *Rivista*, December 1905, 8.

17. Ad for Cora vermouth and Fernet-Branca amaro, *La Nación*, September 20, 1901, 6; ad for Felsina Buton amaro, *La Prensa*, December 2, 1900, 8; ad for Nocera Umbra mineral water, *La Prensa*, December, 1900, 7.

18. Ad for Jannello, *Caras y Caretas*, January 1, 1910.

19. Giuseppe Peretti was a member of the Italian Chamber of Commerce in Buenos Aires. Ad for Cora vermouth, *Caras y Caretas*, January 8, 1910.

20. Ad for La Gran Ciudad de Chicago, *La Patria*, January 1, 1910, 4–5.

21. William Leach, *Land of Desire: Merchants, Power and the Rise of a New American Culture* (New York: Pantheon Books, 1993). On Argentine department stores, see Fernando Rocchi, *Chimneys in the Desert: Industrialization in Argentina during the Export Boom Years, 1870–1930* (Stanford, CA: Stanford University Press, 2006), 71–72, 77, 83.

22. See, for example, "Vida Italiana," *La Nación*, September 4, 1901, 2; "Vida Italiana," *La Nación*, September 4, 1901, 2; "Notas Sociales, Circolo Italiano," *La Nacion*, September 23, 1901, 3.

23. For examples of advertisements for Italian exports in Spanish in *La Patria*, see ad for Fernet-Branca, *La Patria*, January 1, 1910, 6; ad for Malvasia Florio, *La Patria*, January 2, 1900, 7; ad for San Pellegrino, *La Patria*, January 7, 1905, 7; ad for Nocera Umbra, *La Patria*, January 11, 1910, 8.

24. For examples of manifesti in Spanish in *La Patria*, see "Manifesti: Città di Milano," *La Patria*, June 29, 1900, 6; "Manifesto d'arrivo: Italie," *La Patria*, January 6, 1905, 7; "Manifesti d'arrivo: Siena," *La Patria*, January 1, 1910, 5; "Manifesti: Brasile, Principe Umberto," *La Patria*, January 9, 1915, 9; "Manifesto d'arrivo: Tomaso di Savoia," *La Patria*, April 22, 1920, 11; "Manifesto d'arrivo: Ansaldo Savoia I," *La Patria*, February 14, 1925, 10. By 1910, all manifesti in *La Patria* appeared in Spanish rather than in Italian.

25. See, for example, "Bollettino Commerciale," *Il Progresso Italo-Americano* (hereafter *Il Progreso*), September 13, 1900, 2.

26. Alessandro Durante, ed., *A Companion to Linguistic Anthropology* (Malden, MA: Blackwell, 2004). Linguistic anthropologists interested in links between language, identity, and social action have discussed bilingualism and code switching (when two or more languages are employed during the same speech event) as two possible outcomes of language contact between speakers.

27. F. A., "Snazionalizzazione," *Rivista*, November 1909, 10.

28. "Nel nome di Colombo," *La Patria*, April 17, 1910, 7.

29. "Emigrazione," *Bollettino Mensile*, December 1910, 2.

30. Peter D'Agostino, "Craniums, Criminals, and the 'Cursed Race': Italian Anthropology in American Racial Thought, 1861–1924," *Comparative Studies in Society and History* 44, no. 2 (2002): 319–43.

31. Ibid.

32. For an excellent comparative study of racially based immigration policies of countries in the western hemisphere, see David Scott FitzGerald and David Cook-Martín, *Culling the Masses: The Democratic Origins of Racist Immigration Policy in the Americas* (Cambridge: Harvard University Press, 2014).

33. For a sweeping overview of Argentine history, see David Rock, *Argentina, 1516–1987: From Spanish Colonization to Alfonsín* (Berkeley: University of California Press, 1985). *Criollo* was a flexible catagory that changed over time and was shaped by factors such as region and class. For a discussion of the term *criollo* as it relates to Argentine food and identity see Rebekah E. Pite, "*La cocina criolla*: A History of Food and Race in Twentieth-Century Argentina," in *Rethinking Race in Modern Argentina*, ed. Paulina L. Alberto and Eduardo Elena (New York: Cambridge University Press, 2016), 99–125.

34. George Reid Andrews, *The Afro-Argentines of Buenos Aires, 1800–1900* (Madison: University of Wisconsin Press, 1980).

35. In the book, Sarmiento proposed a dichotomy between "barbarianism" embodied by *caudillos* such as the Argentine dictator Juan Manuel de Rosas and "civilization" represented by urban life, European migration, and North America. Domingo F.

Sarmiento, *Facundo; or, Civilization and Barbarism*, trans. Mary Mann (New York: Penguin Books, 1998).

36. Samuel L. Baily, "Sarmiento and Immigration: Changing Views on the Role of Immigration in the Development of Argentina," in *Sarmiento and His Argentina*, ed. Joseph T. Criscenti (Boulder, CO: L. Rienner Publishers, 1993), 131–42. See also Jose C. Moya, *Cousins and Strangers: Spanish Immigrants in Buenos Aires, 1850–1930* (Berkeley: University of California Press, 1998), 48–52.

37. Julia Rodriguez, *Civilizing Argentina: Science, Medicine, and the Modern State* (Chapel Hill: University of North Carolina Press, 2006), 16–18.

38. On scientific racism and immigration in Argentina, see Rodriguez, *Civilizing Argentina*; Eugenia Scarzanella, *Italiani malagente: Immigrazione, criminalità, razzismo in Argentina, 1890–1940* (Milan: F. Angeli, 1999).

39. Baily, *Immigrants in the Lands of Promise*, 80; Scarzanella, *Italiani malagente*.

40. Nancy Leys Stepan, *The Hour of Eugenics: Race, Gender, and Nation in Latin America* (Ithaca, NY: Cornell University Press, 1991), 139–45.

41. Donna R. Gabaccia, "Race, Nation, Hyphen: Italian-American Multiculturalism in Comparative Perspective," in *Are Italians White? How Race Is Made in America*, ed. Jennifer Guglielmo and Salvatore Salerno (New York: Routledge, 2003), 44–59. For a comparative study of racial selection in immigration policy for countries of the western hemisphere, see FitzGerald and Cook-Martín, *Culling the Masses*.

42. Paul R. Spickard, *Almost All Aliens: Immigration, Race, and Colonialism in American History and Identity* (New York: Routledge, 2007); Gary Gerstle, *American Crucible: Race and Nation in the Twentieth Century* (Princeton, NJ: Princeton University Press, 2001).

43. On the racial status of Italians in the United States, see Matthew Frye Jacobson, *Whiteness of a Different Color: European Immigrants and the Alchemy of Race* (Cambridge: Harvard University Press, 1999); James R. Barrett and David Roediger, "Inbetween Peoples: Race, Nationality, and the 'New Immigrant' Working Class," *Journal of American Ethnic History* 16, no. 3 (1997): 3–44; Thomas A. Guglielmo, *White on Arrival: Italians, Race, Color, and Power in Chicago, 1890–1945* (Oxford: Oxford University Press, 2003); Jennifer Guglielmo and Salvatore Salerno, eds., *Are Italians White?: How Race Is Made in America* (New York: Routledge, 2003).

44. Erika Lee, *At America's Gates: Chinese Immigration during the Exclusion Era, 1882–1942* (Chapel Hill: University of North Carolina Press, 2003).

45. On the history of nativism, see John Higham, *Strangers in the Land: Patterns of American Nativism, 1860–1925* (New Brunswick, NJ: Rutgers University Press, 1955). On the Immigration Act of 1924, see Mae M. Ngai, *Impossible Subjects: Illegal Aliens and the Making of Modern America* (Princeton, NJ: Princeton University Press, 2004), 21–55.

46. Gabaccia, "Race, Nation, Hyphen," 56.

47. Over the last two decades, food studies scholars and migration historians have moved beyond exploring food as a means through which migrants forge and sustain ethnic group identity to food as a site for producing racial demarcation and exclusion. This group of scholars, focused mainly on migrants of color in the United States, has provided an important corrective to apolitical and celebratory images of the United States

as "melting pot" of tolerant, multicultural eaters. See especially Anita Mannur, "Asian American Food-Scapes," *Amerasia Journal* 32, no. 2 (2006): 1–5; Robert Ji-Song Ku, Martin F. Manalansan IV, and Anita Mannur, eds., *Eating Asian America: A Food Studies Reader* (New York: New York University Press, 2013); Tanachai Mark Padoongpatt, "Too Hot to Handle: Food, Empire, and Race in Thai Los Angeles," *Radical History Review* 110 (Spring 2011): 83–108; Simone Cinotto, *Soft Soil, Black Grapes: The Birth of Italian Winemaking in California*, trans. Michelle Tarnopoloski (New York: New York University Press, 2012).

48. Istituto Coloniale Italiano, *Italia e Argentina* (Bergamo: Officine dell'Istituto Italiano d'Arti Grafiche, 1910), 9.

49. As Michel Gobat explains, Latin American elites in the mid-nineteenth century used the term "Latin race" to distinguish themselves from U.S. Americans, Europeans, and the masses of racially mixed residents of Latin American countries. Historians interested in "Latin America" as a continental historical invention have pointed to a number of historical actors who contributed to its emergence: French imperialists interested in Mexico; Latin American expats who popularized the term in Europe; and Latin America elites, who invoked the term to unite countries south of the Rio Grande against U.S. and European imperialism while justifying their political, economic, and racial superiority over indigenous and mixed race peoples. Less explored, however, are Italians and other migrant groups who harnessed notions of Latinness and Latin America to build and conserve economic and cultural bridges between themselves and Latin American consumers around global goods. Michel Gobat, "The Invention of Latin America: A Transnational History of Anti-Imperialism, Democracy, and Race," *American Historical Review* 118, no. 5 (2013): 1345–75.

50. "Il Centenario Argentino," *La Patria*, June 22, 1910, 7.

51. M. Gravina, ed., *Almanacco dell'italiano nell'Argentina* (Buenos Aires, 1918), 86.

52. "Solidarietà italo-argentina," *La Patria*, September 16, 1915, 6; "La formazione della razza argentina," *La Patria*, September 3, 1915, 4.

53. "Situazione generale," *Bollettino Mensile*, April 29, 1910, 5.

54. Ibid.; "Il centenario dell'independenza Argentina a Roma," *La Patria*, May 23, 1910, 3.

55. "Voci affettuose," *La Patria*, February 20, 1910, 5.

56. Luigi Einaudi, *Un principe mercante: Studio sulla espansione coloniale italiana* (Turin: Fratelli Bocca, 1900), 11.

57. "L'emigrazione italiana nel 1 semestre 1908," *Bollettino Mensile*, October 15, 1908, 6.

58. Bernardino Frescura, "La Mostra degli Italiani all'Estero, all'Esposizione Internazionale di Milano nel 1906," *Bollettino dell'emigrazione* 18 (1907): 79.

59. "Sull'America Latina," *La Patria*, April 14, 1910, 5.

60. Aldo Visconti, *Emigrazione ed esportazione: Studio dei rapporti che intercedono fra l'emigrazione e le esportazioni italiane per gli Stati Uniti del Nord America e per la Repubblica Argentina* (Turin: Tipografia Baravalle e Falconieri, 1912), 29, 40.

61. Ibid., 16.

62. Emilio Zuccarini, *Il lavoro degli italiani nella Repubblica Argentina dal 1516 al 1910* (Buenos Aires, 1910), 109.

63. Baily, *Immigrants in the Lands of Promise*, 151. On marriage patterns, see Eduardo José Míguez, "Il comportamento matrimoniale degli italiani in Argentina. Un bilancio," in *Identità degli italiani in Argentina: Reti sociali, famiglia, lavoro*, ed. Gianfausto Rosoli (Rome: Edizioni Studium, 1993), 81–105.

64. Conversely, it was not until after World War I that Italians in New York begin to produce a community that included a sizable aging and more gender-balanced population as well as a large second generation of Italian Americans. Baily, *Immigrants in the Lands of Promise*, 63–65.

65. Samuel L. Baily, "Marriage Patterns and Immigrant Assimilation in Buenos Aires, 1882–1923," *Hispanic American Historical Review* 60, no. 1 (1980): 32–48.

66. On eugenics and race in Argentina, see Stepan, *Hour of Eugenics*. For the United States, see Alan M. Kraut, *Silent Travelers: Germs, Genes, and the "Immigrant Menace"* (New York: BasicBooks, 1994); Alexandra Minna Stern, *Eugenic Nation: Faults and Frontiers of Better Breeding in Modern America* (Berkeley: University of California Press, 2005).

67. Comitato della Camera Italiana di Commercio ed Arti, *Gli italiani nella Repubblica Argentina*, 61.

68. "L'espansione italiana," *La Patria*, April 15, 1910, 7.

69. Ad for Luigi Bosca & Sons wines, *La Patria,* May 25, 1910, 12.

70. Ad for Florio Marsala, *La Patria*, April 29, 1911, 13.

71. Comitato della Camera Italiana di Commercio ed Arti, "La Casa di Rappresentanze Francesco Jannello," *Gli italiani nella Repubblica Argentina all'Esposizione di Torino 1911.*

72. Fabio Parasecoli, *Al Dente: A History of Food in Italy* (London: Reaktion Books, 2014), 136–37, 144–45; David Gentilcore, *Pomodoro!: A History of the Tomato in Italy* (New York: Columbia University Press, 2010), 45–68; Vito Teti, *Storia del peperoncino: Un protagonista delle culture mediterranee* (Rome: Donzelli Editore, 2007), 41–57, 83–119.

73. Donna R. Gabaccia, "Making Foods Italian in the Hispanic Atlantic" (unpublished paper, University of Minnesota, November 1, 2006).

74. On the history of Argentina cuisine, see Rebekah E. Pite, *Creating a Common Table in Twentieth-Century Argentina: Doña Petrona, Women, and Food* (Chapel Hill: University of North Carolina Press, 2013); Aníbal B. Arcondo, *Historia de la alimentación en Argentina: Desde los orígenes hasta 1920* (Córdoba: Ferreyra Editor, 2002); Eduardo P. Archetti, "Hibración, pertenencia y localidad en la construcción de una cocina nacional," in *La Argentina en el siglo XX*, ed. Carlos Altamirano (Buenos Aires: Universidad Nacional de Quilmes, 1999), 217–36.

75. Carols M. Urien and Ezio Colombo, *La República Argentina en 1910* (Buenos Aires: Maucci Hermanos, 1910), 63–65. On food's role in shaping ideas about race in colonial Latin America see Rebecca Earle, *The Body of the Conquistador: Food, Race, and the Colonial Experience in Spanish America, 1942–1700* (New York: Cambridge University Press, 2012).

76. On the different meanings of *criolla* in Mexican, Argentine, and Cuban culinary literature, see Jeffrey M. Pilcher, "Eating à la Criolla: Global and Local Foods in Argentina, Cuba, and Mexico," *IdeAs* 3 (Winter 2012): 3–16, and Pite, *"La cocina criolla."* On food policy and eating under Perón, see Natalia Milanesio, "Food Politics and Consumption in Peronist Argentina," *Hispanic American Historical Review* 90, no. 1 (2010): 75–108.

77. Arcondo, *Historia de la alimentación en Argentina*, 25–26.

78. "N. y A. Canale y Ca.," *La Nación*, Numero del centenario, 1810–1910, 1910, 123; "Gran Hotel Italia (Rosario)," *La Nación*, Numero del centenario, 1810–1910, 1910, 150. On La Sonambula, opened by Nicolas Canale in 1863, see Arcondo, *Historia de la alimentación en Argentina*, 280.

79. Pite, *Creating a Common Table*, 72.

80. Quotes come from "Situazione generale," *Bollettino Mensile*, April 1910, 5, and Visconti, *Emigrazione ed esportazione*, 29, 40.

81. E. Mayor des Planches, "Gli Stati Uniti e l'emigrazione italiana," *Rivista Coloniale* (Rome, Italy) 1 (May–August 1906): 75.

82. "Perché gli americani cercano l'amicizia degli italiani," *Il Progresso*, November 8, 1906, 1; "I desiderabili," *Il Progresso,* May 26, 1905, 1.

83. Peter G. Vellon, *A Great Conspiracy against Our Race: Italian Immigrant Newspapers and the Construction of Whiteness in the Early 20th Century* (New York: New York University Press, 2014), 8.

84. Gabaccia, *We Are What We Eat*, 93.

85. Harvey Levenstein, "The American Response to Italian Food, 1880–1930," *Food and Foodways* 1 (1985): 1–24. On the stigmatization of Italian wine-makers and drinkers, see Cinotto, *Soft Soil, Black Grapes.*

86. Robert A. Woods, "Notes on the Italians in Boston," *Charities* 12 (1904): 451.

87. "Chicken with Macaroni," *New York Times,* August 16, 1888, 8.

88. Antonio Mangano, "The Italian Colonies of New York City," MA thesis, Columbia University, 1903, repr. *Italians in the City: Health and Related Social Needs*, ed. Francesco Cordasco (New York: Arno Press, 1975), 51.

89. It would be groups of bohemian writers, artists, and intellectuals in the early twentieth century who would initially explore Italian restaurants in migrant enclaves of cities like New York and San Francisco, seeing in them both tasty food and an exotic "Latin" culture of pleasure seeking and entertainment. Gabaccia, *We Are What We Eat*, 95–102.

90. Levenstein, "American Response to Italian Food," 4–5. Paul Freedman found that macaroni was the most common offered entrée in mid-nineteenth-century upscale restaurants. See Paul Freedman, "American Restaurants and Cuisine in the Mid-Nineteenth Century," *New England Quarterly* 84, no. 1 (2011): 5–59.

91. Emilio Perera, "Pro importazione italiana negli Stati Uniti," *Rivista*, September 1910, 7.

92. "Pure Food Exposition," *Rivista*, September 1911, 14–15; Guido Rossati, "La vendemmia in Italia ed il commercio dei vini italiani cogli S.U. nel 1905," *Rivista*, December, 1905, 41–51.

93. Luigi Solari, "Come sorse, che cosa é e ció che potrebbe essere la Camera di Commercio Italiana in New York," *Rivista,* September 1912, 7.

94. "Pure Food Exposition," 14.

95. "Parte ufficiale," *Rivista*, February 1910, 7; "Pure Food Law," *Rivista*, April 1909, 7.

96. Rossati, "La vendemmia in Italia," 41.

97. "A Few Hints to American Retailers and Consumers of Italian Articles of Food," *Rivista*, February 1911, 15.

98. Guido Rossati, "Olive verdi," *Rivista*, May 1912, 7.

99. "Per l'alimentazione pura," *Rivista*, March 1910, 8.

100. "A Few Hints," 15.

101. The act was part of a nationwide movement by Progressive reformers to protect consumers' safety and health and to enforce labor legislation. On the act, see James Harvey Young, *Pure Food: Securing the Federal Food and Drugs Act of 1906* (Princeton, NJ: Princeton University Press, 1989); Mitchell Okun, *Fair Play in the Marketplace: The First Battle for Pure Food and Drugs* (DeKalb: Northern Illinois University Press, 1986).

102. "Purity of Italian Food Articles," *Rivista*, March 1912, 23.

103. "Condemnations under the Pure Food Law," *Rivista*, September 1912, 23.

104. Italian American Directory Co., *Gli italiani negli Stati Uniti d'America*, 66.

105. "Pure Food Exposition,"14; Gustavo Porges, "'Quality Trust,'" *Rivista*, December 1911, 32.

106. "Prodotti alimentari italiani a Chicago," *Rivista*, February 1911, 13.

107. Perera, "Pro Importazione Italiana negli Stati Uniti," 7.

108. Emilio Longhi, "Italian Products on the Chicago Market," *Italian Chamber of Commerce Bulletin*, May 1908, 73–75.

109. Ibid.

110. Gustavo Porges, "Qualità il segreto del successo," *Rivista*, December 1912, 29.

111. Baily, *Immigrants in the Lands of Promise*, 150–52.

112. John F. Mariani, "Everybody Likes Italian Food," *American Heritage* 40, no. 8 (1989): 127.

113. Velma Phillips and Laura Howell, "Racial and Other Differences in Dietary Customs," *Journal of Home Economics* 12, no. 9 (1920): 411.

114. Gabaccia, *We Are What We Eat*, 122–31; Levenstein, "American Response," 7–10; Stephanie J. Jass, "Recipes for Reform: Americanization and Foodways in Chicago Settlement Houses, 1890–1920," PhD diss., Western Michigan University, 2004; Michael J. Eula, "Failure of American Food Reformers among Italian Immigrants in New York City, 1891–1897," *Italian Americana* 18 (Winter 2000): 86–99; Cinotto, *Italian American Table*, 55, 108–109.

115. Solari, "Come sorse, che cosa é e ció che potrebbe essere la Camera di Commercio Italiana in New York," 7.

116. "Importazione negli Stati Uniti di olio d'oliva dall'Italia," *Rivista*, August 1909, 13.

117. Longhi, "Italian Products on the Chicago Market," 73–75.

118. Gabaccia, *We Are What We Eat*, 99–102; Cinotto, *Italian American Table*, 180–209; Levenstein, "American Response."

119. "Importazione di commestibili e bibite," *Bollettino Mensile,* February 1902, 7.

120. "Importazione di commestibili e bibite," *Bollettino Mensile*, January 1904, 6; "Importazione di commestibili e bibite," *Bollettino Mensile*, February 1904, 6.

121. Donna R. Gabaccia, *Foreign Relations: American Immigration in Global Perspective* (Princeton, NJ: Princeton University Press, 2012), 54–55.

122. Kevin H. O'Rourke and Jeffrey G. Williamson, *Globalization and History: The Evolution of a Nineteenth-Century Atlantic Economy* (Cambridge: MIT Press, 1999), 185, 196.

123. Jeffry A. Frieden, *Global Capitalism: Its Fall and Rise in the Twentieth Century* (New York: Norton, 2006), 64–68. On the history of trusts in the United States, see Naomi R. Lamoreaux, *The Great Merger Movement in American Business, 1895–1904* (New York: Cambridge University Press, 1985); on Argentina, see Yovanna Pineda, *Industrial Development in a Frontier Economy: The Industrialization of Argentina, 1890–1930* (Stanford, CA: Stanford University Press, 2009), 18–39.

124. "Parte ufficiale," *Rivista*, July 1911, 8.

125. Qualcuno, "A chi giova il protezionismo?," *La Patria*, January 15, 1900, 3.

126. "L'immigrazione in pericolo," *La Patria*, March 7, 1900, 4.

127. "Il disagio presente," *La Patria*, April 7, 1900, 3.

128. "Emigrazione," *Bollettino Mensile,* December 5, 1901, 1.

129. "Iniquità fiscali," *La Patria*, June 28, 1900, 3; "La libertá del giuoco," *La Patria*, September 4, 1900, 3.

130. "Un pauroso fenomeno," *La Patria*, April 10, 1900, 3.

131. "Ci danno ragione," *La Patria*, July 17, 1900, 3.

132. "A proposito dello sciopero dei cappellai," *La Patria*, March 3, 1900, 5.

133. "Il plebiscito della Società Rurale," *La Patria*, January 16, 1900, 3.

134. "La conferenza sul regime degli zuccheri," *Bollettino Mensile*, April 5, 1902, 1.

135. "Ci danno ragione," 3.

136. Scholars continue to argue over the extent to which Argentine tariff policy in the late nineteenth and early twentieth centuries protected Argentine industry. Until recently, scholars have blamed the Argentine government not only for favoring agro-export sectors over industry but also for being actively antagonist toward it, pointing to low tariffs as a prime example of Congress' failure to support Argentina's industrialization. Other scholars, however, have argued that tariffs were more effective in encouraging industry than Structuralists writing mainly in the post–World War II era have acknowledged. On these evolving debates, see Carlos F. Díaz Alejandro, *Essays on the Economic History of the Argentine Republic* (New Haven, CT: Yale University Press, 1970), 277–308; María Inés Barbero and Fernando Rocchi, "Industry," in *The New Economic History of Argentina*, ed. Gerardo della Paolera and Alan Taylor (Cambridge: Cambridge University Press, 2003), 261–94.

137. Rocchi, *Chimneys in the Desert*, 204–236; Pineda, *Industrial Development*, 108–123.

138. Carlos F. Díaz Alejandro, "The Argentine Tariff, 1906–1940," *Oxford Economic Papers* 19, no. 1 (1967): 75–98; Carl Solberg, "The Tariff and Politics in Argentina, 1916–1930," *Hispanic American Historical Review* 53, no. 2 (1973): 260–84.

139. "Tariffa doganale argentina," *Bollettino Mensile*, November 1899, 4–6; "La revisione delle tariffe doganali argentine," *Bollettino Mensile*, August 1910, 12.

140. "La 'Liga de Defensa Comercial' e la legge di dogana," *La Patria,* September 9, 1905, 6.

141. "Politica doganale Italo-Argentina," *Bollettino Mensile*, July 1908, 7.

142. "Atti della camera," *Bollettino Mensile*, December 1, 1906, 1–2.

143. "Capitali italiani," *Bollettino Mensile*, January 5, 1901, 1.

144. Gabaccia, *Foreign Relations*, 122–32; Tom E. Terrill, *The Tariff, Politics, and American Foreign Policy, 1874–1901* (Westport, CT: Greenwood Press, 1973); Alfred E. Eckes, *Opening*

America's Market: U.S. Foreign Trade Policy since 1776 (Chapel Hill: University of North Carolina Press, 1995).

145. Mariani, "Tariff Revision," 44.

146. "Tentativo per abrogare la legge Payne-Aldrich fallito," *Il Progresso*, April 3, 1910, 1.

147. "Il discorso del Cav. Uff. L. Solari, presidente della camera," *Rivista*, February 1910, 11.

148. "Condemnations under the Pure Food Law," *Rivista*, September 1912, 23.

149. Gino C. Speranza, "La necessità di un accordo internazionale in riguardo agli emigranti," *Rivista*, December 1905, 21. On the interesting life of Italian American Gino Speranza and his eventual support for immigration restriction and Americanization programs, see Aldo E. Salerno, "America for Americans Only: Gino C. Speranza and the Immigrant Experience," *Italian Americana* 14, no. 2 (1996): 133–47.

150. Mariani, "Tariff Revision," 44.

151. "Per I venditori ambulanti," *Il Progresso*, February 25, 1905, 1; "Imperialismo e monopolismo," *Supplemento al Progresso-Italo-Americano*, October 7, 1900, 3.

152. Francesco Pisani, "Verso l'imperialismo o verso il lavoro?," *Supplemento al Progresso Italo-Americano*, May 7, 1905, 1; "L'America e gli americani giudicati da un europeo," *Il Progresso*, May 11, 1905, 1.

153. "I 'trusts' in America," *Rivista*, December 1905, 38.

154. "Il boicottaggio della carne," *Il Progresso*, April 7, 1910, 1.

155. For example, the paper supported the unsuccessful Democratic candidate William Jennings Bryan in the 1900 presidential election. "Italiani, tutti per Bryan!," *Il Progresso*, October 27, 1900, 1; "Gli italiani sono per Bryan. Il colossale meeting democratico," *Il Progresso*, October 28, 1900, 1.

156. "Ancora le cooperative," *La Patria*, May 7, 1906, 6; "Il rincaro della carne," *La Patria*, July 3, 1906, 6; "Il pane rincara," *La Patria*, May 12, 1906, 6.

157. Rocchi, *Chimneys in the Desert*, 103–115; Pineda, *Industrial Development*, 33–36.

Chapter 3. *Tipo Italiano*: The Production and Sale of Italian-Style Goods, 1880–1914

1. A. Dall'Aste Brandolini, "L'immigrazione e le colonie italiane nella Pennsylvania," *Bollettino dell'emigrazione* (hereafter *Bollettino*) 4 (1902): 57–58.

2. F. A., "Una visita alla Manifattura de Nobili," *Rivista Commerciale* (hereafter *Rivista*), December 1909, 41.

3. Ibid., 37.

4. Luigi Einaudi, *Un principe mercante: Studio sulla espansione coloniale italiana* (Turin: Fratelli Bocca, 1900), 23.

5. Aldo Visconti, *Emigrazione ed esportazione: Studio dei rapporti che intercedono fra l'emigrazione e le esportazioni italiane per gli Stati Uniti del Nord America e per la Repubblica Argentina* (Turin: Tipografia Baravalle e Falconieri, 1912), 30.

6. "L'emigrazione transoceanica, le correnti commerciali, e i servizi marittimi," *Bollettino* 11 (1904): 107.

7. See, for example, F. Prat, "Gli italiani negli Stati Uniti e specialmente nello Stato di New York," *Bollettino* 2 (1902): 26; Obizzo Luigi Maria Orazio Colombano Malaspina, "L'immigrazione nella Repubblica Argentina," *Bollettino* 3 (1902): 16–17.

8. On the wheat and cattle industries in Argentina, see James R. Scobie, *Revolution on the Pampas: A Social History of Argentine Wheat, 1860–1910* (Austin: University of Texas Press, 1964); Peter H. Smith, *Politics and Beef in Argentina: Patterns of Conflict and Change* (New York: Columbia University Press, 1969). For the United States, see William Cronon, *Nature's Metropolis: Chicago and the Great West* (New York: W. W. Norton, 1991), and Jimmy M. Skaggs, *Prime Cut: Livestock Raising and Meatpacking in the United States, 1607–1983* (College Station: Texas A&M University Press, 1986).

9. Silvano Serventi and Françoise Sabban, *Pasta: The Story of a Universal Food* (New York: Columbia University Press, 2002), 3, 192; Scobie, *Revolution on the Pampas*, 87.

10. F. Prat, "Gli italiani negli Stati Uniti e specialmente nello Stato di New York," 26.

11. Ad for Ernesto Bisi, *Il Progresso Italo-Americano* (hereafter *Il Progresso*), December 1, 1905, 3.

12. Ad for Tomba wines, *La Patria degli Italiani* (hereafter *La Patria*), September 29, 2.

13. See, for example, Diane C. Vecchio, *Merchants, Midwives, and Laboring Women: Italian Migrants in Urban America (Urbana: University of Illinois Press, 2006)*; S .P. Breckinridge, *New Homes for Old* (New York: Harper & Brothers Publishing, 1921), 58–60; Mary Sherman, "Manufacturing of Foods in the Tenements," *Charities and the Commons* 15 (1906): 669–73.

14. Simone Cinotto, *Soft Soil, Black Grapes: The Birth of Italian Winemaking in California*, trans. Michelle Tarnopoloski (New York: New York University Press, 2012), 107–113; see also Simone Cinotto, *The Italian American Table: Food, Family, and Community in New York City* (Urbana: University of Illinois Press, 2013), 107–108; see also Hasia R. Diner, *Hungering for America: Italian, Irish, and Jewish Foodways in the Age of Migration* (Cambridge: Harvard University Press, 2001), 69–70.

15. At the "Italians Abroad" exhibit during the 1906 Milan International exhibition, the jury awarded prizes to many tipo italiano companies for the large number of Italian workers they employed. Bernardino Frescura, "La Mostra degli Italiani all'Estero, all'Esposizione Internazionale di Milano nel 1906," *Bollettino* 18 (1907): 219.

16. Comitato della Camera Italiana di Commercio ed Arti, *Gli italiani nell'Repubblica Argentina* (Buenos Aires: Compañia Sud-Americana de Billetes de Banco, 1898), 453.

17. Ettore Patrizi, *Gl'italiani in California, Stati Uniti d'America* (San Francisco: Stabilimento Tipo-Litografico, 1911), 53; Italian American Directory Co., *Gli italiani negli Stati Uniti d'America* (New York: Andrew H. Kellogg Co., 1906), 287.

18. Jennifer Guglielmo, *Living the Revolution: Italian Women's Resistance and Radicalism in New York City, 1880–1945* (Chapel Hill: University of North Carolina Press, 2010), 60–64, 139–75; José Moya, "Italians in Buenos Aires's Anarchist Movement: Gender Ideology and Women's Participation, 1890–1910," in *Women, Gender, and Transnational Lives: Italian Workers of the World*, ed. Donna R. Gabaccia and Franca Iacovetta (Toronto: University of Toronto Press, 2002), 189–216.

19. United States Immigration Commission, *Immigrants in Cities*, vol. 1 (Washington, DC: U.S. Government Printing Office, 1911), 176, 220.

20. Donna J. Guy, "Women, Peonage, and Industrialization: Argentina, 1810–1914," *Latin American Research Review* 16, no. 3 (1981): 78–80.

21. Louise Odencrantz, *Italian Women in Industry* (New York: Russell Sage Foundation, 1919), 17, 48–50. See also Miriam Cohen, *Workshop to Office: Two Generations of Italian Women in New York City, 1900–1950* (Ithaca, NY: Cornell University Press, 1993).

22. Patrizi, *Gl'italiani in California*, 46.

23. Italian American Directory Co., *Gli italiani negli Stati Uniti d'America*, 398–400.

24. Guglielmo, *Living the Revolution*, 69–72.

25. Ibid, 58. See also Donna R. Gabaccia, *From Sicily to Elizabeth Street: Housing and Social Change among Italian Immigrants, 1880–1930* (Albany: State University of New York Press, 1984), 64, 92, 94; Virginia Yans-McLaughlin, *Family and Community: Italian Immigrants in Buffalo, 1880–1930* (Ithaca, NY: Cornell University Press, 1977), 51–53.

26. Frescura, "La Mostra degli Italiani all'Estero," 75–81.

27. Ibid., 219; Lorenzo Faleni and Amedeo Serafini, eds., *La Repubblica Argentina all'Esposizione internazionale di Milano 1906* (Buenos Aires, 1906), 43–47.

28. "Panatteria e fabbrica a vapore di biscotti della Vedova Canale," in Faleni and Serafini, *La Repubblica Argentina*.

29. Italian American Directory Co., *Gli italiani negli Stati Uniti d'America*, 398–400.

30. Guido Rossati, "La nuova tariffa doganale degli Stati Uniti rispetto all'importazione italiana," *Rivista*, August 1909, 10–13; Comitato della Camera Italiana di Commercio ed Arti, *Gli italiani nell'Argentina*, 303.

31. Ad for Ernesto Petrucci, *Il Progresso*, August 23, 1900, 4.

32. Ad for Emilio Longhi & Co., *Italian Chamber of Commerce Bulletin* (Chicago), October 1908, 6.

33. Ad for Francesco Jannello, *La Patria*, April 23, 1905, 4.

34. Samuel L. Baily and Franco Ramella, eds., *One Family, Two Worlds: An Italian Family's Correspondence across the Atlantic, 1901–1922*, trans. John Lenaghan (New Brunswick, NJ: Rutgers University Press, 1988), 90, 157–58.

35. Frescura, "La Mostra degli Italiani all'Estero," 227.

36. Ibid., 227, 229.

37. Italian American Directory Co., *Gli iItaliani negli Stati Uniti d'America*, 264–67.

38. Ibid., 293.

39. "Il commesso viaggiatore," *Bollettino Mensile della Camera Italiana di Commercio in Buenos Aires* (hereafter *Bollettino Mensile*), March 5, 1900, 1.

40. "Progetto di nuova legge sui vini," *Bollettino Mensile*, July 1908, 12.

41. "More about Paste," *Italian Chamber of Commerce Bulletin*(Chicago), January–February 1909, 15, 17.

42. Ad for United States Macaroni Factory, *Il Progresso*, January 13, 1900, 1.

43. Ad for United States Macaroni Factory, *Il Progresso*, December 1, 1905, 3.

44. Ibid.

45. For example, see ad for Grandolini and Company's pasta machine, *La Patria*, July 9, 1910, 4; ad for Grandolini and Company's pasta machine, *Il Progresso*, April 1, 1905, 7.

46. On the industrialization of food production, see Harvey A. Levenstein, *Revolution at the Table: The Transformation of the American Diet* (New York: Oxford University Press, 1988); Amy Bentley, *Inventing Baby Food: Taste, Health, and the Industrialization of the American Diet*

(Berkeley: University of California Press, 2014); Susan Strasser, *Satisfaction Guaranteed: The Making of the American Mass Market* (New York: Pantheon Books, 1989); Helen Zoe Veit, *Modern Food, Moral Food: Self-Control, Science, and the Rise of Modern American Eating in the Early Twentieth Century* (Chapel Hill: University of North Carolina Press, 2013); Roger Horowitz, *Putting Meat on the American Table: Taste, Technology, Transformation* (Baltimore: Johns Hopkins University Press, 2006).

47. Cinotto, *Italian American Table*, 115–30.

48. Donna R. Gabaccia, *We Are What We Eat: Ethnic Food and the Making of Americans* (Cambridge: Harvard University Press, 1998), 55–63, 65–73.

49. Eugenia Scarzanella, *Italiani d'Argentina: Storie di contadini, industriali e missionari italiani in Argentina, 1850–1912* (Venice: Marsilio, 1983), 30–70.

50. Fernando Rocchi, *Chimneys in the Desert: Industrialization in Argentina during the Export Boom Years, 1870–1930* (Stanford, CA: Stanford University Press, 2006), 28–33, 96–115; Yovanna Pineda, *Industrial Development in a Frontier Economy: The Industrialization of Argentina, 1890–1930* (Stanford, CA: Stanford University Press, 2009), especially chapters 4 and 5.

51. Faleni and Serafini, *La Repubblica Argentina*, 92.

52. Rocchi, *Chimneys in the Desert*, 89.

53. Camera di Commercio Italiana in New York, *Nel cinquantenario della Camera di Commercio Italiana in New York, 1887–1937* (New York, 1937), 66–67.

54. "Vino di California," *Rivista*, April 1909, 30.

55. Pineda, *Industrial Development in a Frontier Economy*, 110–15. Oscar Cornblit, "Inmigrantes y empresarios en la politica argentina," *Desarrollo Económico* 6, no. 24 (1967): 641–91.

56. "Pure Food Law," *Rivista*, April 1909, 7.

57. Camera di Commercio Italiana in New York, *Nel cinquantenario della Camera di Commercio Italiana in New York*, 66–67, 132.

58. Frank R. Rutter, *Tariff Systems of South American Countries* (Washington, DC: U.S. Government Printing Office, 1916), 89–90.

59. See, for example, "Atti della camera," *Bollettino Mensile,* April 29, 1910, 3; "Raccomandazioni ai nostri speditori di formaggi," *Bollettino Mensile*, August, 1908, 6.

60. Italian American Directory Co., *Gli italiani negli Stati Uniti d'America*, 67.

61. Visconti, *Emigrazione ed esportazione*, 52.

62. G. Saint-Martin, "Gli italiani nel distretto consolare di Nuova Orleans," *Bollettino* 1 (1903): 16–17.

63. Ads for Castruccio & Sons pasta and B. Piccardo pasta, *Il Progresso*, April 1, 1905, 6.

64. Frescura, "La Mostra degli Italiani all'Estero," 79–80; Cinotto, *Italian American Table*, 135–36.

65. "Il Commercio Italo-Americano," *Rivista*, February 1912, 14–15; "Per le paste alimentari," *Rivista*, May 1906, 4.

66. Leonard Covello, *The Social Background of the Italo-American School Child: A Study of the Southern Italian Family Mores and Their Effect on the School Situation in Italy and America* (Leiden: E. J. Brill, 1967), 295, quoted in Diner, *Hungering for America*, 58.

67. Ministerio de Agricultura, Dirección de Comercio e Industria, *Censo Industrial de la República* (Buenos Aires, Talleres de Publicaciones de la Oficina Meteorológica Argentina, 1910), 12–13.

68. Comitato della Camera Italiana di Commercio ed Arti, *Gli italiani nella Repubblica Argentina*, 169–170.

69. Frescura, "La Mostra degli Italiani all'Estero," 220.

70. Trade statistics for 1916 came from Ufficio Centrale di Stastica, *Annuario statistico italiano* (1917–1918), 247. Trade statistics for 1909 came from Visconti, *Emigrazione ed esportazione*, 51.

71. Ad for United States Macaroni Factory, *Il Progresso*, January 13, 1900, 1.

72. Ad for Italian Cigar & Tobacco Company, *Il Progresso*, June 29, 1900, 3.

73. Elizabeth Zanoni, "'In Italy everyone enjoys it. Why not in America?': Italian Americans and Consumption in Transnational Perspective during the Early Twentieth Century," in *Making Italian America: Consumer Culture and the Production of Ethnic Identities*, ed. Simone Cinotto (New York: Fordham University Press, 2014), 71–82. On the history of marketing and mass adverting, see Strasser, *Satisfaction Guaranteed*; Jackson Lears, *Fables of Abundance: A Cultural History of Advertising in America* (New York: Basic Books, 1994); Stuart Ewen and Elizabeth Ewen, *Channels of Desire: Mass Images and the Shaping of American Consciousness* (New York: McGraw-Hill, 1982).

74. Ad for Razzetti Brothers, *Il Progresso*, December 16, 1910, 8; ad for G. B. Lobravico, *Il Progresso*, April 3, 1910, 3.

75. Ad for American Chianti Wine Company, *Il Progresso*, December 16, 4.

76. Ad for De Nobili, *Il Progresso,* September 2, 1915, 7; ad for De Nobili, *Rivista*, December 1911, 49.

77. John E. Zucchi, "Paesani or Italiani? Local and National Loyalties in an Italian Immigrant Community," in *The Family and Community Life of Italian Americans*, ed. Richard Juliani (New York: American-Italian Historical Association, 1983), 147–60.

78. Zanoni, "In Italy everyone enjoys it," 73–77.

79. Cinotto, *Italian American Table,* 172–76.

80. See, for example, Visconti, *Emigrazione ed esportazione*, 6–7.

81. Frescura, "La Mostra degli Italiani all'Estero," 220–21.

82. Luigi Solari, "Tutela d'interessi legittimi," *Rivista*, September 1909, 7.

83. "Per l'alimentazione pura contro le frodi sugli olii d'oliva," *Rivista*, March 1910, 7.

84. Ramon Ramon-Muñoz, "Modernizing the Mediterranean Olive-Oil Industry," in *The Food Industries of Europe in the Nineteenth and Twentieth Centuries*, ed. Derek J. Oddy and Alain Drouard (Burlington, VT: Ashgate, 2013), 71–88; Leo A. Loubère, *The Red and the White: A History of Wine in France and Italy in the Nineteenth Century* (Albany: State University of New York Press, 1978).

85. See, for example, "Atti della camera," *Rivista*, October 1909, 1; "Atti della camera," *Rivista*, February 1910, 9; "Atti della camera," *Rivista,* May 1910, 1; Solari, "Tutela d'interessi legittimi," 7.

86. "Frode in commercio contraffazione di marche," *Rivista*, October 1910, 6; "Sentenze commerciali," *Rivista*, February 1911, 19.

87. "Sentenze commerciali," *Rivista*, August 1911, 39; "Sentenze commerciali," *Rivista,* July 1911, 16.

88. See, for example, "Frodi e manipolazioni di prodotti italiani," *Bollettino Mensile*, August 31, 1908, 6; "Atti della camera," *Bollettino Mensile*, September 1908, 1.

89. "Assemblea Ordinaria del 15 Luglio 1900," *Bollettino Mensile*, August 5, 1900, 2.

90. "Ai produttori d'Italia," *Bollettino Mensile,* April 5, 1900, 1; "Atti della camera," *Bollettino Mensile,* January 5, 1901, 2; "Il Congresso Ispano-Americano," *Bollettino Mensile*, February 5, 1901.

91. Rocchi, *Chimneys in the Desert*, 90.

92. Comitato della Camera Italiana di Commercio ed Arti, *Gli italiani nella Repubblica Argentina all'Esposizione di Torino 1911* (Buenos Aires: Stabilimento Grafico della Compañia General de Fósforos, 1911), 67–69.

93. "Raccomandazioni ai nostri speditori di formaggi," *Bollettino Mensile*, August 1908, 6.

94. Comitato della Camera Italiana di Commercio ed Arti, *Gli italiani nella Repubblica Argentina all'Esposizione di Torino 1911*.

95. "Concorrenza," *Bollettino Mensile*, April 5, 1901, 1.

96. "Atti della camera," *Bollettino Mensile*, November 1899, 2.

97. "Atti della camera," *Bollettino Mensile*, April 1901, 2.

98. "Frodi e manipolazioni di prodotti italiani,", 6.

99. "Importazione di commestibili e bibite," *Bollettino Mensile*, July 1902, 7; "Importazione di commestibili e bibite," *Bollettino Mensile*, August, 1902, 7; "Notizie varie," *Bollettino Mensile*, February 1904, 5.

100. "Atti della camera," *Bollettino Mensile*, September 30, 1908, 2.

101. "Esportazione vinicola," *Bollettino Mensile*, September 5, 1900, 1–2.

102. "Ministero di Agricoltura, Industria e Commercio, Legge concernente disposizioni per combattere le frodi nella preparazione e nel commercio dei vini," *Bollettino Mensile*, May 5, 1901, 4–5.

103. "Importazione di commestibili e bibite," *Bollettino Mensile*, July 1902, 2; "Importazione di commestibili e bibite," *Bollettino Mensile*, August 1902, 7; "Notizie varie," *Bollettino Mensile*, February 1904, 5.

104. "Atti della camera," *Bollettino Mensile*, April 1901, 2; "Atti della camera," *Bollettino Mensile*, September 1901, 2.

105. "Atti della camera," *Bollettino Mensile*, August 1904, 3; "Atti della camera," *Bollettino Mensile*, January 1, 1905, 1.

106. "Rapporto del R. Enotecnico Italiano al R. Ministero di Agricoltura Industria e Commercio," *Rivista*, January 1909, 7–9.

107. "Concorrenza," 1.

108. "Il Congresso Ispano-Americano," 1–2.

109. Dirección General de Inmigración, *Resumen estadístico del movimiento migratorio en la República Argentina, años 1857–1924* (Buenos Aires: Talleres Gráficos del Ministerio de Agricultura de la Nación, 1925), 8–9.

110. "L'emigrazione italiana nel 1 semestre 1908," *Bollettino Mensile*, October 15, 1908, 5–6.

111. Comitato della Camera Italiana di Commercio ed Arti, *Gli italiani nella Repubblica Argentina all'Esposizione di Torino 1911.* On the competition between Spanish and Italian migrants in Buenos Aires, see Jose C. Moya, *Cousins and Strangers: Spanish Immigrants in Buenos Aires, 1850–1930* (Berkeley: University of California Press, 1998), especially chapter 7.

112. "Il Congresso Ispano-Americano," 1–2.

113. Donna R. Gabaccia, "Making Foods Italian in the Hispanic Atlantic" (unpublished paper, University of Minnesota, November 1, 2006); Vito Teti, ed., *Mangiare meridiano: Culture alimentari del Mediterraneo* (Traversa Cassiodoro, Italy: Abramo, 2002).

114. Visconti, *Emigrazione ed esportazione*, 17–18.

115. "Atto della camera," *Bollettino Mensile*, July 1904, 2–3; "Atto della camera," *Bollettino Mensile*, October 1904, 1.

116. Frescura, "La Mostra degli Italiani all'Estero," 81–86.

117. "Atti della camera," *Bollettino Mensile*, May, 1910, 3.

118. "Atti della camera," *Bollettino Mensile*, June, 1910, 4.

119. Ad for Confiteria los dos Chinos, *La Prensa*, December 16, 1900, 8; ad for Confiteria los dos Chinos, *La Patria*, December 22, 1905, 10.

120. Fabio Parasecoli, *Food Culture in Italy* (Westport, CT: Greenwood Press, 2004), 47.

121. On Mexican foodways, see Jeffrey M. Pilcher, *Que vivan los tamales!: Food and the Making of Mexican Identity* (Albuquerque: University of New Mexico Press, 1998).

122. Rebekah E. Pite discusses a recipe for *pan dulce de Navidad* from Doña Patrona's 1934 cookbook, *El libro de Doña Patrona*. See Rebekah E. Pite, *Creating a Common Table in Twentieth-Century Argentina: Doña Petrona, Women, and Food* (Chapel Hill: University of North Carolina Press, 2013), 91.

123. On Italians in the history of Argentine wine making, see Scarzanella, *Italiani d'Argentina*, 44–45. Aníbal B. Arcondo, *Historia de la alimentación en Argentina: Desde los orígienes hasta 1920* (Córdoba: Ferreyra Editor, 2002), 245–48. On the United States, see Cinotto, *Soft Soil, Black Grapes*; Donna R. Gabaccia, "Ethnicity in the Business World: Italians in American Food Industries," *Italian American Review* 6, no. 2 (1997/1998): 1–19.

124. On Tomba, see Comitato della Camera Italiana di Commercio ed Arti, *Gli italiani nella Repubblica Argentina,* 204–214; "Establecimiento vitivínicola Domingo Tomba, *La Nación*, Numero del centenario, 1810–1910, 1910, 241–43.

125. On the Italian-Swiss Colony, see Patrizi, *Gl'italiani in California,* 35–37; Cinotto, *Soft Soil, Black Grapes*, 4, 19; Gabaccia, *We Are What We Eat*, 73, 87.

126. Comitato della Camera Italiana di Commercio ed Arti, *Gli italiani nella Repubblica Argentina*, 213.

127. "Crisi vinicola," *Bollettino Mensile*, March 1902, 1–2.

128. Ad for Tomba wine, *La Patria*, September 20, 1910, 2.

129. "Establecimiento vitivícola Domingo Tomba," 241.

130. Ad for Tomba wine, *La Patria,* September 29, 1905, 2.

131. Marni Davis, *Jews and Booze: Becoming American in the Age of Prohibition* (New York: New York University Press, 2012).

132. Cinotto, *Soft Soil, Black Grapes*.

133. Patrizi, *Gl'italiani in California*, 35–37.

134. "Vino di California," *Rivista*, April 1909, 29–30; Rossati, "La vendammia in Italia ed il commercio dei vini italiani cogli S.U. nel 1905," 41.

135. "Perche non progredisce l'esportazione dei vini italiani," *La Patria*, January 20, 1910, 7.

136. Eugenia Scarzanella, *Italiani malagente: Immigrazione, criminalità, razzismo in Argentina, 1890–1940* (Milan: F. Angeli, 1999), 25; Arcondo, *Historia de la alimentación en Argentina*, 100–105.

Chapter 4. "Pro Patria": Women and the Normalization of Migrant Consumption during World War I

1. B. A., "Quel che insegna il presente dissidio tra l'Argentina e l'Italia," *Rivista Commerciale* (hereafter *Rivista*), October 1911, 6.

2. *Gli italiani nel Sud America ed il loro contributo alla guerra: 1915–1918* (Buenos Aires: Arigoni & Barbieri, 1922), 21–53; "Comitato Generale Italiano di Soccorso Pro Croce Rossa e famiglie dei richiamati," *Il Progresso Italo-Americano* (hereafter *Il Progresso*), September 8, 1915, 2; "Pro patria," *Il Progresso*, November 9, 1915, 1; E. Zuccarini, "Virtù italica," *La Patria degli* Italiani (hereafter *La Patria*), September 18, 1915, 4. On Italians' war efforts in Buenos Aires and Montevideo, see John Starosta Galante, "The 'Great War' in *Il Plata*: Italian Immigrants in Buenos Aires and Montevideo during the First World War," *Journal of Migration History* 2, no. 1 (2016): 57–92.

3. See, for example, "Vita sociale," *La Patria*, September 26, 1915, 5; "Vita sociale," *La Patria*, October 6, 1915, 5; *Gli italiani nel Sud America*, 199; "Sottocomitato della Boca," *La Patria*, September 15, 1915, 5; "Il gran concerto al Century per le famiglie dei richiamati," *Il Progresso*, July 29, 1915, 2.

4. For Argentina, see "Pro patria, la festa pro lana," *La Patria*, January 6, 1916, 5; "L'iniziativa della 'Patria degli Italiani,'" *La Patria*, September 1, 1915, 6; Un italiano, "Per il freddo dei nostri soldati," *La Patria*, September 4, 1915, 4; "Vita sociale," *La Patria*, October 8, 1915, 4. For the United States, see, "Il cuore delle nostre colonie per i prodi soldati d'Italia," *Il Progresso*, September 9, 1915, 1; "Proteggiamo i nostri soldati," *Il Progresso*, September 12, 1915, 1.

5. "Cronaca rosarino," *La Patria,* September 6, 1915, 7.

6. "Sezione Femminile del Comitato di Guerra," *La Patria*, January 8, 1916, 5.

7. "Per i nostri combattenti," *Il Progresso*, October 19, 1915, 2.

8. "Vita sociale," *La Patria*, September 2, 1915, 5.

9. Gianfausto Rosoli, ed., *Un secolo di emigrazione italiana, 1876–1976* (Rome: Centro Studi Emigrazione, 1978), 354.

10. Mark I. Choate, *Emigrant Nation: The Making of Italy Abroad* (Cambridge: Harvard University Press, 2008), 208.

11. For Argentina, see "Notizie statistiche sui movimenti migratori," *Bollettino dell'emigrazione* (hereafter *Bollettino*) 1 (1910): 77. For the United States, see "Emigrazione dall'Europa e immigrazione in America e in Australia," *Bollettino* 14 (1907): 127.

12. Commissariato Generale dell'Emigrazione, *Annuario statistico della emigrazione italiana dal 1876 al 1925* (Rome: Edizione del Commissariato Generale dell'Emigrazione, 1926), 432.

13. See, for example, "La moda," *Il Progresso*, November 3, 1915, 5; "Per le signore," *Il Progresso*, November 10, 1915, 9.

14. Examples include "Cronache femminile," *La Patria*, April 14, 1915, 4; "Cronache femminile," *La Patria*, May 23, 1915, 4. By the end of World War I the column had expanded slightly and changed its name to "Vita femminile" (feminine life). "Vita femminile," *La Patria*, June 1, 1917, 4; "Vita femminile," *La Patria*, January 1, 1920, 8.

15. "Per i nostri soldati," *Il Progresso*, September 4, 1915, 1.

16. "L'offerta del 'Costurero Italo-Argentino,'" *La Patria*, September 18, 1915, 5.

17. "Appunti. . . . lana, lana, lana!," *La Patria*, August 29, 1915, 4.

18. "Sezione Femminile del Comitato Italiano di Guerra," *La Patria*, August 15, 1915, 7.

19. S. Magnani-Tedeschi, "La donna italiana e la guerra," *La Patria*, September 21, 1915, 3.

20. "Appunti. . . . lana, lana, lana!," 4; "Date lana," *La Patria*, November 10, 1915, 4.

21. Magnani-Tedeschi, "La donna italiana e la guerra."

22. "L'offerta del 'Costurero Italo-Argentino,'" *La Patria*, September 18, 1915, 5.

23. "La raccolta per la lana del Comitato delle Donne Italiane sotto gli auspicii del Progresso Italo-Americano," *Il Progresso*, September 17, 1915, 2.

24. "Proteggiamo i nostri soldati," *Il Progresso*, September 12, 1915, 1.

25. "Date lana," *La Patria*, 4.

26. "Pro patria," *La Patria*, January 6, 1916, 5.

27. Ibid., 6.

28. *Gli italiani nel sud America*, 38.

29. "Italian-American Ladies Association for Italian Soldiers Relief Fund," *Rivista*, January 17, 1916, 1.

30. "Il fascio delle forze italiane," *La Patria*, July 13, 1915, 3. For the United States, see Camera di Commercio Italiana in New York, *Nel cinquantenario della Camera di Commercio Italiana in New York, 1887–1937* (New York, 1937), 142.

31. "Il dovere della nostra colonia," *Rivista*, July 19, 1915, 1.

32. Luigi Solari, "Appello ai soci," *Rivista*, August 23, 1915, 1. Emphasis is in the original.

33. "Comitato Generale Italiano di Soccorso pro Croce Rossa e famiglie dei richiamati," *Il Progresso*, July 10, 1915, 2.

34. "Agli abbonati, ai lettori, agli amici," *La Patria*, May 24, 1915, 5.

35. "Per l'album al Principe di Udine," *Il Progresso*, June 7, 1917, 2.

36. *Gli italiani nel Sud America*, 32, 199.

37. "Per la celebrazione del XX Settembre," *La Patria*, September 18, 1915, 5; "Attraverso i rioni della città durante la distribuzione dei cesti-regalo," *La Patria*, September 24, 1915, 4. See also M. Gravina, *Almanacco dell'Italiano nell'Argentina (Buenos Aires, 1918)*, 189; *Gli italiani nel Sud America*, 34–35.

38. "Per la celebrazione del XX Settembre," *La Patria*, September 18, 1915, 5.

39. "Appunti. . . . lana, lana, lana!," *La Patria*, August 29, 1915, 4.

40. "Vita sociale," *La patria*, September 7, 1916, 5.

41. See ad for Gath & Chaves, *La Patria*, September 10, 1915, 6; ad for Gath & Chaves, *La Patria*, August 13, 1915.

42. "Le assemblee della Camera Italiana di Commercio," *La Patria*, September 12, 1915, 6.

43. "Agli italiani residenti all'estero," *La Patria*, September 20, 1916, 27.

44. Camera di Commercio Italiana in New York, *Nel cinquantenario della Camera di Commercio Italiana in New York*, 142.

45. "Cronaca rosarina," *La Patria*, December 15, 1915, 6.

46. "Agli italiani residenti all'estero," 27.

47. "Un grande dovere patriottico," *La Patria*, September 26, 1915, 7.

48. Ad for San Pellegrino sparking water, *La Patria*, September 20, 1917.

49. Ad for Florio Marsala and Malvasia, *La Patria*, July 8, 1917, 7.

50. D. Bórea, "Comitato Agrario Italiano nell'Argentina," *La Patria*, September 5, 1915, 6.

51. Ibid.

52. *Gli italiani nel Sud America*, 39–40.

53. "Pro patria," *La Patria*, January 13, 1916, 7.

54. "L'offerta del 'Costurero Italo-Argentino,'" *La Patria*, September 18, 1915, 5. See also "Solidarietà italo-argentina," *La Patria*, September 16, 1915.

55. "Costurero Privado Italo-Argentino," *La Patria*, September 12, 1915, 6.

56. *Gli italiani nel Sud America*, 252–53.

57. See, for example, "Vita sociale," *La Patria*, September 26, 1915, 5; "Per la Croce Rossa Alleata," *La Patria*, September 26, 1915, 8.

58. Ad for Harrods, *La Patria,* September 26, 1915, 8.

59. On Piccardo, see Fernando Rocchi, *Chimneys in the Desert: Industrialization in Argentina during the Export Boom Years, 1870–1930* (Stanford, CA: Stanford University Press, 2006), 55, 79–81.

60. "Italian Wines in Competition with French and German Wines," *Rivista*, October 5, 1914, 1–2.

61. "Italian Sparkling Wines Successfully Competing with the Best Champagne Brands," *Rivista*, October 12, 1914, 1–2.

62. Ministero delle Finanze, *Movimento commerciale del Regno d'Italia nell'anno 1919*, Parte Seconda (Rome: Stabilimento Poligrafico per l'Amministrazione della Guerra, 1921), 1046, 1065. A quintal is one hundred kilograms.

63. Ibid., 1288, 1301, 1302.

64. Ibid., 1064, 1065, 1301.

65. Camera di Commercio Italiana in New York, *Nel cinquantenario della Camera di Commercia Italiana in New York*, 141–43. In response to the chambers' pleas, Italy allowed Italian ships to devote a higher quantity of tonnage to the transport of Italian merchandise to the United States. For a list of items banned from exportation by the Italian government, see "Elenco completo delle merci delle quali il governo italiano ha vietata l'esportazione," *Rivista*, March 15, 1915, 1.

66. On La Rosa, see Simone Cinotto, *The Italian American Table: Food, Family, and Community in New York City* (Urbana: University of Illinois Press, 2013), 137.

67. Ibid., 130–41.

68. Camera di Commercio Italiana in New York, *Nel cinquantenario della Camera di Commercio Italiana in New York*, 142.

69. "Una nostra nuova iniziativa pro patria," *Il Progresso*, November 20, 1915, 1. See also "Pro patria," *Il Progresso*, November 21, 1915, 2.

70. "Per la patria," *Il Progresso*, November 25, 1915, 3.

71. "Sigari per i nostri soldati!," *Il Progresso*, September 2, 1917, 3.

72. Ibid.

73. Ibid. See ad for De Nobili cigars, *Il Progresso*, September 2, 1915, 7.

74. See, for example, "Per la salute dei nostri soldati," *Il Progresso*, October 15, 1917, 3.

75. "La cassetta di Natale per i nostri trionfatori sulle Alpi," *Il Progresso*, September 2, 1917, 2; emphasis is mine.

76. "La cassetta di Natale per i nostri trionfatori sulle Alpi," *Il Progresso*, September 3, 1917, 2.

77. "La casette di Natale per i nostri trionfatori sulle Alpi," *Il Progresso*, November 15, 1917, 3.

78. "Una nostra nuova iniziativa pro patria," 1.

79. "Sigari e sigarette pei nostri combattenti," *Il Progresso*, September 1, 1917, 3.

80. Gary Gerstle, *American Crucible: Race and Nation in the Twentieth Century* (Princeton, NJ: Princeton University Press, 2001), 81–127; John Higham, *Strangers in the Land: Patterns of American Nativism, 1860–1925* (New Brunswick, NJ: Rutgers University Press, 1955), 194–221; Donna R. Gabaccia, *Foreign Relations: American Immigration in Global Perspective* (Princeton, NJ: Princeton University Press, 2012), 122–25.

81. Camera di Commercio Italiana in New York, *Nel cinquantenario della Camera di Commercio Italiana in New York*, 142; "Prestito del Governo Italiano negli Stati Uniti," *Il Progresso*, October 30, 1915, 9; "Gli italiani nell'esercito dei Uncle Sam," *Il Progresso*, September 18, 1917, 1; "Il secondo prestito della libertà," *Il Progresso*, October 5, 1917, 2; "Il Progresso a disposizione dei suoi lettori per l'acquisto dei liberty-bonds," *Il Progresso*, October 5, 1917, 2. On foreign-born soldiers in the U.S. military, see Nancy Gentile Ford, *Americans All!: Foreign-Born Soldiers in World War I* (College Station: Texas A&M University Press, 2001).

82. "La raccolta per la lana e la nostra iniziativa," *Il Progresso*, October 16, 1915, 2.

83. Gravina, *Almanacco dell'italiano nell'Argentina*, 190; "Per le desolate famiglie dei nostri combattenti," *Il Progresso*, November 7, 1915, 2.

84. For additional examples, see ads for Cinzano, *La Patria*, December 11, 1915, 6; *La Patria*, December 15, 1915, 6; *La Patria*, December 17, 1915, 6; *La Patria*, November 27, 1915, 6.

85. Ad for Cinzano vermouth, *La Patria*, February 4, 1905, 3; ad for Cinzano, *La Patria*, January 23, 1910, 4.

86. Donna R. Gabaccia, "Women of the Mass Migrations: From Minority to Majority, 1820–1930," in *European Migrants: Global and Local Perspectives*, ed. Dirk Hoerder and Leslie Page Moch (Boston: Northeastern University Press, 1996), 90–111.

87. David Cook-Martín, "Soldiers and Wayward Women: Gendered Citizenship, and Migration Policy in Argentina, Italy, and Spain since 1850," *Citizenship Studies* 10, no. 5 (2006): 571–90; Donna J. Guy, *Sex and Danger in Buenos Aires: Prostitution, Family, and Nation in Argentina* (Lincoln: University of Nebraska Press, 1991), 16–35; Martha Gardner, *The Qualities of a Citizen: Women, Immigration, and Citizenship, 1870–1965* (Princeton, NJ: Prince-

ton University Press, 2005); Eithne Luibhéid, *Entry Denied: Controlling Sexuality at the Border* (Minneapolis: University of Minnesota Press, 2002).

88. Donna R. Gabaccia, "When the Migrants Are Men: Italy's Women and Transnationalism as a Working-Class Way of Life," in *American Dreaming, Global Realities: Rethinking U.S. Immigration History*, ed. Donna R. Gabaccia and Vicki L. Ruiz (Urbana: University of Illinois Press, 2006), 199.

89. Bernardino Frescura, "La Mostra degli Italiani all'Estero, all'Esposizione Internazionale di Milano nel 1906," *Bollettino* 18 (1907): 171.

90. Leopoldo Corinaldi, "L'emigrazione italiana negli Stati Uniti," *Bollettino* 2 (1902): 5.

91. José María Ramos Mejía, *The Argentine Masses* (1899), in *Darwinism in Argentina: Major Texts, 1845–1909*, ed. Leila Gómez, trans. Nicholas Ford Callaway (Lanham, MD: Bucknell University Press), 209–210.

92. "Instruzioni per chi emigra negli Stati Uniti dell'America del Nord," *Bollettino* 15 (1904): 52.

93. G. Naselli, "Gli Italiani nel distretto consolare di Filadelfia," *Bollettino* 10 (1903): 33.

94. On the gendering of consumption globally, see especially Victoria de Grazia, with Ellen Furlough, eds., *The Sex of Things: Gender and Consumption in Historical Perspective* (Berkeley: University of California Press, 1996).

95. For classic treatment of this gendered process, see Louise A. Tilly and Joan W. Scott, *Women, Work, and Family* (New York: Holt, Rinehart, and Winston, 1978).

96. Leonore Davidoff and Catherine Hall, *Family Fortunes: Men and Women of the English Middle Class, 1780–1850* (Chicago: University of Chicago Press, 1987); Erika D. Rappaport, *Shopping for Pleasure: Women in the Making of London's West End* (Princeton, NJ: Princeton University Press, 2000); Lisa Tiersten, *Marianne in the Market: Envisioning Consumer Society in Fin-de-Siècle France* (Berkeley: University of California Press, 2001).

97. For the United States, see Kristin L. Hoganson, *Consumers' Imperium: The Global Production of American Domesticity, 1865–1920* (Chapel Hill: University of North Carolina Press, 2007). For Argentina, see Cecilia Tossounian, "Images of the Modern Girl: From the Flapper to the Joven Moderna (Buenos Aires, 1920–1940)," *Forum for Inter-American Research* 6, no. 2 (2013): 41–70.

98. Jennifer Scanlon, *Inarticulate Longings: The* Ladies' Home Journal*, Gender, and the Promises of Consumer Culture* (New York: Routledge, 1995).

99. Carolyn M. Goldstein, *Creating Consumers: Home Economists in Twentieth-Century America* (Chapel Hill: University of North Carolina Press, 2012); Rebekah E. Pite, *Creating a Common Table in Twentieth-Century Argentina: Doña Petrona, Women, and Food* (Chapel Hill: University of North Carolina Press, 2013), 27–29.

100. For an overview of this literature, see de Grazia and Furlough, *Sex of Things*.

101. Elizabeth Zanoni, "'Per Voi, Signore': Gendered Representations of Fashion, Food, and Fascism in *Il Progresso Italo-Americano* during the 1930s," *Journal of American Ethnic History* 31, no. 3 (2012): 33–71; Simone Cinotto, "All Things Italian: Italian American Consumers, the Transnational Formation of Taste, and the Commodification of Difference," in *Making Italian America: Consumer Culture and the Production of Ethnic Identities*, ed. Simone Cinotto (New York: Fordham University Press, 2014), 1–31.

102. For the United States, see Lizabeth Cohen, *Making a New Deal: Industrial Workers in Chicago, 1919–1939*, 2nd ed. (Cambridge: Cambridge University Press, 2008); Meg Jacobs, *Pocketbook Politics: Economic Citizenship in Twentieth-American America* (Princeton, NJ: Princeton University Press, 2005). For Argentina, see Eduardo Elena, *Dignifying Argentina: Peronism, Citizenship, and Mass Consumption* (Pittsburgh: University of Pittsburgh Press, 2011); Natalia Milanesio, *Workers Go Shopping in Argentina: The Rise of Popular Consumer Culture* (Albuquerque: University of New Mexico Press, 2013).

103. Ad for Banfi Products Co., *Il Progresso*, December 21, 1927, 2; emphasis is mine.

104. Ad for Cirio tomato extract, *La Patria*, September 4, 1925, 3. On Cirio and the early development of the Italian tomato production, see David Gentilcore, *Pomodoro!: A History of the Tomato in Italy* (New York: Columbia University Press, 2010), 83–89, 109–115, 136–42.

105. On migration's influence on the tomato industry in Italy, see Gentilcore, *Pomodoro!*, 109–119.

106. Ad for Florio Marsala, *Il Progresso*, November 21, 1937, 9.

107. Ad for Bertolli olive oil, *L'Italia* (Chicago), August 14, 1932.

108. Notice the only man in the ad is in a chef outfit, showing continued gendered divisions between the informal, non-remunerative food work done by women in the domestic realm and the professional, paid work of male chefs in restaurants and public eating cultures.

109. Elizabeth Zanoni, "'In Italy everyone enjoys it. Why not in America?': Italian Americans and Consumption in Transnational Perspective during the Early Twentieth Century," in Cinotto, *Making Italian America*, 79–81.

110. Ad for Bertolli, *L'Italia* (Chicago), August 14, 1932.

111. Ad for Sasso, *La Patria*, January 3, 1916, 2.

112. On the rising standards of living in Italy, see Emanuela Scarpellini, *Material Nation: A Consumer's History of Modern Italy*, trans. Daphne Hughes and Andrew Newton (New York: Oxford University Press, 2011); Francesco Chiapparino, Renato Covino, and Gianni Bovini, *Consumi e industria alimentare in Italia dall'unità a oggi: Lineamenti per una storia*, 2nd ed. (Narni, Italy: Giada, 2002), 81–90.

113. Linda Reeder, *Widows in White: Migration and the Transformation of Rural Italian Women, Sicily, 1880–1920* (Toronto: University of Toronto Press, 2003), 14.

114. On gendered messages about food and domesticity intended for middle-class, female readers in the United States, see Scanlon, *Inarticulate Longings*; Laura Shapiro, *Perfection Salad: Women and Cooking at the Turn of the Century* (New York: Farrar, Straus, and Giroux, 1986); Amy Bentley, *Inventing Baby Food: Taste, Health, and the Industrialization of the American Diet* (Berkeley: University of California Press, 2014). For Argentina, see Pite, *Creating a Common Table*, 45–52, 58–65, 76–84. See also Sherrie A. Inness, ed., *Kitchen Culture in America: Popular Representations of Food, Gender, and Race* (Philadelphia: University of Pennsylvania Press, 2001); Sherrie A. Inness, *Cooking Lessons: The Politics of Gender and Food* (Lanham, MD: Rowman & Littlefield, 2001).

115. Ad for Sasso olive oil, *La Patria*, November 29, 1915, 2.

116. Ad for Brioschi digestive aid, *Il Progresso*, February 9, 1936, illustrated section.

117. Hasia Diner, among others, discusses these generational networks of food culture within the Italian American context. Hasia R. Diner, *Hungering for America: Italian, Irish, and Jewish Foodways in the Age of Migration* (Cambridge: Harvard University Press, 2001), 79.

118. Cohen, *Making a New Deal*, 110–16; Donna R. Gabaccia, *We Are What We Eat: Ethnic Food and the Making of Americans* (Cambridge: Harvard University Press, 1998), 45–63.

119. For the history of advertising, see Pamela Walker Laird, *Advertising Progress: American Business and the Rise of Consumer Marketing* (Baltimore: Johns Hopkins University Press, 1998); Jackson Lears, *Fables of Abundance: A Cultural History of Advertising in America* (New York: Basic Books, 1994); Roland Marchand, *Advertising the American Dream: Making Way for Modernity, 1920–1940* (Berkeley: University of California Press, 1985); Stuart Ewen, *Captains of Consciousness: Advertising and the Social Roots of the Consumer Culture* (New York: McGraw-Hill, 1976). For Argentina, see Rocchi, *Chimneys in the Desert*, 77–85; Milanesio, *Workers Go Shopping in Argentina*, 83–122. This trend also coincided with the slow move away from selling food in bulk, instigated in part by food safety laws requiring the labeling of foods by origin, weight, and contents; the smaller food units sold in boxes, cans, jars, and bottles were more easily branded than foodstuff transported and sold in bulk. Susan Strasser, *Satisfaction Guaranteed: The Making of the American Mass Market* (New York: Pantheon Books, 1989); Aníbal B. Arcondo, *Historia de la alimentación en Argentina: Desde los orígenes hasta 1920* (Córdoba, Argentina: Ferreyra Editor, 2002).

120. Ad for Locatelli cheese, *Il Progresso*, December 16, 1934, 14.

121. Ad for Pastene coffee, *Il Progresso*, October 29, 1939, illustrated section.

122. Tracy N. Poe, "The Labour and Leisure of Food Production as a Mode of Ethnic Identity Building among Italians in Chicago, 1890–1940," *Rethinking History* 5, no. 1 (2001): 131–48; John E. Zucchi, "Paesani or Italiani? Local and National Loyalties in an Italian Immigrant Community," in *The Family and Community Life of Italian Americans*, ed. Richard Juliani (New York: American-Italian Historical Association, 1983), 147–60.

Chapter 5. Reorienting Migrant Marketplaces in *le due Americhe* during the Interwar Years

1. "Commentarii," *La Patria degli Italiani* (hereafter *La Patria*), July 4, 1917, 3.

2. F. Filippini, "Camera Italo-Argentina di Industria e Commercio di Genova," *La Patria*, May 12, 1915, 6.

3. I follow the lead of Lok Siu, who argues that the migration of Chinese to Latin America should be understood within the context of migrants' enduring ties to their homeland and to the larger history of U.S. economic, political, and military interventions in Latin America. Lok C. D. Siu, *Memories of a Future Home: Diasporic Citizenship of Chinese in Panama* (Stanford, CA: Stanford University Press, 2005).

4. On Garibaldi, see especially Lucy Riall, *Garibaldi: Invention of a Hero* (New Haven, CT: Yale University Press, 2007); Alfonso Scirocco, *Garibaldi: Citizen of the World*, trans. Allan Cameron (Princeton, NJ: Princeton University Press, 2007).

5. "Il Congresso ispano-americano," *Bollettino Mensile della Camera Italiana di Commercio in Buenos Aires*, February 5, 1901, 1–2.

6. On U.S. economic and cultural expansion in Latin America, see especially Emily S. Rosenberg, *Financial Missionaries to the World: The Politics and Culture of Dollar Diplomacy, 1900–1930* (Cambridge: Harvard University Press, 1999); María I. Barbero and Andrés M. Regalsky, eds. *Americanización: Estados Unidos y América Latina en el siglo XX. Transferencias económicas, tecnológicas y culturales* (Buenos Aires: EDUNTREF, 2014); Gilbert M. Joseph, Catherine C. LeGrand, and Ricardo D. Salvatore, eds., *Close Encounters of Empire: Writing the Cultural History of U.S.-Latin American Relations* (Durham, NC: Duke University Press, 1998).

7. "Lotta fra due mondi," *La Patria,* January 26, 1910, 5; "La crisi del Monroismo," *La Patria*, April 8, 1906, 3.

8. "Psicologia Pan-americana," *La Patria*, July 17, 1910, 5.

9. "La politica yankee," *La Patria,* July 21, 1910, 6.

10. "Lo spauracchio yankee," *La Patria*, July 24, 1910, 5.

11. "La lacuna del Congresso Pan-americano," *La Patria*, July 19, 1910, 5.

12. "Panamericanismo," *La Patria*, October 18, 1910, 7.

13. "Tocchi in penna," *La Patria*, March 3, 1910, 5.

14. Vico Mantegazza, "Italia e Argentina," *La Patria*, May 16, 1910, 3.

15. "Il verbo della disillusione," *La Patria*, June 8, 1910, 5.

16. "Italia e Argentina," *La Patria*, July 9, 1910, 7.

17. "Il Centenario Argentino," *La Patria*, June 22, 1910, 7.

18. "L'America Latina per Enrico Piccione," *La Patria*, August 3, 1906, 3.

19. S. Magnani Tedeschi, "Argentini e Italiani per il trionfo della latinità," *La Patria*, August 18, 1925, 4.

20. Paulina L. Alberto and Eduardo Elena, "Introduction: Shades of the Nation," in *Rethinking Race in Modern Argentina*, eds. Paulina L. Alberto and Eduardo Elena (New York: Cambridge University Press, 2016), 1–22.

21. "Sull'America Latina," *La Patria*, April 14 1910, 4.

22. R. Lucente, "Vita americana," *La Patria*, May 2, 1906, 4.

23. Michel Gobat, "The Invention of Latin America: A Transnational History of Anti-Imperialism, Democracy, and Race," *American Historical Review* 118, no. 5 (2013): 1345–75.

24. Alfredo Malaurie and Juan M. Gazzano, *La Industria Argentina y la Exposición del Paraná* (Buenos Aires: De Juan M. Gazzano y Cia., 1888), 9, 64, 145.

25. Andrés Regalsky and Aníbal Jáuregui, "Americanización, proyecto económico y las ideas de Alejandro Bunge en la década de 1920," in Barbero and Regalsky, *Americanización*, 85–117.

26. Samuel Flagg Bemis, *The Latin American Policy of the United States* (New York: Harcourt, Brace and Co., 1943), 147, 230.

27. Donna R. Gabaccia, *Foreign Relations: American Immigration in Global Perspective* (Princeton, NJ: Princeton University Press, 2012), 84.

28. From the mid-nineteenth century to World War I, the British invested heavily in Argentina's agro-export industries and in railroads, ports, and communication networks. Donna J. Guy, "Dependency, the Credit Market, and Argentine Industrialization, 1860–1940," *Business History Review* 58, no. 4 (1984): 532–61.

29. Direzione Generale di Statistica e del Lavoro, *Annuario statistico italiano* (Rome, 1915), 178; Direzione Generale di Statistica, *Annuario statistico italiano* (Rome, 1919–1921), 268.

30. Department of Commerce, Bureau of Foreign and Domestic Commerce, *Statistical Abstract of the United States: 1920* (Washington, DC: U.S. Government Printing Office, 1921), 411.

31. Department of Commerce, Bureau of Foreign and Domestic Commerce, *Statistical Abstract of the United States: 1931* (Washington, DC: U.S. Government Printing Office, 1931), 503.

32. Ibid., 504.

33. Department of Commerce, Bureau of Foreign and Domestic Commerce, *Statistical Abstract of the United States: 1929* (Washington, DC: U.S. Government Printing Office, 1929), 487.

34. James A. Farrell, "South Americans Can Gain by Use of Own Money Locally," National Foreign Trade Council Bulletin #55 (1938), box 25, folder bulletins B-1B-68 1938, National Foreign Trade Council Records, Hagley Museum and Library, Wilmington, Delaware (hereafter Hagley).

35. Dudley Maynard Phelps, *Migration of Industry to South America* (New York: McGraw-Hill, 1936), 13, 15, 18–21.

36. Ibid., 128. M. J. French, "The Emergence of a U.S. Multinational Enterprise: The Goodyear Tire and Rubber Company, 1910–1939," *Economic History Review* 40, no. 1 (1987): 69, 72.

37. Phelps, *Migration of Industry*, 11.

38. Ibid., 46–47. On Armour, see "L'inaugurazione del frigorifico Armour," *La Patria*, July 3, 1915, 6. See also Swift's biography: Louis F. Swift and Arthur Van Vlissingen Jr., *The Yankee of the Yards: The Biography of Gustavus Franklin Swift* (Chicago: A. W. Shaw, 1927).

39. On the history of the meatpacking industry, see especially Peter H. Smith, *Politics and Beef in Argentina: Patterns of Conflict and Change* (New York: Columbia University Press, 1969). See also Pepé Treviño, *La carne podrida: El caso Swift-Deltec*, 2nd ed. (Buenos Aires: A. Peña Lillo, 1972); Diego P. Roldán, *Chimeneas de carne: Una historia del frigorífico Swift de Rosario, 1907–1943* (Rosario, Argentina: Prohistoria Ediciones, 2008).

40. Samuel L. Baily and Franco Ramella, eds., *One Family, Two Worlds: An Italian Family's Correspondence across the Atlantic, 1901–1922*, trans. John Lenaghan (New Brunswick, NJ: Rutgers University Press, 1988), 195,198; Daniel James, *Doña Maria's Story: Life History, Memory, and Political Identity* (Durham, NC: Duke University Press, 2000).

41. Erika Lee, *At America's Gates: Chinese Immigration during the Exclusion Era, 1882–1943* (Chapel Hill: University of North Carolina, 2003); Robert F. Zeidel, *Immigrants, Progressives, and Exclusion Politics: The Dillingham Commission, 1900–1927* (DeKalb: Northern Illinois University Press, 2004); Aristide R. Zolberg, *A Nation by Design: Immigration Policy in the Fashioning of America* (New York: Russell Sage Foundation, 2006).

42. Gianfausto Rosoli, ed., *Un secolo di emigrazione italiana, 1876–1976* (Rome: Centro Studi Emigrazione, 1978), 354.

43. Gardner, *The Qualities of a Citizen: Women, Immigration, and Citizenship, 1870–1965* (Princeton, NJ: Princeton University Press, 2005).

44. See Eduardo José Míguez, "Introduction: Foreign Mass Migration to Latin America in the Nineteenth and Twentieth Centuries—an Overview," in *Mass Migration to Modern Latin America*, ed. by Samuel Baily and Eduardo José Míguez (Wilmington, DE: Scholarly Resources, 2003), xiii.

45. Erika Lee, "Orientalisms in the Americas: A Hemispheric Approach to Asian American History," *Journal of Asian American Studies* 8, no. 3 (2005): 235–65. The United States pressured countries in the western hemisphere to enact harsh immigration laws, especially Canada and Mexico. See Adam M. McKeown, *Melancholy Order: Asian Migration and the Globalization of Borders* (New York: Columbia University Press, 2008); Erika Lee, "Enforcing the Borders: Chinese Exclusion along the U.S. Borders with Canada and Mexico, 1882–1924," *Journal of American History* 89, no. 1 (2002): 54–86.

46. David Scott FitzGerald and David Cook-Martín, *Culling the Masses: The Democratic Origins of Racist Immigration Policy in the Americas* (Cambridge: Harvard University Press, 2014), 299–332; Samuel L. Baily and Eduardo José Míguez, ed., *Mass Migration to Modern Latin America* (Wilmington, DE: SR Books, 2003).

47. After about 1930 the U.S. would again become the more popular destination for western hemisphere–bound Italians but only by a couple thousand migrants annually. Rosoli, *Un secolo di emigrazione italiana*, 353–54.

48. Donna R. Gabaccia, *Italy's Many Diasporas* (Seattle: University of Washington Press, 2000), 129–52; Philip V. Cannistraro and Gianfausto Rosoli, "Fascist Emigration Policy in the 1920s: An Interpretive Framework," *International Migration Review* 13, no. 4 (1979): 673–92.

49. Kevin H. O'Rourke and Jeffrey G. Williamson, *Globalization and History: The Evolution of a Nineteenth-Century Atlantic Economy* (Cambridge: MIT Press, 1999), 186–206.

50. On immigration and ethnicity in Brazil, see Jeffrey Lesser, *Immigration, Ethnicity, and National Identity in Brazil, 1808 to the Present* (Cambridge: Cambridge University Press, 2013).

51. William Ricketts, "What Does South America Offer the American Advertiser," 31–37, box MN5, folder 1928, March-1929, Dec., JWT Newsletter Collection, 1910–1986, J. Walter Thompson Company Collections, Hartman Center for Sales, Advertising and Marketing History, David. M. Rubenstein Rare Book & Manuscript Library, Duke University, Durham, North Carolina (hereafter JWT Collections).

52. Phelps, *Migration of Industry*, 104–106.

53. Ibid., 239.

54. Ibid., 241.

55. Julie Greene, *The Canal Builders: Making America's Empire at the Panama Canal* (New York: Penguin Press, 2009); John Soluri, *Banana Cultures: Agriculture, Consumption, and Environmental Change in Honduras and the United States* (Austin: University of Texas Press, 2005); Steve Striffler and Mark Moberg, *Banana Wars: Power, Production, and History in the Americas* (Durham, NC: Duke University Press, 2003).

56. Phelps, *Migration of Industry*, 243. On U.S. racial attitudes toward Italian and other working-class migrant groups, see Matthew Frye Jacobson, *Whiteness of a Different Color: European Immigrants and the Alchemy of Race* (Cambridge: Harvard University Press, 1999);

James R. Barrett and David Roediger, "Inbetween Peoples: Race, Nationality, and the 'New Immigrant' Working Class," *Journal of American Ethnic History* 16, no. 3 (1997): 3–44.

57. "Le correnti di emigrazione latina negli Stati Uniti e le restrizioni legali nordamericane," *La Patria*, February 6, 1925, 1.

58. Donna R. Gabaccia, "Nations of Immigrants: Do Words Matter?," *Pluralist* 5, no. 3 (2010): 5–31.

59. Russell Pierce, "See How We've Grown in South America!," *News Letter*, September 15, 1929; 3–4, box MN8, folder 1929 Jan. 1–Dec. 15, JWT Newsletter Collection, 1910–1986, JWT Collections; emphasis in the original.

60. Mark I. Choate, *Emigrant Nation: The Making of Italy Abroad* (Cambridge: Harvard University Press, 2008), 92–97.

61. *Printers' Ink*, June 5, 1914, cited in Stuart Ewen, *Captains of Consciousness: Advertising and the Social Roots of the Consumer Culture* (New York: McGraw-Hill, 1976), 65.

62. "Cities within a City—And Each One a Worth-While Market," *JWT News Letter* no. 13, February 7, 1924, 4, box MN6, folder 1924, Jan. 3–Feb. 28 Newsletters, JWT Newsletter Collection, 1910–1986, JWT Collections.

63. *JWT News Letter* no. 19, March 20, 1924, 6, box MN6, folder 1924, Mar. 6–April 7 Newsletters, JWT Newsletter Collection, 1910–1986, JWT Collections.

64. "Unrecognized Cities in the United States, No. 4: The Czechoslovak City of Chicago," *News Letter*, no. 28, May 2, 1924, 3, box MN6, folder 1924, May 22–June 24, JWT Newsletter Collection, 1910–1986, JWT Collections; "Unrecognized Cities in the United States, No. 7: The Swedish City of Chicago," *News Letter*, no. 45, September 16, 1924, box MN6, folder 1924 Aug.–Oct. 9, JWT Newsletter Collection, 1910–1986, JWT Collections; "Unrecognized Cities in the United States, No. 9: The Polish City of Milwaukee," *News Letter*, no. 50, October 13, 1924, 3, box MN6, folder 1924, Oct. 16–Nov. 13, JWT Newsletter Collection, 1910–1986, JWT Collections; "Unrecognized Cities in the United States: The Policy City of Chicago," *News Letter*, no. 75, April 9, 1925, 2, box MN6, folder 1925 March 26–May 7, JWT Newsletter Collection, 1910–1986, JWT Collections.

65. Lizabeth Cohen, *Making a New Deal: Industrial Workers in Chicago, 1919–1939*, 2nd ed. (Cambridge: Cambridge University Press, 2008), 99–203.

66. Ibid., 100–158; Simone Cinotto, *The Italian American Table: Food, Family, and Community in New York City* (Urbana: University of Illinois Press, 2013); Hasia R. Diner, *Hungering for America: Italian, Irish, and Jewish Foodways in the Age of Migration* (Cambridge: Harvard University Press, 2001).

67. "Jewish Papers and New York Coverage," News Letter, no. 192, Nov. 15, 1927, 485, box MN8, folder 1927: Nov. 1–Dec. 15, JWT Newsletter Collection, 1910–1986, JWT Collections.

68. "Unrecognized Cities in the United States, No. 5: The German City of Philadelphia," *News Letter*, no. 31, June 12, 1924, box MN6, folder 1924, May 22–June 24, JWT Newsletter Collection, 1910–1986, JWT Collections.

69. "Unrecognized Cities in the United States, No. 3: The Jewish City of New York," *News Letter*, no. 24, April 24, 1924, 6, box MN6, folder 1924, April 17–May 15, JWT Newsletter Collection, 1910–1986, JWT Collections.

70. Ad for Helmar cigarettes, *Il Progresso Italo-Americano* (hereafter *Il Progresso*), November 4, 1924, 8.

71. See, for example, ad for Helmar cigarettes, *Il Progresso*, April 10, 1917, 7.

72. Ad for Heckers flour, *Il Progresso*, January 8, 1935, 5.

73. Cinotto, *Italian American Table*, 57–58, 64–69; Sherrie A. Inness, ed., *Kitchen Culture in America: Popular Representations of Food, Gender, and Race* (Philadelphia: University of Pennsylvania Press, 2001); Laura Shapiro, *Perfection Salad: Women and Cooking at the Turn of the Century* (New York: Farrar, Straus, and Giroux, 1986).

74. Ad for Armour prosciutto, *Il Progresso*, December 6, 1937, 3.

75. Armour and Company, *Catalogue of Products Manufactured by Armour and Company* (Chicago: Armour and Company, 1916), 64–69.

76. Ibid., 56.

77. Similarly, in the 1930s U.S. food companies such as Quaker Oats, Hershey's, and Procter and Gamble sold kosher pancake mix, chocolate, and Crisco to Jewish consumers. Jenna Weissman Joselit, *The Wonders of America: Reinventing Jewish Culture 1880–1950* (New York: Henry Holt, 1994), 171–218; Roger Horowitz, *Kosher USA: How Coke Became Kosher and Other Tales of Modern Food* (New York: Columbia University Press, 2016).

78. "Spanish Advertising Department," in *A Series of Talks on Advertising* (New York, 1909), box DG6, folder 1909, Publications Collection, 1887–2005, Domestic Publications Series, JWT Collections.

79. A second Brazil branch opened in Rio de Janeiro in 1931. The Montevideo office closed in 1930 but reopened again in 1937. See "Our South American Forces," *News Letter*, March 15, 1929, 3, box MN8, folder 1929 Jan. 1–Dec. 15, JWT Newsletter Collection, 1910–1986, JWT Collections; Arthur Farlow, "J.W.T. Pioneers in South America," *News Letter*, July 1, 1929, box MN8, folder 1929 Jan. 1–Dec. 15, JWT Newsletter Collection, 1910–1986, JWT Collections; Pierce, "See How We've Grown in South America!"

80. There has been debate over the size of Argentina's middle class and, relatedly, the extent to which working-class Argentines could participate in a national consumer culture before World War II. Fernando Rocchi sees a rising middle class assisting in the large-scale formation of a consumer society already in the early twentieth century. Natalia Milanesio argues that it was not until the mid-nineteenth century, after Juan Domingo Perón's government took power, that most lower-income Argentines could fully participate in mass consumption. Fernando Rocchi, *Chimneys in the Desert: Industrialization in Argentina during the Export Boom Years, 1870–1930* (Stanford, CA: Stanford University Press, 2006), 51, 61–62. Natalia Milanesio, *Workers Go Shopping in Argentina: The Rise of Popular Consumer Culture* (Albuquerque: University of New Mexico Press, 2013), 2.

81. Rocchi, *Chimneys in the Desert*, 49–85.

82. Clement H. Watson, "Markets Are People—Not Places," JWT *News Bulletin*, July 1938, 3–4, JWT Newsletter Collections, 1910–1986, box MN5, JWT Newsletter Collection, 1910–1986, JWT Collection.

83. Ibid., 6. On J. Walter Thompson and advertising in South America, see Jennifer Scanlon, "Mediators in the International Marketplace: U.S. Advertising in Latin America in the Early Twentieth Century," *Business History Review* 77, no. 3 (2003): 387–415; Ricardo

D. Salvatore, "Yankee Advertising in Buenos Aires: Reflections on Americanization," *Interventions* 7, no. 2 (2005): 216–35; Milanesio, *Workers Go Shopping*, 54–60, 115.

84. See, for example, ad for Remington typewriter, *La Patria*, January 18, 1900, 7; ad for American Light Company, *La Patria*, April 20, 1905, 2; ad for Victor phonograph, *La Patria*, June 15, 1911, 4; ad for Singer sewing machines, *La Patria*, March 4, 1900, 2.

85. See, for example, ad for Michelin, *La Patria*, September 20, 1917; ad for United States Rubber Export Co., *La Patria*, January 8, 1920, 6; ad for Overland, *La Patria*, May 9, 1920, 12; ad for Studebaker Corporation of America, *La Patria*, September 20, 1920, 106; ad for Harley-Davidson, *La Patria*, September 20, 1920, 8; ad for Goodyear, *La Patria*, September 8, 1925, 3; ad for Ford Motor Co., *La Patria*, August 17, 1930, 11; ad for Chevrolet, *La Patria*, January 26, 1930, 11; ad for Case, *La Patria*, September 3, 1930, 8; ad for International Harvester Corporation, *La Patria*, January 1, 1920, 11; ad for Fairbanks, Morse & Co., *La Patria*, April 1, 1920, 7.

86. Ad for Kodak, *La Patria*, May 15, 1920, 6; ad for Frederick Gee Watches Company, *La Patria*, January 9, 1915, 1; ad for Westinghouse Electric International, *La Patria*, December 20, 1920, 7; ad for B.V.D. Company, *La Patria*, November 18, 1920, 9; ad for Keds shoes, *La Patria*, November 4, 1925, 4; ad for Bayer, *La Patria*, January 4, 1925, 6.

87. On the evolution of advertising in the U.S., see Jackson Lears, *Fables of Abundance: A Cultural History of Advertising in America* (New York: Basic Books, 1994); Pamela Walker Laird, *Advertising Progress: American Business and the Rise of Consumer Marketing* (Baltimore: Johns Hopkins University Press, 1998). For Argentina, see Rocchi, *Chimneys in the Desert*, 77–85; Milanesio, *Workers Go Shopping*, 83–122; Noemí M. Girbal-Blacha and María Silvia Ospital, "'Vivir con lo nuestro': Publicidad y política en la Argentina de los años 1930," *European Review of Latin American and Caribbean Studies* 78 (April 2005): 49–66.

88. *Armour and Company: Containing Facts about Business and Organization* (Armour & Company, 1917).

89. Ad for Armour tripe, *Il Mattino d'Italia* (Buenos Aires, Argentina; hereafter *Il Mattino*), November 10, 1935, 5.

90. Ad for Veedol oil, *La Patria*, August 5, 1925, 2.

91. Ad for Veedol oil, *La Patria*, August 6, 1925, 2; ad for Veedol, *La Patria*, August 5, 1925, 2.

92. Carina Frid de Silberstein, "Migrants, Farmers, and Workers: Italians in the Land of Ceres," in *Italian Workers of the World: Labor Migration and the Formation of Multiethnic States*, ed. Donna R. Gabaccia and Fraser M. Ottanelli, 79–101 (Urbana: University of Illinois Press, 2001).

93. Ad for Case threshing machine, *La Patria*, January 29, 1920, 8.

94. See, for example, ad for Elvea, Vitelli & Company canned tomatoes, *Il Progresso*, March 3, 1935, 3; and ad for L. Gandolfi & Company, *Il Progresso*, December 5, 1926, 9.

95. Ad for Joseph Personeni Inc., *Il Progresso,* December 1, 1925, 12.

96. Ad for Aguila Saint chocolate, *La Patria*, September 20, 1917; ad for Barilá Turrochole, *Il Mattino*, July 7, 1936, 3; ad for Quilmes beer, *Il Mattino*, July 7, 1936, 4.

97. See, for example, ad for Terrabussi cookies, *Il Mattino*, December 15, 1935, 7; ad for Confetteria del gas, *Il Mattino*, December 25, 1935, 9. Confetteria del gas, while calling out to connazionali, also described its baked goods as an "Argentine tradition."

98. Ricardo D. Salvatore, "The Enterprise of Knowledge: Representational Machines of Informal Empire," in *Close Encounters of Empire: Writing the History of U.S.–Latin American Culture Relations*, ed. Gilbert M. Joseph, Catherine C. LeGrand, and Ricardo D. Salvatore (Durham, NC: Duke University Press, 1998), 94.

Chapter 6. Fascism and the Competition for Migrant Consumers, 1922–1940

1. Ad for Motta panettone, *Il Progresso Italo-Americano* (hereafter *Il Progresso*), November 28, 1937, 8.

2. Ad for Confetería del Molino panettone, *Il Mattino*, November 7, 1936, 3.

3. Jeffry A. Frieden, *Global Capitalism: Its Fall and Rise in the Twentieth Century* (New York: Norton, 2006), 177–81.

4. Yovanna Pineda, *Industrial Development in a Frontier Economy: The Industrialization of Argentina, 1890–1930* (Stanford, CA: Stanford University Press, 2009), 115–23; María Inés Barbero and Fernando Rocchi, "Industry," in *The New Economic History of Argentina*, ed. Gerardo della Paolera and Alan Taylor (Cambridge: Cambridge University Press, 2003), 261–94.

5. Fernando Rocchi, *Chimneys in the Desert: Industrialization in Argentina during the Export Boom Years, 1870–1930* (Stanford, CA: Stanford University Press, 2006), 204–207.

6. Peter H. Smith, *Politics and Beef in Argentina: Patterns of Conflict and Change* (New York: Columbia University Press, 1969), 16.

7. Vera Zamagni, *The Economic History of Italy, 1860–1990* (Oxford: Oxford University Press, 1993), 243–71.

8. Istituto Centrale di Statistica del Regno d'Italia, *Annuario statistico italiano* (1931, publisher varies), 271; Istituto Nazionale di Statistica, *Commercio di importazione e di esportazione del Regno d'Italia* (1939, publisher varies), 638, 640.

9. Carol Helstosky, *Garlic and Oil: Politics and Food in Italy* (Oxford: Berg, 2004), 63–126.

10. Donna R. Gabaccia, *Italy's Many Diasporas* (Seattle: University of Washington Press, 2000), 136, 141–44; Philip V. Cannistraro and Gianfausto Rosoli, "Fascist Emigration Policy in the 1920s: An Interpretive Framework," *International Migration Review* 13, no. 4 (1979): 673–92.

11. Matteo Pretelli, "Culture or Propaganda? Fascism and Italian Culture in the United States," *Studi Emigrazione* 43, no. 161 (2006): 171–92; Stefano Luconi and Guido Tintori, *L'ombra lunga del fascio: Canali di propaganda fascista per gli "italiani d'America"* (Milan: M&B Publishing, 2004); Stefano Luconi, *La "diplomazia parallela": Il regime fascista e la mobilitazione politica degli italo-americani* (Milan: F. Angeli, 2000).

12. Stefano Luconi, "Etnia e patriottismo nella pubblicità per gli italo-americani durante la guerra d'Etiopia," *Italia Contemporanea* 241 (2005): 514–22; Simone Cinotto, "'Buy Italiano!': Italian American Food Importers and Ethnic Consumption in 1930s New York," in *Italian Americans: A Retrospective on the Twentieth Century*, edited by Paola A. Sensi-Isolani and Anthony Julian Tamburri (Chicago Heights, IL: American Italian Historical Association, 2001), 167–78.

13. Federico Finchelstein, *Transatlantic Fascism: Ideology, Violence, and the Sacred in Argentina and Italy, 1919–1945* (Durham, NC: Duke University Press, 2010), 35–41; Vanni Blengino and Eugenia Scarzanella, ed., *Fascisti in Sud America* (Florence: Le Lettere, 2005).

14. República Argentina, *Censo Industrial de 1935* (Buenos Aires: Ministerio de Hacienda, 1938), 35.

15. María Inés Barbero, "Empresas y empresarios Italianos en la Argentina (1900–30)," in *Studi sull'emigrazione: Un'analisi comparata*. Atti del Convegno storico internazionale sull'emigrazione, ed. Maria Rosaria Ostuni (Biella, Italy: Fondazione Sella, 1989), 303–313. On Pirelli, see María Inés Barbero, "Grupos empresarios, intercambio commercial e inversiones italianas en la Argentina. El caso de Pirelli (1910–1920)," *Estudios Migratorios Latinoamericanos* 5, nos. 15–16 (1990): 311–41.

16. Federico Finchelstein, *Fascismo, liturgia e imaginario. El mito del General Uriburu y la Argentina nacionalista* (Buenos Aires: Fondo de Cultura Económica, 2002).

17. Finchelstein, *Transatlantic Fascism*, 80–89, 104–107.

18. On Pope and pro-Mussolini support, see Philip V. Cannistraro, "The Duce and the Prominenti: Fascism and the Crisis of Italian American Leadership," *Altreitalie* 31 (July–December, 2005): 82–83; Philip V. Cannistraro, "Generoso Pope and the Rise of Italian-American Politics, 1925–1936," in *Italian Americans: New Perspectives in Italian Immigration and Ethnicity*, ed. Lydio F. Tomasi (Staten Island: Center for Migration Studies of New York Inc., 1985), 265–88.

19. David Aliano, *Mussolini's National Project in Argentina* (Madison, WI: Fairleigh Dickinson University Press, 2012), 64–71, 120; Ronald C. Newton, "Ducini, Prominenti, Antifascisti: Italian Fascism and the Italo-Argentine Collectivity, 1922–1945," *Americas* 51, no. 1 (1994): 48–49; Pantaleone Sergi, "Fascismo e antifascismo nella stampa italiana in Argentina: così fu spenta 'La Patria degli Italiani,'" *Altreitalie* 35 (July–December 2007): 4–43.

20. Luca de Caprariis, "'Fascism for Export'? The Rise and Eclipse of the Fasci Italiani all'Estero," *Journal of Contemporary History* 35, no. 2 (2000): 151–183; Emilio Gentile, "La politica estera del partito fascista: Ideologia e organizzazione dei fasci italiani all'estero (1920–1930)," *Storia Contemporanea* 26, no. 2 (1995): 897–956.

21. Two good places to start for antifascist activities of Italians in the United States and Argentina are Pietro Rinaldo Fanesi, "Italian Antifascism and the Garibaldine Tradition in Latin America," trans. Michael Rocke, and Fraser M. Ottanelli, "'If Fascism Comes to America We Will Push It Back into the Ocean': Italian American Antifascism in the 1920s and 1930s," both in *Italian Workers of the World: Labor Migration and the Formation of Multiethnic States*, ed. Donna R. Gabaccia and Fraser M. Ottanelli (Urbana: University of Illinois Press, 2001), 163–77 and 178–95 respectively.

22. Pretelli, "Culture or Propaganda?"; Matteo Pretelli, "Tra estremismo e moderazione. Il ruolo dei circoli fascisti italo-americani nella politica estera italiana degli anni Trenta," *Studi Emigrazione* 40, no. 150 (2003): 315–23; Luconi, *La "diplomazia parallela"*; Stefano Luconi, "The Italian-Language Press, Italian American Voters, and Political Intermediation in Pennsylvania in the Interwar Years," *International Migration Review* 33, no. 4 (1999): 1031–61.

23. Cannistraro, "The Duce and the Prominenti," 77–78.

24. Leo V. Kanawada Jr., *Franklin D. Roosevelt's Diplomacy and American Catholics, Italians, and Jews* (Ann Arbor, MI: UMI Research Press, 1982), 75–89.

25. Finchelstein, *Transatlantic Fascism*, 53–57.

26. "Il peso attivo dell'Italia nella bilancia del commercio mondiale," *Il Progresso*, November 29, 1935, 6.

27. Angelo Flavio Guidi, "Italia e le sanzioni," *Il Progresso*, November 28, 1935, 6.

28. Generoso Pope, "Gli Stati Uniti e le sanzioni," *Il Progresso*, November 26, 1935, 1; Generoso Pope, "Rispettare la neutralità," *Il Progresso*, November 27, 1935, 1; "Note del giorno," *Il Progresso*, November 6, 1935, 6. On the pressure applied to Congress to stay neutral by Italian migrants in Philadelphia, see Stefano Luconi, *From* Paesani *to White Ethnics: The Italian Experience in Philadelphia* (Albany: State University of New York Press, 2001), 87–89.

29. I. C. Fablo, "L'embargo alle materie prime," *Il Progresso*, November 23, 1935, 6; I. C. Falbo, "Stati Uniti e Italia," *Il Progresso*, November 20, 1935, 6.

30. "Autorevoli voci contro la politica sanzionista," *Il Progresso*, November 2, 1935, 2; I. C. Falbo, "La neutralità americana," *Il Progresso*, November 2, 1935, 6.

31. Department of Commerce, Bureau of Foreign and Domestic Commerce, *Statistical Abstract of the United States 1937*, no. 59 (Washington, D.C.: U.S. Government Printing Office, 1938), 455, 457.

32. Generoso Pope, "Vigili e attivi," *Il Progresso*, December 7, 1935, 1.

33. Falbo, "Stati Uniti e Italia," 6. On Italians and the New Deal, see Stefano Luconi, "Italian Americans, the New Deal State, and the Making of Citizen Consumers," in *Making Italian America: Consumer Culture and the Production of Ethnic Identities,* ed. Simone Cinotto (New York: Fordham University Press, 2014), 137–47.

34. "Il problema delle sanzioni," *La Nuova Patria degli Italiani* (hereafter *La Nuova Patria*), October 6, 1935, 2.

35. "I doveri degli italiani in Argentina," *La Nuova Patria*, December 15, 1935, 1.

36. "Nervi a posto, connazionali!," *La Nuova Patria*, October 20, 1935, 1.

37. "Il boicottaggio agli inglesi," *La Nuova Patria*, August 25, 1935, 1.

38. "L'Argentina e le sanzioni," *La Nuova Patria*, October 13, 1935, 1.

39. "Dati e impressioni dall'italia alla vigilia della guerra in Africa," *La Nuova Patria*, August 18, 1935, 3.

40. G. Chiummiento, "Di fronte alle sanzioni," *La Nuova Patria*, November 24, 1935, 1.

41. Generoso Pope, "L'aggressione economica," *Il Progresso*, November 19, 1935, 1; I. C. Falbo, "Fronte e retrofronte," *Il Progresso*, November 10, 1935, 6.

42. "Note del giorno," *Il Progresso*, November 23, 1935, 6.

43. "Il malvagio esperimento," *Il Progresso*, November 24, 1935, 6.

44. Luconi, "Etnia e patriottismo nella pubblicità per gli italo-americani durante la guerra d'Etiopia."

45. Simone Cinotto, *The Italian American Table: Food, Family, and Community in New York City* (Urbana: University of Illinois Press, 2013), 168.

46. Ad for WOV radio, "Per la diffusione dei prodotti italiani," *Il Progresso*, November 24, 1935, 2.

47. "'Comprate prodotti italiani,'" *Il Progresso*, December 6, 1935, 11.

48. Ibid.

49. Ibid.

50. "Non dimenticate," *Il Progresso*, December 8, 1935, 6.

51. Ad for Negroni salami, *Il Progresso*, December 12, 1935, 10; ad for Luigi Vitelli–Elvea, *Il Progresso*, November 10, 1935, 3.

52. Ad for De Nobili, *Il Progresso*, November 17, 1935, 3.

53. Ad for Planters Edible Oil Company, *Il Progresso*, November 3, 1935, illustrated section; Cinotto, *Italian American Table*, 166.

54. "Le sanzioni dell'Italia," *Il Mattino*, November, 14, 1935, 1; "Ricordate," *Il Mattino*, December 11, 1935, 3.

55. "Nervi a posto, connazionali!," *La Nuova Patria*, October 20, 1935, 1.

56. "L'accademia d'Italia contro le infami sanzioni," *Il Mattino*, November 19, 1935, 1.

57. "Il mirabile slancio della collettivitá per l'assistenza ai volontari e la solidarietá nazionale," *Il Mattino*, November 10, 1935, 3. On roast beef, see "Spaghetti o roastbeef," *Il Mattino*, December 24, 1935, 2.

58. "La data dell'iniquitá e dell'ignominia nella storia del mondo," *Il Mattino*, November 18, 1935, 1.

59. "Il rassegna del 'Commercio Italiano' in Argentina," *Il Mattino*, November 18, 1935, 5.

60. "Le donne italiane contro le sanzioni," *Il Mattino*, November 2, 1935, 1; "La parola del Duce alle donne italiane," *Il Mattino*, December 2, 1935, 1; "Costituzione del Comitato Femminile 'Pro Patria,'" *Il Mattino*, November 14, 1935, 5.

61. Ad for Società Anonima Tabacchi Italiani, *Il Mattino*, November 11, 1935, 5.

62. Ad for Martini & Rossi, *Il Mattino*, April 26, 1936, 28. See also ad for Ferro China Bisleri, *Il Mattino*, May 6, 1936, 2. For the United States, see ad for Fernet-Branca amaro, *Il Progresso*, January 12, 1936, illustrated section.

63. Cinotto discusses the accelerated growth of the tipo italiano food industry in the United States after, and as a result of, World War I. Cinotto, *Italian American Table*, 105–148.

64. Smith, *Politics and Beef in Argentina*, 16–17.

65. For examples of tipo italiano businesses, see ad for Fraschini tobacco, *La Patria degli Italiani* (hereafter *La Patria*), September 20, 1916, 16; ad for Alfredo Canonico's pasta factory, *Il Progresso*, September 20, 1930, 30.

66. On Chiummiento and *La Patria,* see Sergi, "Fascismo e antifascismo nella stampa italiana in Argentina."

67. G. Chiummiento, "Dove entrano in scena perfino i salami, con rispetto parlando," *La Nuova Patria*, September 13, 1936, 1.

68. G. Chiummiento, "Dove si parla di presunti anti italiani fabbricanti di formaggio," *La Nuova Patria*, September 6, 1936, 1.

69. G. Chiummiento, "Dove si parla anche di traditori fabbricanti di carta," *La Nuova Patria*, August 30, 1936, 1, 2.

70. Girbal-Blacha, Noemí M., and María Silvia Ospital. "'Vivir con lo nuestro': Publicidad y política en la Argentina de los años 1930." *European Review of Latin American and Caribbean Studies* 78 (April 2005): 49–66.

71. Chiummiento, "Dove entrano in scena perfino i salami, con rispetto parlando," 1; Chiummiento, "Dove si parla anche di traditori fabbricanti di carta," 1, 2.

72. Chiummiento, "Dove si parla di presunti anti italiani fabbricanti di formaggio," 1.

73. G. Chiummiento, "Il diritto di fabbricare prodotti italiani all'estero," *La Nuova Patria*, August 23, 1936, 1; Chiummiento, "Dove si parla anche di traditori fabbricanti di carta," 1, 2.

74. Ad for Vinos Ruiseñor, *Il Mattino*, December 8, 1935, 7. See also, for example, ad for Vino de Calidad, *Il Mattino*, November 2, 1935, 6; ad for Tampieri & Cia. pasta, *Il Mattino*, February 16, 1936, 12; ad for prodotti Barilà, *Il Mattino*, November 15, 1935, 17.

75. Ad for Fox ham, *Il Mattino*, December 25, 1935, 15; ad for Fox ham, *Il Mattino*, January 1, 1931, 12, 14.

76. Ad for Società Anonima Tabacchi Italiani, *Il Mattino*, November 11, 1935, 5.

77. "Una grande iniziativa del 'Mattino d'Italia' per la resistenza interna," *Il Mattino*, November 18, 1935, 3. See also "Il grandioso successo del 'pacco tricolore,'" *Il Mattino*, November 19, 1935, 3; "I pacchi tricolori per le nostre famiglie," *Il Mattino*, December 12, 1935, 9.

78. "Una grande iniziativa del 'Mattino d'Italia' per la resistenza interna," 3. On packing house workers before 1940, see Smith, *Politics and Beef*, 52–53, 233–36. Daniel James presents the life story of Doña María, a meatpacking worker from Berisso, Argentina. Daniel James, *Doña Maria's Story: Life History, Memory, and Political Identity* (Durham, NC: Duke University Press, 2000).

79. Gilbert M. Joseph, "Close Encounters: Toward a New Cultural History of U.S.–Latin American Relations," in *Close Encounters of Empire: Writing the Cultural History of U.S.–Latin American Relations*, ed. Gilbert M. Joseph, Catherine C. LeGrand, and Ricardo D. Salvatore (Durham, NC: Duke University Press, 1998), 5.

80. "Il mirabile slancio della collettivitá per l'assistenza ai volontari e la solidarietá nazionale," 3.

81. "Nobile incitamento agli italiani d'America," *Il Mattino*, November 4, 1935, 5.

82. Finchelstein, *Transatlantic Fascism*, 40–41, 86–90.

83. On the Italian Commercial Mission, see "Il ricevimento di S. E. Asquini alla Camera di Commercio Italiana," *Il Mattino*, November 7, 1935, 5; "S. E. Asquini e i membri della missione hanno visitato gli stabilimenti dell CIAE e della Pirelli," *Il Mattino*, November 21, 1935, 5; "La missione commerciale italiana accompagnata da S. E. l'Ambasciatore visita lo stabilimento della S.A.T.I.," *Il Mattino*, November 22, 1935, 4; "S. E. Asquini e i membri della missione commerciale hanno visitato la 'Cinzano,'" *Il Mattino*, December 20, 1935, 5.

84. "Cerimonie in onore della missione commerciale italiana," *Il Mattino*, November 14, 1935, 5.

85. Ad for Martini vermouth, *Il Mattino*, May 25, 1936, 2.

86. "L'Argentina e le sanzioni," *Il Mattino*, December 13, 1935, 1.

87. Michele Intaglietta, "Italia e Argentina," *Il Mattino*, May 25, 1936, 2.

88. Michele Intaglietta, "Amici lettori," *Il Mattino*, December 1, 1935, 1.

89. "Semana de Italia" propaganda, *Il Mattino*, December 15, 1935, 9.

90. T.S., "Bandiere! Bandiere! Bandiere!," *Il Mattino*, December 15 1935, 9; "Fraternitá italo-argentina," *Il Mattino*, December 16, 1935, 1.

91. For another example, see ad for Ferro-Quina Bisleri, *Il Mattino*, October 18, 1936, 7.

92. This section builds off a previous article on portrayals of gendered consumption under fascism in *Il Progresso*. See Elizabeth Zanoni, "'Per Voi Signore': Gendered Representations of Fashion, Food, and Fascism in *Il Progresso Italo-Ameriano* during the 1930s," *Journal of American Ethnic History* 31, no. 3 (Spring 2012): 33–71.

93. Examples include "Vita femminile," *La Patria*, January 4, 1920, 6; "Vita femminile," *La Patria*, February 1, 1920, 4; "La donna in casa e fuori," *La Patria*, March 1925, 9; "La donna in casa e fuori," *La Patria*, May 10, 1925, 9.

94. Examples include "Per voi, signore," *Il Progresso*, January 16, 1925; "Per voi, signore," *Il Progresso*, July 31, 1932, 6-S. In 1933 the column changed to "Per voi, signore e signorine" (For you ladies and misses), reflecting the paper's growing interest in targeting a younger group of second-generation Italians.

95. See, for example, "Vita femminile," *Il Progresso*, October 9, 1937, 9.

96. Nancy C. Carnevale, *A New Language, A New World: Italian Immigrants in the United States, 1890–1945* (Urbana: University of Illinois Press, 2009); Donna Gabaccia, *From the Other Side: Women, Gender, and Immigrant Life in the U.S., 1820–1990* (Bloomington: Indiana University Press, 1994), 27–41; Samuel L. Baily, *Immigrants in the Lands of Promise: Italians in Buenos Aires and New York City, 1870–1914* (Ithaca, NY: Cornell University Press, 1999), 63–66.

97. Ad for Gath y Chaves, *La Patria*, January 20, 1910, 8; ad for Gath y Chaves, *La Patria*, April 18, 1915, 7; ad for Harrods, *La Patria*, January 4, 1920, 11; ad for Gath y Chaves, *La Patria*, August 30, 1925, 10; ad for Harrods, *La Patria*, March 16, 1930, 4.

98. Ad for 43-brand cigarettes, *La Patria*, September 20, 1920, 6; ad for Diadema cooking oil, *Il Mattino*, December 6, 1936, 5.

99. Ad for Alexander's, *Il Progresso*, March 31, 1935, 4; ad for May's, *Il Progresso*, March 3, 1935, 4; ad for May's, *Il Progresso*, October 8, 1939, 2; ad for Kaye's Studio Shop, *Il Progresso*, October 11, 1939, 3.

100. See, for example, ad for General Electric, *Il Progresso*, November 7, 1937, 21; ad for Lucky Strike, *Il Progresso*, June 19, 1932, illustrated section; ad for Pontiac, *Il Progresso*, October 31, 1937, 4-S; ad for Ford, *Il Mattino*, September 6, 1935, 9; ad for Armour, *Il Mattino*, November 24, 1935, 5.

101. Ad for Florio Marsala, *Il Progresso*, December 21, 1924, 9-S; ad for Fernet-Branca amaro, *Il Progresso*, January 12, 1926, illustrated section; ad for Fiat cars, *Il Mattino*, July 15, 1936, 2.

102. On Mussolini and masculinity, see Gigliola Gori, "Model of Masculinity: Mussolini, the 'New Italian' of the Fascist Era," in *Superman Supreme: Fascist Body as Political Icon—Global Fascism*, ed. J. A. Mangan (London: Frank Cass Publishers, 1999), 27–61; Sandro Bellassai, "The Masculine Mystique: Anti-Modernism and Virility in Fascist Italy," *Journal of Modern Italian Studies* 10, no. 3 (September 2005): 314–35.

103. Ad for Florio Marsala, *Il Progresso*, April 21, 1935, 3; ad for Buitoni pasta, *Il Progresso*, December 3, 1926; ad for Caffè Pastene, *Il Progresso*, October 15, 1939, illustrated section; ad for Caffè Pastene, *Il Progresso*, October 29, 1939, illustrated section; ad for Bertolli olive oil, *L'Italia*, August 14, 1932.

104. Ad for Ferro-China, *Il Mattino*, January 11, 1931, 13; ad for Olio Sasso, *La Patria*, April 27, 1924, 4; ad for Cirio, *La Patria*, November 12, 1925, 3; ad for Cinzano, *La Patria*, November 12, 1925, 4; ad for Spumante Margherita, *La Patria*, December 5, 1925, 4; ad for Toscano cigarettes, *Il Mattino*, December 27, 1935, 5.

105. Helstosky, *Garlic and Oil*, 63–89.

106. Ibid., 81–85.

107. Victoria de Grazia, *How Fascism Ruled Women, Italy, 1922–1945* (Berkeley: University of California Press, 1992), 211–24. On femininity, beauty, and consumption under fascism, see also Stephen Gundle, *Bellissima: Feminine Beauty and the Idea of Italy* (New Haven, CT: Yale University Press, 2007), 80–106; Eugenia Paulicelli, *Fashion under Fascism: Beyond the Black Shirt* (Oxford: Berg, 2004).

108. On fascism's treatment of Italian women, see Perry Willson, *Peasant Women and Politics in Fascist Italy: The Massaie Rurali* (New York: Routledge, 2002); Perry Willson, *The Clockwork Factory: Women and Work in Fascist Italy* (New York: Oxford University Press, 1993); Piero Meldini, *Sposa e madre esemplare: Ideologia e politica della donna e della famiglia durante il fascismo* (Rimini, Italy: Guaraldi, 1975); Robin Pickering-Iazzi, ed., *Mothers of Invention: Women, Italian Fascism, and Culture* (Minneapolis: University of Minneapolis Press, 1995); Maura E. Hametz, *In the Name of Italy: Nation, Family, and Patriotism in a Fascist Court* (New York: Fordham University Press, 2012).

109. Ad for Florio Marsala, *Il Progresso*, April 21, 1935, 3.

110. Ad for Pirelli shoes, *Il Mattino*, October 28, 1936, 19. Kathy Peiss, "Making Up, Making Over: Cosmetics, Consumer Culture, and Women's Identity," in *The Sex of Things: Gender and Consumption in Historical Perspective*, ed. Victoria de Grazia, with Ellen Furlough (Berkeley: University of California Press, 1996), 311–36.

111. Ad for Pirelli tires, *La Patria*, August 29, 1920, back page; ad for Ferro-China liqueur, *Il Mattino*, January 11, 1931, 13.

112. Hasia R. Diner, *Hungering for America: Italian, Irish, and Jewish Foodways in the Age of Migration* (Cambridge: Harvard University Press, 2001), 20–35; Cinotto, *Italian American Table*; Vito Teti, "Emigrazione, alimentazione, culture popolari," in *Storia dell'emigrazione Italiana. Partenze*, ed. Piero Bevilacqua, Andrina De Clementi, and Emilio Franzina (Rome: Donzelli Editore, 2001), 575–600; Paola Corti, "Emigrazione e consuetudini alimentari. L'esperienza di una catena migratoria," in *Storia d'Italia*, Annali 13. *L'alimentazione nella storia dell'Italia contemporanea*, ed. Alberto Capatti, Albero De Bernardi, and Angelo Varni (Turin, Italy: Einaudi, 1998), 683–719.

113. For a discussion of recipes on women's pages in *Il Progresso*, see Zanoni, "'Per Voi, Signore," 50–52. For Thanksgiving dinner, see "Pel pranzo di Thanksgiving," *Il Progresso*, November 19, 1939, 7-S. For empanadas, see ad for Confetería del Molino empanadas, *Il Mattino*, April 5, 1936, 12. For yerba mate, see ad for Carbador yerba, *La Patria*, January 15, 1910, 4. Yerba is the plant from which *mate* tea is made; the tea is a staple in many South American countries. For examples of recipes from *Il Mattino* that incorporate animal proteins into traditional foods from Italy, see "La ricetta gastronomica: Pizza rustica," *Il Mattino*, February 10, 1931, 8; "La ricetta gastronomica: Tagliatelle alla messinese," *Il Mattino*, January 24, 1931, 8; "La ricetta gastronomica: Agnello alla cacciatora," *Il Mattino*, January 20, 1931, 6.

114. Finchelstein, *Transatlantic Fascism*, 50–52, 70–78.

115. On the birth and growth of the Argentine female consumer, as well as on Argentines' embrace of and challenge to U.S. consumer models, see Cecilia Tossounian, "Images of the Modern Girl: From the Flapper to the Joven Moderna (Buenos Aires, 1920–1940)," *Forum for Inter-American Research* 6, no. 2 (2013): 41–70. See also Fernando Rocchi, "La americanización del consumo: Las batallas por el mercado argentino, 1920–1945," in *Americanización: Estados Unidos y América Latina en el siglo XX*, ed. María Barbero and Andrés Regalsky (Buenos Aires: EDUNTREF, 2014), 150–216; Fernando Rocchi, "Inventando la soberanía del consumidor," in *Historia de la vida privada en la Argentina: La Argentina plural 1870–1930*, ed. Fernando Devoto and Marta Madero (Buenos Aires: Taurus, 1999), 301–321; Natalia Milanesio, *Workers Go Shopping in Argentina: The Rise of Popular Consumer Culture* (Albuquerque: University of New Mexico Press, 2013), 70–71, 101–109, 177–78.

116. On sporting events and college life, see, for example, "Ragazze di collegio," *Il Progresso*, September 3, 1939, 7-S; "Lavori estivi," *Il Progresso*, July 31, 1932, 6-3. For suggestions on clothing for automobile rides, see "Per voi, signore e signorine," *Il Progresso*, April 21, 1935, 7-S. On Hollywood, see "Quel che si porta ad Hollywood," *Il Progresso*, October 19, 1937, 9; "Le sosia delle stelle del cinema," *Il Progresso*, June 26, 1932, 4; "Illusioni del trucco," *Il Progresso*, March 31, 1935, 7-S. On ideas of individuality and modernity in mainstream U.S. women's fashion literature, see Kathy Peiss, *Hope in a Jar: The Making of America's Beauty Industry* (Philadelphia: University of Pennsylvania Press, 2011).

117. See, for example, "Tra cinematografi e 'films,'" *La Patria*, July 14, 1925, 5; "Tra cinematografi e films," *La Patria*, January 2, 1930, 12; "Il cinematografo," *Il Mattino*, January 1, 1931, 17; "Il cinematografo," *Il Mattino*, January 9, 1931, 12. On cinema in Argentina, see Matthew B. Karush, *Culture of Class: Radio and Cinema in the Making of a Divided Argentina, 1920–1946* (Durham, NC: Duke University Press, 2012).

118. Ad for Coleman lamps, *La Patria*, September 20, 1920, 12; ad for Kent cigarettes, *Il Mattino*, December 14, 1925, 7; ad for General Electric vacuum, *La Patria*, May 16, 1930, 5; ad for DUO electric water heater, *La Patria*, July 15, 1930, 4.

119. Ad for United States Rubber Company, *La Patria*, January 22, 1920, 8; ad for Ford, *Il Mattino*, September 6, 1936, 9. On women, femininity, and the development of the U.S. automobile industry, see Virginia Scharff, *Taking the Wheel: Women and the Coming of the Motor Age* (Albuquerque: University of New Mexico Press, 1992).

120. "Donne dell'Italia fascista," *Il Mattino*, February 23, 1936, 1.

121. See, for example, "Comitato Femminile Italiano Pro-patria. Raccolta dell'oro," *Il Mattino*, December 5, 1935, 10; "La 'Giornata della Fede' e la raccolta dell'oro," *Il Mattino*, December 19, 1935, 4; "L'imponente celebrazione del 'Giorno della Fede," *Il Mattino*, December 19, 1935, 6; "Benefica attività del fascio femminile," *Il Mattino*, July 18, 1936, 5.

122. The commandments were originally published in the Italian daily *Il Giornale d'Italia*. "I diece comandamenti per le donne italiane durante la guerra d'Africa," *Il Mattino*, November 3, 1935, 9. See also "Le donne italiane contro le sanzioni," *Il Mattino*, November 2, 1935, 1.

123. Filippo Tommaso Marinetti, *The Futurist Cookbook* (1932; repr., San Francisco: Bedford Arts, 1989). On futurist cooking, see Carol Helstosky, "Recipe for the Nation:

Reading Italian History through *La scienza in cucina* and *La cucina futurista*," *Food and Food-ways* 11, nos. 2/3 (2003): 113–40.

124. On *El Hogar* and Argentine cuisine, see Rebekah E. Pite, *Creating a Common Table in Twentieth-Century Argentina: Doña Petrona, Women, and Food* (Chapel Hill: University of North Carolina Press, 2013), 58–64, 76–79.

125. "Marinetti cuoco futurista," *Il Mattino*, September 2, 1936, 3.

126. Ad for Kent cigarettes, *Il Mattino*, December 13, 1925, 3.

127. Ad for Armour pasta sauce, *Il Mattino*, November 17, 1935, 5.

Epilogue

1. Robert B. Sherman and Richard M. Sherman, "A Spoonful of Sugar," performed by Julie Andrews, *Mary Poppins*, Walt Disney Productions, 1964, CD.

2. Mario "Pájaro" Gomez and Jorge Risso, "Fernet con coca," performed by Vilma Palma e Vampiros, *Fondo Profundo*, Barca Discos, 1994, CD. The best popular piece I have found on Fernet con Coca is Diego Vecino, "Fernet: Una historia de amor argentina," *Brando*, http://www.conexionbrando.com/1387961-fernet-una-historia-de-amor-argentina. See also Jonathan Gilbert, "How One Company Turned Grandpa's Booze into Argentina's National Drink, *Fortune*, March 18, 2016, http://fortune.com/2016/03/18/fernet-branca -argentina.

3. For example, see ad for Fernet-Branca, *La Patria*, January 15, 1900, 7; ad for Fernet-Branca, *La Patria*, April 2, 1905, 2; ad for Fernet-Branca, *La Patria*, January 19, 1910, 16; ad for Fernet-Branca, *La Patria*, December 8, 1915, 10; ad for Fernet-Branca, *La Patria*, January 20, 1920, 8. See also "Che cosa é il Fernet-Branca," *La Patria*, September 20, 1911.

4. For example, see ad for Fernet-Branca, *La Nación*, September 22, 1901, 8; ad for Fernet-Branca, *La Prensa*, December 9, 1915, 15.

5. Mark Pendergrast, *For God, Country, and Coca-Cola: The Unauthorized History of the Great American Soft Drink and the Company that Makes It* (New York: Scribner's, 1993), 230.

6. Coca-Cola Argentina, "Nuestra historia," http://www.cocacoladeargentina.com.ar/ nuestra-compania/nuestra-historia.

7. F.lli Branca Destilerias S.A., Branca International S.p.A., http://www.brancainter national.com/en/THEGROUP/companies/Distillerie/Destilerias/index.html.

8. Paul H. Lewis, *The Crisis of Argentine Capitalism* (Chapel Hill: University of North Carolina Press, 1990), 289–328.

9. Vecino, "Fernet." Fernet has followers in the United States as well, especially among Californians, where in San Francisco, fernet is often followed by a chaser of ginger ale. Nate Cavalleri, "The Myth of Fernet," *SF Weekly*, December 7, 2005, http://archives.sf weekly.com/sanfrancisco/the-myth-of-fernet/Content?oid=2158526.

10. By 1959 only 10,806 Italians entered the United States and only 7,549 entered Argentina. Gianfausto Rosoli, ed., *Un secolo di emigrazione italiana, 1876–1976* (Rome: Centro Studi Emigrazione, 1978), 355.

11. On postwar Italian emigration, see Donna R. Gabaccia, *Italy's Many Diasporas* (Seattle: University of Washington Press, 2000), 153–73.

12. Andrea Leonardi, Alberto Cova, and Pasquale Galea, *Il Novecento economico italiano: Dalla grande guerra al "miracolo economico" (1914–1962)* (Bologna, Italy: Monduzzi, 1997); Guido Crainz, *Storia del miracolo italiano: Culture, identità, trasformazioni fra anni cinquanta e sessanta* (Roma: Donzelli, 1996).

13. Ugo Ascoli, *Movimenti migratori in Italia* (Bologna, Italy: Il Mulino, 1979); Giovanni Pellicciari and Gianfranco Albertelli, ed., *L'immigrazione nel triangolo industriale* (Milan: Angeli, 1970); Gabaccia, *Italy's Many Diasporas*, 168–70.

14. Fabio Parasecoli, *Al Dente: A History of Food in Italy* (London: Reaktion Books, 2014), 271.

15. Gino C. Speranza, "La necessità di un accordo internazionale in riguardo agli emigranti," *Rivista Commerciale*, December 1905, 21.

16. Silvia Lepore, "Economic Profile of Italian Argentines in the 1980s," in *The Columbus People: Perspectives in Italian Immigration to the Americas and Australia*, ed. Lydio F. Tomasi, Piero Gastaldo, and Thomas Row (New York: Center for Migration Studies, 1994), 125–51; Joel Perlmann, *Italians Then, Mexicans Now: Immigrant Origins and Second-Generation Progress, 1890 to 2000* (New York: Russell Sage Foundation, 2007).

17. On the globalization of Italian cuisines, see especially Parasecoli, *Al Dente*, 225–47; John F. Mariani, *How Italian Food Conquered the World* (New York: Palgrave Macmillian, 2011).

18. Carlo Petrini, *Slow Food: The Case for Taste*, trans. William McCuaig (New York: Columbia University Press, 2003).

19. Jeffrey M. Pilcher, "'Old Stock' Tamales and Migrant Tacos: Taste, Authenticity, and the Naturalization of Mexican Food," *Social Research* 81, no. 2 (2014): 441–62; Krishnendu Ray, *The Ethnic Restaurateur* (New York: Bloomsbury Academic, 2016).

20. Eataly, "Stores," https://www.eataly.com/us_en/stores. On geographical indicators, see Parasecoli, *Al Dente*, 253–59.

21. Marilyn Halter, *Shopping for Identity: The Marketing of Ethnicity* (New York: Schocken Books, 2000); Matthew Frye Jacobson, *Roots Too: White Ethnic Revival in Post–Civil Rights America* (Cambridge: Harvard University Press, 2006).

22. David Gentilcore, *Pomodoro!: A History of the Tomato in Italy* (New York: Columbia University Press, 2010),130.

23. Simone Cinotto, *The Italian American Table: Food, Family, and Community in New York City* (Urbana: University of Illinois Press, 2013), 211–17.

24. Fabio Parasecoli, "We Are Family: Ethnic Food Marketing and the Consumption of Authenticity in Italian-Themed Chain Restaurants," *Making Italian America: Consumer Culture and the Production of Ethnic Identities,* ed. Simone Cinotto (New York: Fordham University Press, 2014), 244–55; Davide Girardelli, "Commodified Identities: The Myth of Italian Food in the United States," *Journal of Communication Inquiry* 28, no. 4 (2004): 307–324.

25. Mark I. Choate, *Emigrant Nation: The Making of Italy Abroad* (Cambridge: Harvard University Press, 2008).

26. On migration to Italy, see, for example, Russell King, "Recent Immigration to Italy: Character, Causes and Consequences," *GeoJournal* 30, no. 3 (July 1993): 283–92; Paul M. Sniderman et al., *The Outsider: Prejudice and Politics in Italy* (Princeton, NJ: Princeton

University Press, 2000); Graziella Parati, *Migration Italy: The Art of Talking Back in a Destination Culture* (Toronto: University of Toronto Press, 2005); Hans Lucht, *Darkness before Daybreak: African Migrants Living on the Margins in Southern Italy Today* (Berkeley: University of California Press, 2012); Elisabetta Zontini, *Transnational Families, Migration and Gender: Moroccan and Filipino Women in Bologna and Barcelona* (New York: Berghahn Books, 2010).

27. Over 450,000 Argentines with Italian ancestry acquired Italian citizenship from 1998 to 2010. See Guido Tintori, "More than One Million Individuals Got Italian Citizenship Abroad in Twelve Years (1998–2010)," European Union Democracy, November 21, 2012, http://eudo-citizenship.eu/news/citizenship-news/748-more-than-one-million-individuals-got-italian-citizenship-abroad-in-the-twelve-years-1998-2010%3E. On Italian-Argentines and dual nationality, see David Cook-Martín, *The Scramble for Citizens: Dual Nationality and State Competition for Immigrants* (Stanford, CA: Stanford University Press, 2012).

28. Parasecoli, *Al Dente*, 246–47.

29. Kitty Calavita, *Immigrants at the Margins: Law, Race, and Exclusion in Southern Europe* (New York: Cambridge University Press, 2005); Demetrios G. Papademetriou and Kimberly A. Hamilton, *Converging Paths to Restriction: French, Italian, and British Responses to Immigration* (Washington, DC: Carnegie Endowment for International Peace, 1996).

30. Parasecoli, *Al Dente*, 238–40.

31. Donna R. Gabaccia, "Food, Mobility, and World History," in *The Oxford Handbook of Food History*, ed. Jeffrey M. Pilcher (New York: Oxford University Press, 2012), 305–323.

32. Jeffrey M. Pilcher, *Planet Taco: A Global History of Mexican Food* (New York: Oxford University Press, 2012), 211–20. On NAFTA and Mexican migration, see Alejandro Portes, "NAFTA and Mexican Immigration," Border Battles: The U.S. Immigration Debates, Social Science Research Council, 2006, http://borderbattles.ssrc.org/Portes; Deborah Barndt, *Women Working the NAFTA Food Chain: Women, Food, and Globalization* (Toronto: Second Story Press, 1999); Douglas S. Massey, Jorge Durand and Nolan J. Malone, *Beyond Smoke and Mirrors: Mexican Immigration in an Era of Economic Integration* (New York: Russell Sage Foundation, 2002).

33. Peter Andreas, *Border Games: Policing the U.S.-Mexico Divide* (Ithaca, NY: Cornell University Press, 2000).

34. Ibid., 115–39. See also Christina Boswell and Andrew Geddes, *Migration and Mobility in the European Union* (New York: Palgrave Macmillan, 2011).

Bibliography

Archives

ITALY

Archivio Centrale dello Stato, Rome
Biblioteca Nazionale Centrale di Roma, Rome
Biblioteca Storica Nazionale dell'Agricoltura, Rome
Istituto Nazionale di Statistica, Rome

ARGENTINA

Biblioteca Nacional de la República Argentina, Buenos Aires
Centro de Documentación e Información, Ministerio de Hacienda y Finanzas Públicas, Buenos Aires
Centro de Estudios Migratorios Latinoamericanos, Buenos Aires

UNITED STATES

David M. Rubenstein Rare Book and Manuscript Library, Duke University, Durham, North Carolina
Hagley Museum and Library, Wilmington, Delaware
Immigration History Research Center, University of Minnesota
New York Public Library, New York

Newspapers and Journals

Bollettino Mensile della Camera Italiana di Commercio in Buenos Aires (Buenos Aires)
Il Mattino d'Italia (Buenos Aires)

Il Progresso Italo-Americano (New York)
Italian Chamber of Commerce Bulletin (Chicago)
L'Italia (Chicago)
La Nación (Buenos Aires, Argentina)
La Nuova Patria degli Italiani (Buenos Aires)
La Patria degli Italiani (Buenos Aires)
La Prensa (Buenos Aires)
New York Times (New York)
Rivista Commerciale (New York)

Government Publications

ITALY

Atti della giunta per la inchiesta agraria e sulle condizioni della classe agricola. 15 vols. Rome: Forzani e C., 1881–1886.

Bordiga, Oreste. *Inchiesta parlamentare sulle condizioni dei contadini nelle provincie meridionali e nella Sicilia.* Vol. 4. *Campania*, Tomo 1. Rome: Tipografia nazionale di Giovanni Bertero e C., 1909.

Commissariato Generale dell'Emigrazione. *Annuario statistico della emigrazione italiana dal 1876 al 1925.* Rome: Edizione del Commissariato Generale dell'Emigrazione, 1926.

———. *Bollettino dell'emigrazione.* Rome: Tipografia Società Cartiere Centrali, 1902–1927.

Direzione della Statistica Generale (1881–1885); Direzione Generale della Statistica (1886–1907; 1919–1921); Direzione Generale della Statistica e del Lavoro (1911–1915); Ufficio Centrale di Statistica (1916–1918); Istituto Centrale di Statistica del Regno d'Italia (1927–1943). *Annuario statistico italiano.* Publisher varies.

Istituto Centrale di Statistica. *Commercio di importazione e di esportazione del Regno d'Italia.* Publisher varies,1934–1951.

Jarach, Cesare. *Inchiesta parlamentare sulle condizioni dei contadini nelle provincie meridionali e nella Sicilia.* Vol. 2. *Abruzzi e Molise*, Tomo 1. Rome: Tipografia nazionale di Giovanni Bertero e C., 1909.

Lorenzoni, Giovanni. *Inchiesta parlamentare sulle condizione dei contadini nelle provincie meridionali e nella Sicilia.* Vol. 6. *Sicilia*, Tomo 1. Rome: Tipografia nazionale di Giovanni Bertero e C., 1910.

Marenghi, Ernesto. *Inchiesta parlamentare sulle condizioni dei contadini nelle provincie meridionali e nella Sicilia.* Vol. 5. *Basilicata e Calabrie*, Tomo 2 Calabrie. Rome: Tipografia nazionale di Giovanni Bertero e C., 1909.

Ministero delle Finanze. Direzione Generale della Gabelle. *Movimento commerciale del Regno d'Italia.* Publisher varies,1880–1904.

ARGENTINA

Dirección General de Inmigración. *Resumen estadistico del movimiento migratorio en la Republica Argentina, años 1857–1924.* Buenos Aires: Talleres Gráficos del Ministerio de Agricultura de la Nación, 1925.

Ministerio de Agricultura, Dirección de Comercio e Industria. *Censo Industrial de la República*. Buenos Aires, Talleres de Publicaciones de la Oficina Meteorológica Argentina, 1910.

República Argentina. *Censo Industrial de 1935*. Buenos Aires: Ministerio de Hacienda, 1938.

República Argentina. *Tercer Censo Nacional*, Tomo 8, *Censo del Comercio*. Buenos Aires: Talleres Gráficos de L. J. Rosso y Cía., 1917.

UNITED STATES

Department of Commerce. Bureau of Foreign and Domestic Commerce. *Statistical Abstract of the United States*. Washington, DC: U.S. Government Printing Office, 1921–1938.

United States Immigration Commission. *Immigrants in Cities*. Vol. 1. Washington, DC: U.S. Government Printing Office, 1911.

Willcox, Walter F., and Imre Ferenczi. *International Migrations*. Vol. 1. New York: Gordon and Breach Science Publishers, 1969.

Additional Primary Sources

Amedeo, Luigi and Umberto Cagni. *On the* Polar Star *in the Arctic Sea*. Translated by William Le Queux. London: Hutchinson & Co., 1903.

Armour and Company. *Catalogue of Products Manufactured by Armour and Company*. Chicago: Armour & Company, 1916.

Armour and Company: Containing Facts about Business and Organization. Armour & Company, 1917.

Bernardy, Amy A. "L'etnografia della 'piccole italie.'" In *Atti del primo congresso di etnografia italiana*, edited by Società di Etnografia Italiana, 173–79. Perugia, Italy: Unione Tipografica Cooperativa, 1912.

Breckinridge, S. P. *New Homes for Old*. New York: Harper & Brothers Publishing, 1921.

Camera di Commercio Italiana in New York. *Nel cinquantenario della Camera di Commercio Italiana in New York, 1887–1937*. New York, 1937.

Comitato della Camera Italiana di Commercio ed Arti. *Gli italiani nella Repubblica Argentina*. Buenos Aires: Compañia Sud-Americana de Billetes de Banco, 1898.

———. *Gli italiani nella Repubblica Argentina all'Esposizione di Torino 1911*. Buenos Aires: Stabilimento Grafico della Compañia General de Fósforos, 1911.

Covello, Leonard. *The Social Background of the Italo-American School Child: A Study of the Southern Italian Family Mores and Their Effect on the School Situation in Italy and America*. Leiden, Netherlands: E. J. Brill, 1967.

Dall'Italia all'Argentina: Guida practica per gli italiani che si recano nell'Argentina. Genoa: Libreria R. Istituto Sordo-Muti, 1888.

Einaudi, Luigi. *Un principe mercante: Studio sulla espansione coloniale italiana*. Turin: Fratelli Bocca, 1900.

Faleni, Lorenzo, and Amedeo Serafini, eds. *La Repubblica Argentina all'Esposizione internazionale di Milano 1906*. Buenos Aires, 1906.

Fontana-Russo, Luigi. "Emigrazione d'uomini ed esportazione di merci." *Rivista Coloniale* (1906): 26–40.

Gli italiani nel Sud America ed il loro contributo alla guerra: 1915–1918. Buenos Aires: Arigoni & Barbieri, 1922.

Godio, Guglielmo. *L'America ne' suoi primi fattori: La colonizzazione e l'emigrazione*. Florence: Tipografia di G. Barbèra, 1893.

Gravina, M. ed. *Almanacco dell'italiano nell'Argentina*. Buenos Aires, 1918.

Istituto Coloniale Italiano. *Italia e Argentina*. Bergamo, Italy: Officine dell'Istituto Italiano d'Arti Grafiche, 1910.

L'Italia nell'America Latina: Per l'incremento dei rapporti industriali e commerciali fra l'Italia e l'America del Sud. Milan: Società Tipografica Editrice Popolare, 1906.

Italian American Directory Co. *Gli italiani negli Stati Uniti d'America*. New York: Andrew H. Kellogg Co., 1906.

Malaurie, Alfredo, and Juan M. Gazzano. *La Industria Argentina y la Exposición del Paraná*. Buenos Aires: De Juan M. Gazzano y Cia., 1888.

Mangano, Antonio. "The Italian Colonies of New York City." MA thesis, Columbia University, 1903. Reprint, *Italians in the City: Health and Related Social Needs*, edited by Francesco Cordasco, 1–57. New York: Arno Press, 1975.

Marinetti, Filippo Tommaso. *The Futurist Cookbook*. 1932. Reprint, San Francisco: Bedford Arts, 1989.

Mayor des Planches, E. "Gli Stati Uniti e l'emigrazione italiana." *Rivista Coloniale* 1 (May–August 1906).

Odencrantz, Louise. *Italian Women in Industry*. New York: Russell Sage Foundation, 1919.

Patrizi, Ettore. *Gl'italiani in California, Stati Uniti d'America*. San Francisco: Stabilimento Tipo-Litografico, 1911.

Phelps, Dudley Maynard. *Migration of Industry to South America*. New York: McGraw-Hill, 1936.

Phillips, Velma, and Laura Howell. "Racial and Other Differences in Dietary Customs." *Journal of Home Economics* 12, no. 9 (1920): 396–411.

Rutter, Frank R. *Tariff Systems of South American Countries*. Washington, DC: U.S. Government Printing Office, 1916.

Sarmiento, Domingo F. *Facundo; or, Civilization and Barbarism*. Translated by Mary Mann. New York: Penguin Books, 1998.

Sherman, Mary. "Manufacturing of Foods in the Tenements." *Charities and the Commons* 15 (1906): 669–73.

Swift, Louis F., and Arthur Van Vlissingen Jr. *The Yankee of the Yards: The Biography of Gustavus Franklin Swift*. Chicago: A. W. Shaw, 1927.

Tonissi, Luigi. "Progetto per un banco del commercio italo-americano." In *L'esplorazione commerciale e L'esploratore*, 383–401. Milan: Premiato stabilimento tipografico P. B. Bellini, 1896.

Urien, Carols M., and Ezio Colombo. *La República Argentina en 1910*. Buenos Aires: Maucci Hermanos, 1910.

Visconti, Aldo. *Emigrazione ed esportazione: Studio dei rapporti che intercedono fra l'emigrazione e le esportazioni italiane per gli Stati Uniti del Nord America e per la Repubblica Argentina*. Turin: Tipografia Baravalle e Falconieri, 1912.

Woods, Robert A. "Notes on the Italians in Boston." *Charities* 12 (1904): 451–52.

Zuccarini, Emilio. *Il lavoro degli italiani nella Repubblica Argentina dal 1516 al 1910.* Buenos Aires, 1910.

Scholarly Books and Articles

Abarca, Meredith E. *Voices in the Kitchen: Views of Food and the World from Working-Class Mexican and Mexican American Women.* College Station: Texas A&M University Press, 2006.

Adelman, Jeremy. *Sovereignty and Revolution in the Iberian Atlantic.* Princeton, NJ: Princeton University Press, 2006.

Alberto, Paulina L., and Eduardo Elena. "Introduction: Shades of the Nation." In *Rethinking Race in Modern Argentina,* edited by Paulina L. Alberto and Eduardo Elena, 1–21. New York: Cambridge University Press, 2016.

Aliano, David. *Mussolini's National Project in Argentina.* Madison, WI: Fairleigh Dickinson University Press, 2012.

Andreas, Peter. *Border Games: Policing the U.S.-Mexico Divide.* Ithaca, NY: Cornell University Press, 2000.

Andrews, George Reid. *The Afro-Argentines of Buenos Aires, 1800–1900.* Madison: University of Wisconsin Press, 1980.

Appadurai, Arjun. *The Social Life of Things: Commodities in Cultural Perspective.* Cambridge: Cambridge University Press, 1988.

Archetti, Eduardo P. "Hibración, pertenencia y localidad en la construcción de una cocina nacional." In *La Argentina en el siglo XX,* edited by Carlos Altamirano, 217–36. Buenos Aires: Universidad Nacional de Quilmes, 1999.

Arcondo, Aníbal B. *Historia de la alimentación en Argentina: Desde los orígienes hasta 1920.* Córdoba, Argentina: Ferreyra Editor, 2002.

Ascoli, Ugo. *Movimenti migratori in Italia.* Bologna, Italy: Il Mulino, 1979.

Baily, Samuel L. *Immigrants in the Lands of Promise: Italians in Buenos Aires and New York City, 1870–1914.* Ithaca, NY: Cornell University Press, 1999.

———. "Marriage Patterns and Immigrant Assimilation in Buenos Aires, 1882–1923." *Hispanic American Historical Review* 60, no. 1 (1980): 32–48.

———. "Sarmiento and Immigration: Changing Views on the Role of Immigration in the Development of Argentina." In *Sarmiento and His Argentina,* edited by Joseph T. Criscenti, 131–42. Boulder, CO: L. Rienner Publishers, 1993.

Baily, Samuel L., and Eduardo José Míguez, eds. *Mass Migration to Modern Latin America.* Wilmington, DE: SR Books, 2003.

Baily, Samuel L., and Franco Ramella, eds. *One Family, Two Worlds: An Italian Family's Correspondence across the Atlantic, 1901–1922.* Translated by John Lenaghan. New Brunswick, NJ: Rutgers University Press, 1988.

Baldassar, Loretta, and Donna R. Gabaccia. "Home, Family, and the Italian Nation in a Mobile World: The Domestic and the National among Italy's Migrants." In *Intimacy and Italian Migration: Gender and Domestic Lives in a Mobile World,* edited by Loretta Baldassar and Donna R. Gabaccia, 1–24. New York: Fordham University Press, 2011.

Balletta, Francesco. *Il Banco di Napoli e le rimesse degli emigrati (1914–1925)*. Naples: ISTOB, 1972.

Barbero, María Inés. "Empresas y empresarios Italianos en la Argentina (1900–30)." In *Studi sull'emigrazione: Un'analisi comparata*. Atti del Convegno storico internazionale sull'emigrazione, ed. Maria Rosaria Ostuni, 303–313. Biella, Italy: Fondazione Sella, 1989.

———. "Grupos empresarios, intercambio commercial e inversiones italianas en la Argentina. El caso de Pirelli (1910–1920)." *Estudios Migratorios Latinoamericanos* 5, nos. 15–16 (1990): 311–41.

Barbero, María I., and Andrés M. Regalsky, eds. *Americanización: Estados Unidos y América Latina en el siglo XX. Transferencias económicas, tecnológicas y culturales*. Buenos Aires: EDUNTREF, 2014.

Barbero, María Inés, and Fernando Rocchi. "Industry." In *The New Economic History of Argentina*, edited by Gerardo della Paolera and Alan Taylor, 261–94. Cambridge: Cambridge University Press, 2003.

Barndt, Deborah. *Women Working the NAFTA Food Chain: Women, Food, and Globalization*. Toronto: Second Story Press, 1999.

Barrett, James R., and David Roediger. "Inbetween Peoples: Race, Nationality, and the 'New Immigrant' Working Class." *Journal of American Ethnic History* 16, no. 3 (1997): 3–44.

Barthes, Roland. "Toward a Psychosociology of Contemporary Food Consumption." In *Food and Culture: A Reader*, 3rd ed., edited by Carole Counihan and Penny Van Esterik, 23–30. New York: Routledge, 2013.

Bellassai, Sandro. "The Masculine Mystique: Anti-Modernism and Virility in Fascist Italy." *Journal of Modern Italian Studies* 10, no. 3 (2005): 314–35.

Bemis, Samuel Flagg. *The Latin American Policy of the United States*. New York: Harcourt, Brace and Co., 1943.

Bentley, Amy. *Inventing Baby Food: Taste, Health, and the Industrialization of the American Diet*. Berkeley: University of California Press, 2014.

Bernardy, Amy Allemand, and Maddalena Tirabassi. *Ripensare la patria grande: Gli scritti di Amy Allemand Bernardy sulle migrazioni italiane*. Isernia, Italy: C. Iannone, 2005.

Bevilacqua, Piero. "Emigrazione transoceanica e mutamenti dell'alimentazione contadina calabrese fra Otto e Novecento." *Quaderni storici* 47 (August 1981): 520–55.

Bevilacqua, Piero, Andreina De Clementi, and Emilio Franzina, eds. *Storia dell'emigrazione italiana. Arrivi*. Rome: Donzelli, 2002.

———. *Storia dell'emigrazione italiana. Partenze*. Rome: Donzelli, 2001.

Blengino, Vanni, and Eugenia Scarzanella, eds. *Fascisti in Sud America*. Florence: Le Lettere, 2005.

Boswell, Christina, and Andrew Geddes. *Migration and Mobility in the European Union*. New York: Palgrave Macmillan, 2011.

Bosworth, Richard J. *Italy, the Least of the Great Powers: Italian Foreign Policy before the First World War*. New York: Cambridge University Press, 1979.

Bourdieu, Pierre. *Distinction: A Social Critique of the Judgment of Taste*. Translated by Richard Nice. Cambridge: Harvard University Press, 1984.

Briggs, John W. *An Italian Passage: Immigrants to Three American Cities, 1890–1930*. New Haven, CT: Yale University Press, 1978.

Bugiardini, Sergio. "La Camera di commercio italiani di New York." In *Profili di Camera di Commercio Italiane all'estero*, vol. 1, edited by Giovanni Luigi Fontana and Emilio Franzina, 105–121. Soveria Manelli, Italy: Rubbettino Editore, 2001.

Calavita, Kitty. *Immigrants at the Margins: Law, Race, and Exclusion in Southern Europe*. New York: Cambridge University Press, 2005.

Cancian, Sonia. "'Tutti a Tavola!' Feeding the Family in Two Generations of Italian Immigrant Households in Montreal." In *Edible Histories, Cultural Politics: Towards a Canadian Food History*, edited by Franca Iacovetta, Valerie J. Korinek, and Marlene Epp, 209–221. Toronto: University of Toronto Press, 2012.

Cannistraro, Philip V. "The Duce and the Prominenti: Fascism and the Crisis of Italian American Leadership." *Altreitalie* 31 (July–December, 2005): 76–86.

———. "Generoso Pope and the Rise of Italian-American Politics, 1925–1936." In *Italian Americans: New Perspectives in Italian Immigration and Ethnicity*, edited by Lydio F. Tomasi, 265–88. Staten Island: Center for Migration Studies of New York Inc., 1985.

Cannistraro, Philip V., and Gianfausto Rosoli. "Fascist Emigration Policy in the 1920s: An Interpretive Framework." *International Migration Review* 13, no. 4 (1979): 673–92.

Carnevale, Nancy C. *A New Language, A New World: Italian Immigrants in the United States, 1890–1945*. Urbana: University of Illinois Press, 2009.

Cavalleri, Nate. "The Myth of Fernet." *SF Weekly*, December 7, 2005. http://archives.sf weekly.com/sanfrancisco/the-myth-of-fernet/Content?oid=2158526.

Ceserani, Gian Paolo. *Storia della pubblicità in Italia*. Rome: Laterza, 1988.

Chen, Yong. *Chop Suey USA: The Story of Chinese Food in America*. New York: Columbia University Press, 2014.

Chiapparino, Francesco. "Industrialization and Food Consumption in United Italy." In *Food Technology, Science, and Marketing: European Diet in the Twentieth Century*, edited by Adel P. den Hartog, 139–55. East Linton, Scotland: Tuckwell Press, 1995.

Chiapparino, Francesco, Renato Covino, and Gianni Bovini. *Consumi e industria alimentare in Italia dall'unità a oggi: Lineamenti per una storia*. 2nd ed. Narni, Italy: Giada, 2002.

Choate, Mark I. *Emigrant Nation: The Making of Italy Abroad*. Cambridge: Harvard University Press, 2008.

———. "Sending States' Transnational Interventions in Politics, Culture, and Economics: The Historical Example of Italy." *International Migration Review* 41, no. 3 (2007): 728–68.

Cinel, Dino. *From Italy to San Francisco: The Immigrant Experience*. Stanford, CA: Stanford University Press, 1982.

———. *National Integration of Italian Return Migration, 1870–1929*. New York: Cambridge University Press, 1991.

Cinotto, Simone. "All Things Italian: Italian American Consumers, the Transnational Formation of Taste, and the Commodification of Difference." In *Making Italian America: Consumer Culture and the Production of Ethnic Identities*, edited by Simone Cinotto, 1–31. New York: Fordham University Press, 2014.

———. "'Buy Italiano!': Italian American Food Importers and Ethnic Consumption in 1930s New York." In *Italian Americans: A Retrospective on the Twentieth Century*, edited by Paola A. Sensi-Isolani and Anthony Julian Tamburri, 167–78. Chicago Heights, IL: American Italian Historical Association, 2001.

———. *The Italian American Table: Food, Family, and Community in New York City*. Urbana: University of Illinois Press, 2013.

———, ed., *Making Italian America: Consumer Culture and the Production of Ethnic Identities*. New York: Fordham University Press, 2014.

———. *Soft Soil, Black Grapes: The Birth of Italian Winemaking in California*. Translated by Michelle Tarnopoloski. New York: New York University Press, 2012.

Cohen, Lizabeth. *Making a New Deal: Industrial Workers in Chicago, 1919–1939*. 2nd ed. Cambridge: Cambridge University Press, 2008.

Cohen, Miriam. *Workshop to Office: Two Generations of Italian Women in New York City, 1900–1950*. Ithaca, NY: Cornell University Press, 1993.

Cook-Martín, David. *The Scramble for Citizens: Dual Nationality and State Competition for Immigrants*. Stanford, CA: Stanford University Press, 2012.

———. "Soldiers and Wayward Women: Gendered Citizenship, and Migration Policy in Argentina, Italy, and Spain since 1850." *Citizenship Studies* 10, no. 5 (2006): 571–90.

Coppa, Frank J. *Planning, Protectionism, and Politics in Liberal Italy: Economics and Politics in the Giolittian Age*. Washington, DC: Catholic University of America Press, 1971.

Cornblit, Oscar. "Inmigrantes y empresarios en la politica argentina." *Desarrollo Económico* 6, no. 24 (1967): 641–91.

Corti, Paola. "Emigrazione e consuetudini alimentari. L'esperienza di una catena migratoria." In *Storia d'Italia*, Annali 13. *L'alimentazione nella storia dell'Italia contemporanea*, edited by Alberto Capatti, Albero De Bernardi, and Angelo Varni, 683–719. Turin: Einaudi, 1998.

Crainz, Guido. *Storia del miracolo italiano: Culture, identità, trasformazioni fra anni cinquanta e sessanta*. Roma: Donzelli, 1996.

Cronon, William. *Nature's Metropolis: Chicago and the Great West*. New York: W. W. Norton, 1991.

D'Agostino, Peter. "Craniums, Criminals, and the 'Cursed Race': Italian Anthropology in American Racial Thought, 1861–1924." *Comparative Studies in Society and History* 44, no. 2 (2002): 319–43.

Davidoff, Leonore, and Catherine Hall. *Family Fortunes: Men and Women of the English Middle Class, 1780–1850*. Chicago: University of Chicago Press, 1987.

Davis, John. "Changing Perspectives on Italy's 'Southern Problem.'" In *Italian Regionalism: History, Identity, and Politics*, edited by Carl Levy, 53–68. Oxford: Berg, 1996.

Davis, Marni. *Jews and Booze: Becoming American in the Age of Prohibition*. New York: New York University Press, 2012.

de Caprariis, Luca. "'Fascism for Export'? The Rise and Eclipse of the Fasci Italiani all'Estero." *Journal of Contemporary History* 35, no. 2 (2000): 151–83.

De Grand, Alexander. *The Hunchback's Tailor: Giovanni Giolitti and Liberal Italy from the Challenge of Mass Politics to the Rise of Fascism, 1882–1922*. Westport, CT: Praeger, 2001.

de Grazia, Victoria. *How Fascism Ruled Women, Italy, 1922–1945*. Berkeley: University of California Press, 1992.

de Grazia, Victoria, with Ellen Furlough, eds. *The Sex of Things: Gender and Consumption in Historical Perspective*. Berkeley: University of California Press, 1996.

Devoto, Fernando J. *Historia de los italianos en la Argentina*. Buenos Aires: Editorial Biblos, 2006.

———. "Programs and Politics of the First Italian Elite of Buenos Aires, 1852–80." In *Italian Workers of the World: Labor Migration and the Formation of Multiethnic States*, edited by Donna R. Gabaccia and Fraser M. Ottanelli, 41–59. Urbana: University of Illinois Press, 2001.

Devoto, Fernando J., and Gianfausto Rosoli, eds. *L'Italia nella società argentina: Contributi sull'emigrazione italiana in Argentina*. Rome: Centro Studi Emigrazione, 1988.

Díaz Alejandro, Carlos F. "The Argentine Tariff, 1906–1940." *Oxford Economic Papers* 19, no. 1 (1967): 75–98.

———. *Essays on the Economic History of the Argentine Republic*. New Haven, CT: Yale University Press, 1970.

Diner, Hasia R. *Hungering for America: Italian, Irish, and Jewish Foodways in the Age of Migration*. Cambridge: Harvard University Press, 2001.

———. *Roads Taken: The Great Jewish Migrations to the New World and the Peddlers Who Forged the Way*. New Haven, CT: Yale University Press, 2015.

Donato, Katharine M., Donna Gabaccia, Jennifer Holdaway, Martin Manalansan, and Patricia R. Pessar. "A Glass Half Full? Gender in Migration Studies." *International Migration Review* 40, no. 1 (2006): 3–26.

Douglas, Mary. "Deciphering a Meal." *Daedalus* 101, no. 1 (1971): 61–81.

Douglas, Mary, and Baron Isherwood. *The World of Goods: Towards an Anthropology of Consumption*. London: A. Lane, 1979.

Douki, Caroline. "The Liberal Italian State and Mass Emigration, 1860–1914." In *Citizenship and Those Who Leave: The Politics of Emigration and Expatriation*, edited by Nancy L. Green and François Weil, 91–113. Urbana: University of Illinois Press, 2007.

Durante, Alessandro, ed., *A Companion to Linguistic Anthropology*. Malden, MA: Blackwell, 2004.

Earle, Rebecca. *The Body of the Conquistador: Food, Race, and the Colonial Experience in Spanish America, 1942–1700*. New York: Cambridge University Press, 2012.

Eckes, Alfred E. *Opening America's Market: U.S. Foreign Trade Policy since 1776*. Chapel Hill: University of North Carolina Press, 1995.

Elena, Eduardo. *Dignifying Argentina: Peronism, Citizenship, and Mass Consumption*. Pittsburgh: University of Pittsburgh Press, 2011.

Enstad, Nan. *Ladies of Labor, Girls of Adventure: Working Women, Popular Culture, and Labor Politics at the Turn of the Twentieth Century*. New York: Columbia University Press, 1999.

Estudios Migratorios Latinoamericanos. Special Issue: *Las cadenas migratorias italianas a la Argentina* 3, no. 8 (1988).

Eula, Michael J. "Failure of American Food Reformers among Italian Immigrants in New York City, 1891–1897." *Italian Americana* 18 (Winter 2000): 86–99.

Ewen, Elizabeth. *Immigrant Women in the Land of Dollars: Life and Culture on the Lower East Side, 1890–1925*. New York: Monthly Review Press, 1985.

Ewen, Stuart. *Captains of Consciousness: Advertising and the Social Roots of the Consumer Culture*. New York: McGraw-Hill, 1976.

Ewen, Stuart, and Elizabeth Ewen. *Channels of Desire: Mass Images and the Shaping of American Consciousness*. New York: McGraw-Hill, 1982.

Falasca-Zamponi, Simonetta. *Fascist Spectacle: The Aesthetics of Power in Mussolini's Italy*. Berkeley: University of California Press, 1997.

Fanesi, Pietro Rinaldo. "Italian Antifascism and the Garibaldine Tradition in Latin America." In *Italian Workers of the World: Labor Migration and the Formation of Multiethnic States*, translated by Michael Rocke, edited by Donna R. Gabaccia and Fraser M. Ottanelli, 163–77. Urbana: University of Illinois Press, 2001.

Fernández, Alejandro. *Un "mercado étnico" en la Plata: Emigración y exportaciones españolas a la Argentina, 1880–1935*. Madrid: Consejo Superior de Investigaciones Científicas, 2004.

Feys, Torsten. *The Battle for the Migrants: The Introduction of Steamshipping on the North Atlantic and Its Impact on the European Exodus*. St. John's, Newfoundland: International Maritime Economic History Association, 2013.

Finchelstein, Federico. *Fascismo, liturgia e imaginario. El mito del General Uriburu y la Argentina nacionalista*. Buenos Aires: Fondo de Cultura Económica, 2002.

———. *Transatlantic Fascism: Ideology, Violence, and the Sacred in Argentina and Italy, 1919–1945*. Durham, NC: Duke University Press, 2010.

FitzGerald, David Scott, and David Cook-Martín. *Culling the Masses: The Democratic Origins of Racist Immigration Policy in the Americas*. Cambridge: Harvard University Press, 2014.

Foerster, Robert F. *The Italian Emigration of Our Times*. Cambridge: Harvard University Press, 1919.

Fontana, Giovanni Luigi, and Emilio Franzina, eds. *Profili di Camere di commercio italiane all'estero*. Vol. 1. Soveria Mannelli, Italy: Rubbettino Editore, 2001.

Ford, Nancy Gentile. *Americans All!: Foreign-Born Soldiers in World War I*. College Station: Texas A&M University Press, 2001.

Freedman, Paul. "American Restaurants and Cuisine in the Mid-Nineteenth Century." *New England Quarterly* 84, no. 1 (2011): 5–59.

French, M. J. "The Emergence of a U.S. Multinational Enterprise: The Goodyear Tire and Rubber Company, 1910–1939." *Economic History Review* 40, no. 1 (1987): 64–79.

Frid de Silberstein, Carina. "Migrants, Farmers, and Workers: Italians in the Land of Ceres." In *Italian Workers of the World: Labor Migration and the Formation of Multiethnic States*, edited by Donna R. Gabaccia and Fraser M. Ottanelli, 79–101. Urbana: University of Illinois Press, 2001.

Frieden, Jeffry A. *Global Capitalism: Its Fall and Rise in the Twentieth Century*. New York: Norton, 2006.

Gabaccia, Donna R. "Ethnicity in the Business World: Italians in American Food Industries." *Italian American Review* 6, no. 2 (1997/1998): 1–19.

———. "Food, Mobility, and World History." In *The Oxford Handbook of Food History*, edited by Jeffrey M. Pilcher, 305–323. New York: Oxford University Press, 2012.

———. *Foreign Relations: American Immigration in Global Perspective*. Princeton, NJ: Princeton University Press, 2012.

———. *From Sicily to Elizabeth Street: Housing and Social Change among Italian Immigrants, 1880–1930*. Albany: State University of New York Press, 1984.

——. *From the Other Side: Women, Gender, and Immigrant Life in the U.S., 1820–1990*. Bloomington: Indiana University Press, 1994.

——. "In the Shadows of the Periphery: Italian Women in the Nineteenth Century." In *Connecting Spheres: Women in the Western World, 1500 to Present*, edited by Marilyn J. Boxer and Jean H. Quataert, 166–76. New York: Oxford University Press, 1987.

——. *Italy's Many Diasporas*. Seattle: University of Washington Press, 2000.

——. "Making Foods Italian in the Hispanic Atlantic." Unpublished paper, University of Minnesota, November 1, 2006.

——. *Militants and Migrants: Rural Sicilians Become American Workers*. New Brunswick, NJ: Rutgers University Press 1988.

——. "Nations of Immigrants: Do Words Matter?" *Pluralist* 5, no. 3 (2010): 5–31.

——. "Race, Nation, Hyphen: Italian-American Multiculturalism in Comparative Perspective." In *Are Italians White? How Race Is Made in America*, edited by Jennifer Guglielmo and Salvatore Salerno, 44–59. New York: Routledge, 2003.

——. *We Are What We Eat: Ethnic Food and the Making of Americans*. Cambridge: Harvard University Press, 1998.

——. "When the Migrants Are Men: Italy's Women and Transnationalism as a Working-Class Way of Life." In *American Dreaming, Global Realities: Rethinking U.S. Immigration History*, edited by Donna R. Gabaccia and Vicki L. Ruiz, 190–206. Urbana: University of Illinois Press, 2006.

——. "Women of the Mass Migrations: From Minority to Majority, 1820–1930." In *European Migrants: Global and Local Perspectives*, edited by Dirk Hoerder and Leslie Page Moch, 91–111. Boston: Northeastern University Press, 1996.

Gabaccia, Donna R., and Franca Iacovetta, eds. *Women, Gender, and Transnational Lives: Italian Workers of the World*. Toronto: University of Toronto Press, 2002.

Gabaccia, Donna R., and Fraser M. Ottanelli, eds. *Italian Workers of the World: Labor Migration and the Formation of Multiethnic States*. Urbana: University of Illinois Press, 2001.

Gabaccia, Donna R., and Jeffrey M. Pilcher. "'Chili Queens' and Checkered Tablecloths: Public Dining Cultures of Italians in New York City and Mexicans in San Antonio, Texas, 1870s–1940s." *Radical History Review* 110 (Spring 2011): 109–126.

Galante, John Starosta. "The 'Great War' in *Il Plata*: Italian Immigrants in Buenos Aires and Montevideo during the First World War." *Journal of Migration History* 2, no. 1 (2016): 57–92.

Gardner, Martha. *The Qualities of a Citizen: Women, Immigration, and Citizenship, 1870–1965*. Princeton, NJ: Princeton University Press, 2005.

Gentilcore, David. *Pomodoro!: A History of the Tomato in Italy*. New York: Columbia University Press, 2010.

Gentile, Emilio. "La politica estera del partito fascista: Ideologia e organizzazione dei fasci italiani all'estero (1920–1930)." *Storia Contemporanea* 26, no. 2 (1995): 897–956.

Gerstle, Gary. *American Crucible: Race and Nation in the Twentieth Century*. Princeton, NJ: Princeton University Press, 2001.

Gilbert, Jonathan. "How One Company Turned Grandpa's Booze into Argentina's National Drink." *Fortune*, March 18, 2016. http://fortune.com/2016/03/18/fernet-branca-argentina.

Girardelli, Davide. "Commodified Identities: The Myth of Italian Food in the United States." *Journal of Communication Inquiry* 28, no. 4 (2004): 307–324.

Girbal-Blacha, Noemí M., and María Silvia Ospital. "'Vivir con lo nuestro': Publicidad y política en la Argentina de los años 1930." *European Review of Latin American and Caribbean Studies* 78 (April 2005): 49–66.

Gobat, Michel. "The Invention of Latin America: A Transnational History of Anti-Imperialism, Democracy, and Race." *American Historical Review* 118, no. 5 (2013): 1345–75.

Goldstein, Carolyn M. *Creating Consumers: Home Economists in Twentieth-Century America.* Chapel Hill: University of North Carolina Press, 2012.

Gomez, Mario, and Jorge Risso. "Fernet con Coca." Performed by Vilma Palma e Vampiros. *Fondo Profundo.* Barca Discos, 1994. CD.

Goody, Jack. *Cooking, Cuisine, and Class: A Study in Comparative Sociology.* Cambridge: Cambridge University Press, 1982.

Gori, Gigliola. "Model of Masculinity: Mussolini, the 'New Italian' of the Fascist Era." In *Superman Supreme: Fascist Body as Political Icon—Global Fascism,* edited by J. A. Mangan, 27–61. London: Frank Cass Publishers, 1999.

Greene, Julie. *The Canal Builders: Making America's Empire at the Panama Canal.* New York: Penguin Press, 2009.

Guglielmo, Jennifer. *Living the Revolution: Italian Women's Resistance and Radicalism in New York City, 1880–1945.* Chapel Hill: University of North Carolina Press, 2010.

Guglielmo, Jennifer, and Salvatore Salerno, eds. *Are Italians White?: How Race Is Made in America.* New York: Routledge, 2003.

Guglielmo, Thomas A. *White on Arrival: Italians, Race, Color, and Power in Chicago, 1890–1945.* Oxford: Oxford University Press, 2003.

Gundle, Stephen. *Bellissima: Feminine Beauty and the Idea of Italy.* New Haven, CT: Yale University Press, 2007.

Guy, Donna J. "Dependency, the Credit Market, and Argentine Industrialization, 1860–1940." *Business History Review* 58, no. 4 (1984): 532–61.

———. *Sex and Danger in Buenos Aires: Prostitution, Family, and Nation in Argentina.* Lincoln: University of Nebraska Press, 1991.

———. "Women, Peonage, and Industrialization: Argentina, 1810–1914." *Latin American Research Review* 16, no. 3 (1981): 65–89.

Halter, Marilyn. *Shopping for Identity: The Marketing of Ethnicity.* New York: Schocken Books, 2000.

Hametz, Maura E. *In the Name of Italy: Nation, Family, and Patriotism in a Fascist Court.* New York: Fordham University Press, 2012.

Harzig, Christiane, and Dirk Hoerder, with Donna Gabaccia. *What Is Migration History?* Cambridge, MA: Polity Press, 2009.

Heinze, Andrew R. *Adapting to Abundance: Jewish Immigrants, Mass Consumption, and the Search for American Identity.* New York: Columbia University Press, 1990.

Helstosky, Carol. *Garlic and Oil: Politics and Food in Italy.* Oxford: Berg, 2004.

———. "Recipe for the Nation: Reading Italian History through *La scienza in cucina* and *La cucina futurista.*" *Food and Foodways* 11, nos. 2/3 (2003): 113–40.

Higham, John. *Strangers in the Land: Patterns of American Nativism, 1860–1925*. New Brunswick, NJ: Rutgers University Press, 1955.

Hoerder, Dirk. *Cultures in Contact: World Migrations in the Second Millennium*. Durham, NC: Duke University Press, 2002.

———, ed. *Labor Migration in the Atlantic Economies: The European and North American Working Classes during the Period of Industrialization*. Westport, CT: Greenwood Press, 1985.

Hoffenberg, Peter H. *An Empire on Display: English, Indian, and Australian Exhibitions from the Crystal Palace to the Great War*. Berkeley: University of California Press, 2001.

Hoganson, Kristin L. *Consumers' Imperium: The Global Production of American Domesticity, 1865–1920*. Chapel Hill: University of North Carolina Press, 2007.

Horowitz, Roger. *Kosher USA: How Coke Became Kosher and Other Tales of Modern Food*. New York: Columbia University Press, 2016.

———. *Putting Meat on the American Table: Taste, Technology, Transformation*. Baltimore: Johns Hopkins University Press, 2006.

Inness, Sherrie A., ed. *Cooking Lessons: The Politics of Gender and Food*. Lanham, MD: Rowman & Littlefield, 2001.

———. *Kitchen Culture in America: Popular Representations of Food, Gender, and Race*. Philadelphia: University of Pennsylvania Press, 2001.

Jacobs, Meg. *Pocketbook Politics: Economic Citizenship in Twentieth-American America*. Princeton, NJ: Princeton University Press, 2005.

Jacobson, Matthew Frye. *Roots Too: White Ethnic Revival in Post–Civil Rights America*. Cambridge: Harvard University Press, 2006.

———. *Whiteness of a Different Color: European Immigrants and the Alchemy of Race*. Cambridge: Harvard University Press, 1999.

James, Daniel. *Doña Maria's Story: Life History, Memory, and Political Identity*. Durham, NC: Duke University Press, 2000.

Jass, Stephanie J. "Recipes for Reform: Americanization and Foodways in Chicago Settlement Houses, 1890–1920." PhD diss., Western Michigan University, 2004.

Jones, Jennifer. "*Coquettes* and *Grisettes*: Women Buying and Selling in Ancien Régime Paris." In *The Sex of Things: Gender and Consumption in Historical Perspective*, edited by Victoria de Grazia, with Ellen Furlough, 25–53. Berkeley: University of California Press, 1996.

Joselit, Jenna Weissman. *The Wonders of America: Reinventing Jewish Culture 1880–1950*. New York: Henry Holt, 1994.

Joseph, Gilbert M. "Close Encounters: Toward a New Cultural History of U.S.–Latin American Relations." In *Close Encounters of Empire: Writing the Cultural History of U.S.–Latin American Relations*, edited by Gilbert M. Joseph, Catherine C. LeGrand, and Ricardo D. Salvatore, 3–46. Durham, NC: Duke University Press, 1998.

Joseph, Gilbert M., Catherine C. LeGrand, and Ricardo D. Salvatore, eds. *Close Encounters of Empire: Writing the Cultural History of U.S.–Latin American Relations*. Durham, NC: Duke University Press, 1998.

Kanawada, Leo V., Jr. *Franklin D. Roosevelt's Diplomacy and American Catholics, Italians, and Jews*. Ann Arbor, MI: UMI Research Press, 1982.

Karush, Matthew B. *Culture of Class: Radio and Cinema in the Making of a Divided Argentina, 1920–1946*. Durham, NC: Duke University Press, 2012.

Keeling, Drew. *The Business of Transatlantic Migration between Europe and the United States, 1900–1914*. Zurich: Chronos, 2012.

Kertzer, David I. *Family Life in Central Italy, 1880–1910: Sharecropping, Wage Labor, and Coresidence*. New Brunswick, NJ: Rutgers University Press, 1984.

King, Russell. "Recent Immigration to Italy: Character, Causes and Consequences." *Geo-Journal* 30, no. 3 (1993): 283–92.

Kinsbruner, Jay. *Independence in Spanish America: Civil Wars, Revolutions, and Underdevelopment*, 2nd rev. ed. Albuquerque: University of New Mexico Press, 2000.

Kraut, Alan M. *Silent Travelers: Germs, Genes, and the "Immigrant Menace."* New York: Basic-Books, 1994.

Ku, Robert Ji-Song, Martin F. Manalansan IV, and Anita Mannur, eds., *Eating Asian America: A Food Studies Reader*. New York: New York University Press, 2013.

Laird, Pamela Walker. *Advertising Progress: American Business and the Rise of Consumer Marketing*. Baltimore: Johns Hopkins University Press, 1998.

Lamoreaux, Naomi R. *The Great Merger Movement in American Business, 1895–1904*. New York: Cambridge University Press, 1985.

Leach, William. *Land of Desire: Merchants, Power and the Rise of a New American Culture*. New York: Pantheon Books, 1993.

Lears, Jackson. *Fables of Abundance: A Cultural History of Advertising in America*. New York: Basic Books, 1994.

Lee, Erika. *At America's Gates: Chinese Immigration during the Exclusion Era, 1882–1943*. Chapel Hill: University of North Carolina, 2003.

———. "Enforcing the Borders: Chinese Exclusion along the U.S. Borders with Canada and Mexico, 1882–1924." *Journal of American History* 89, no. 1 (2002): 54–86.

———. "Orientalisms in the Americas: A Hemispheric Approach to Asian American History." *Journal of Asian American Studies* 8, no. 3 (2005): 235–56.

Leonardi, Andrea, Alberto Cova, and Pasquale Galea. *Il Novecento economico italiano: Dalla grande guerra al "miracolo economico" (1914–1962)*. Bologna, Italy: Monduzzi, 1997.

Lepore, Silvia. "Economic Profile of Italian Argentines in the 1980s." In *The Columbus People: Perspectives in Italian Immigration to the Americas and Australia*, edited by Lydio F. Tomasi, Piero Gastaldo, and Thomas Row, 125–51. New York: Center for Migration Studies, 1994.

Lesser, Jeffrey. *Immigration, Ethnicity, and National Identity in Brazil, 1808 to the Present*. Cambridge: Cambridge University Press, 2013.

Levenstein, Harvey. "The American Response to Italian Food, 1880–1930." *Food and Foodways* 1 (1985): 1–24.

———. *Revolution at the Table: The Transformation of the American Diet*. New York: Oxford University Press, 1988.

Lévi-Strauss, Claude. "The Culinary Triangle." In *Food and Culture: A Reader*. 3rd ed., edited by Carole Counihan and Penny Van Esterik, 40–47. New York: Routledge, 2013.

Levy, Carl, ed. *Italian Regionalism: History, Identity, and Politics*. Oxford: Berg, 1996.

Lewis, Paul H. *The Crisis of Argentine Capitalism*. Chapel Hill: University of North Carolina Press, 1990.

Lobato, Mirta Zaida. "*La Patria degli Italiani* and Social Conflict in Early Twentieth-Century Argentina," translated by Amy Ferlazzo. In *Italian Workers of the World: Labor Migration and the Formation of Multiethnic States*, edited by Donna R. Gabaccia and Fraser M. Ottanelli, 63–78. Urbana: University of Illinois Press, 2001.

Loubère, Leo A. *The Red and the White: A History of Wine in France and Italy in the Nineteenth Century*. Albany: State University of New York Press, 1978.

Lucht, Hans. *Darkness before Daybreak: African Migrants Living on the Margins in Southern Italy Today*. Berkeley: University of California Press, 2012.

Luconi, Stefano. *La 'diplomazia parallela': Il regime fascista e la mobilitazione politica degli italo-americani*. Milan: F. Angeli, 2000.

——. "Etnia e patriottismo nella pubblicità per gli italo-americani durante la guerra d'Etiopia." *Italia Contemporanea* 241 (2005): 514–22.

——. *From* Paesani *to White Ethnics: The Italian Experience in Philadelphia*. Albany: State University of New York Press, 2001.

——. "Italian Americans, the New Deal State, and the Making of Citizen Consumers." In *Making Italian America: Consumer Culture and the Production of Ethnic Identities,* edited by Simone Cinotto, 137–47. New York: Fordham University Press, 2014.

——. "The Italian-Language Press, Italian American Voters, and Political Intermediation in Pennsylvania in the Interwar Years." *International Migration Review* 33, no. 4 (1999): 1031–61.

Luconi, Stefano, and Guido Tintori. *L'ombra lunga del fascio: Canali di propaganda fascista per gli "italiani d'America."* Milan: M&B Publishing, 2004.

Luibhéid, Eithne. *Entry Denied: Controlling Sexuality at the Border*. Minneapolis: University of Minnesota Press, 2002.

Lumley, Robert, and Jonathan Morris. *The New History of the Italian South: The Mezzogiorno Revisited*. Exeter: University of Exeter Press, 1997.

Mahler, Sarah J., and Patricia R. Pessar. "Gendered Geographies of Power: Analyzing Gender across Transnational Spaces." *Identities* 7, no. 4 (2001): 441–59.

Manning, Patrick. *Migration in World History*, 2nd ed. New York: Routledge, 2013.

Mannur, Anita. "Asian American Food-Scapes." *Amerasia Journal* 32, no. 2 (2006): 1–5.

Marchand, Roland. *Advertising the American Dream: Making Way for Modernity, 1920–1940*. Berkeley: University of California Press, 1985.

Mariani, John F. "Everybody Likes Italian Food." *American Heritage* 40, no. 8 (1989): 122–131.

——. *How Italian Food Conquered the World*. New York: Palgrave Macmillian, 2011.

Massey, Douglas S., Jorge Durand, and Nolan J. Malone. *Beyond Smoke and Mirrors: Mexican Immigration in an Era of Economic Integration*. New York: Russell Sage Foundation, 2002.

Matsumoto, Valerie J. "Apple Pie and *Makizushi*: Japanese American Women Sustaining Family and Community." In *Eating Asian America: A Food Studies Reader*, edited by Robert Ji-Song Ku, Martin F. Manalansan IV, and Anita Mannur, 255–73. New York: New York University Press, 2013.

McKeown, Adam M. *Melancholy Order: Asian Migration and the Globalization of Borders*. New York: Columbia University Press, 2008.

McKibben, Carol Lynn. *Beyond Cannery Row: Sicilian Women, Immigration, and Community in Monterey, California, 1915–99*. Urbana: University of Illinois Press, 2006.

Meldini, Piero. *Sposa e madre esemplare: Ideologia e politica della donna e della famiglia durante il fascismo*. Rimini, Italy: Guaraldi, 1975.

Míguez, Eduardo José. "Il comportamento matrimoniale degli italiani in Argentina. Un bilancio." In *Identità degli italiani in Argentina: Reti sociali, famiglia, lavoro*, edited by Gianfausto Rosoli, 81–105. Rome: Edizioni Studium, 1993.

———. "Introduction: Foreign Mass Migration to Latin America in the Nineteenth and Twentieth Centuries—an Overview." In *Mass Migration to Modern Latin America*, edited by Samuel Baily and Eduardo José Míguez, xiii–xxv. Wilmington, DE: Scholarly Resources, 2003.

Milanesio, Natalia. "Food Politics and Consumption in Peronist Argentina." *Hispanic American Historical Review* 90, no. 1 (2010): 75–108.

———. *Workers Go Shopping in Argentina: The Rise of Popular Consumer Culture*. Albuquerque: University of New Mexico Press, 2013.

Moya, Jose C. *Cousins and Strangers: Spanish Immigrants in Buenos Aires, 1850–1930*. Berkeley: University of California Press, 1998.

———. "Italians in Buenos Aires's Anarchist Movement: Gender Ideology and Women's Participation, 1890–1910." In *Women, Gender, and Transnational Lives: Italian Workers of the World*, edited by Donna R. Gabaccia and Franca Iacovetta, 189–216. Toronto: University of Toronto Press, 2002.

Newton, Ronald C. "Ducini, Prominenti, Antifascisti: Italian Fascism and the Italo-Argentine Collectivity, 1922–1945." *Americas* 51, no. 1 (1994): 41–66.

Ngai, Mae M. *Impossible Subjects: Illegal Aliens and the Making of Modern America*. Princeton, NJ: Princeton University Press, 2004.

Okun, Mitchell. *Fair Play in the Marketplace: The First Battle for Pure Food and Drugs*. DeKalb: Northern Illinois University Press, 1986.

O'Rourke, Kevin H., and Jeffrey G. Williamson. *Globalization and History: The Evolution of a Nineteenth-Century Atlantic Economy*. Cambridge: MIT Press, 1999.

Ottanelli, Fraser M. "'If Fascism Comes to America We Will Push It Back into the Ocean': Italian American Antifascism in the 1920s and 1930s." In *Italian Workers of the World: Labor Migration and the Formation of Multiethnic States*, edited by Donna R. Gabaccia and Fraser M. Ottanelli, 178–95. Urbana: University of Illinois Press, 2001.

Padoongpatt, Tanachai Mark. "Too Hot to Handle: Food, Empire, and Race in Thai Los Angeles." *Radical History Review* 110 (Spring 2011): 83–108.

Papademetriou, Demetrios G., and Kimberly A. Hamilton. *Converging Paths to Restriction: French, Italian, and British Responses to Immigration*. Washington, DC: Carnegie Endowment for International Peace, 1996.

Parasecoli, Fabio. *Al Dente: A History of Food in Italy*. London: Reaktion Books, 2014.

———. *Food Culture in Italy*. Westport, CT: Greenwood Press, 2004.

———. "We Are Family: Ethnic Food Marketing and the Consumption of Authenticity in Italian-Themed Chain Restaurants." In *Making Italian America: Consumer Culture and the*

Production of Ethnic Identities, edited by Simone Cinotto, 244–55. New York: Fordham University Press, 2014.

Parati, Graziella. *Migration Italy: The Art of Talking Back in a Destination Culture*. Toronto: University of Toronto Press, 2005.

Paulicelli, Eugenia. *Fashion under Fascism: Beyond the Black Shirt*. Oxford: Berg, 2004.

Pedrocco, Giorgio. "La conservazione del cibo: Dal sale all'industria agro-alimentare." In *Storia d'Italia, Annali 13, L'alimentazione*, edited by Alberto Capatti, Albero De Bernardi, and Angelo Varni, 419–52. Turin: Einaudi, 1998.

Peiss, Kathy. *Cheap Amusements: Working Women and Leisure in Turn-of-the-Century New York*. Philadelphia: Temple University Press, 1986.

———. *Hope in a Jar: The Making of America's Beauty Industry*. Philadelphia: University of Pennsylvania Press, 2011.

———. "Making Up, Making Over: Cosmetics, Consumer Culture, and Women's Identity." In *The Sex of Things: Gender and Consumption in Historical Perspective*, edited by Victoria de Grazia, with Ellen Furlough, 311–36. Berkeley: University of California Press, 1996.

Pellicciari, Giovanni, and Gianfranco Albertelli, ed. *L'immigrazione nel triangolo industriale*. Milan: Angeli, 1970.

Pendergrast, Mark. *For God, Country, and Coca-Cola: The Unauthorized History of the Great American Soft Drink and the Company That Makes It*. New York: Scribner's, 1993.

Perlmann, Joel. *Italians Then, Mexicans Now: Immigrant Origins and Second-Generation Progress, 1890 to 2000*. New York: Russell Sage Foundation, 2007.

Petrini, Carlo. *Slow Food: The Case for Taste*. Translated by William McCuaig. New York: Columbia University Press, 2003.

Pickering-Iazzi, Robin, ed. *Mothers of Invention: Women, Italian Fascism, and Culture*. Minneapolis: University of Minneapolis Press, 1995.

Pilcher, Jeffrey M. "Eating à la Criolla: Global and Local Foods in Argentina, Cuba, and Mexico." *IdeAs* 3 (Winter 2012): 3–16.

———. "'Old Stock' Tamales and Migrant Tacos: Taste, Authenticity, and the Naturalization of Mexican Food." *Social Research* 81, no. 2 (2014): 441–62.

———. *Planet Taco: A Global History of Mexican Food*. New York: Oxford University Press, 2012.

———. *Que vivan los tamales!: Food and the Making of Mexican Identity*. Albuquerque: University of New Mexico Press, 1998.

Pineda, Yovanna. *Industrial Development in a Frontier Economy: The Industrialization of Argentina, 1890–1930*. Stanford, CA: Stanford University Press, 2009.

Pite, Rebekah E. "*La cocina criolla*: A History of Food and Race in Twentieth-Century Argentina." In *Rethinking Race in Modern Argentina*, edited by Paulina L. Alberto and Eduardo Elena, 99–125. New York: Cambridge University Press, 2016.

———. *Creating a Common Table in Twentieth-Century Argentina: Doña Petrona, Women, and Food*. Chapel Hill: University of North Carolina Press, 2013.

Poe, Tracy N. "The Labour and Leisure of Food Production as a Mode of Ethnic Identity Building among Italians in Chicago, 1890–1940." *Rethinking History* 5, no. 1 (2001): 131–48.

Portes, Alejandro, ed. *The Economic Sociology of Immigration: Essays on Networks, Ethnicity, and Entrepreneurship*. New York: Russell Sage Foundation, 1995.

———. "NAFTA and Mexican Immigration." Border Battles: The U.S. Immigration Debates. Social Science Research Council, 2006. http://borderbattles.ssrc.org/Portes.

Portes, Alejandro, Luis Eduardo Guarnizo, and William J. Haller. "Transnational Entrepreneurs: An Alternative Form of Immigration Economic Adaption." *American Sociological Review* 67, no. 2 (2002): 278–98.

Pozzetta, George E. "The Italian Immigrant Press of New York City: The Early Years, 1880–1915." *Journal of Ethnic Studies* 1 (Fall 1973): 32–46.

Pretelli, Matteo. "Culture or Propaganda? Fascism and Italian Culture in the United States." *Studi Emigrazione* 43, no. 161 (2006): 171–92.

———. "Tra estremismo e moderazione. Il ruolo dei circoli fascisti italo-americani nella politica estera italiana degli anni Trenta." *Studi Emigrazione* 40, no. 150 (2003): 315–23.

Ragionieri, Ernesto. "Italiani all'estero ed emigrazione di lavoratori italiani: Un tema di storia del movimento operaio." *Belfagor, Rassegna di varia umanità* 17, no. 6 (1962): 640–69.

Ramon-Muñoz, Ramon. "Modernizing the Mediterranean Olive-Oil Industry." In *The Food Industries of Europe in the Nineteenth and Twentieth Centuries*, edited by Derek J. Oddy and Alain Drouard, 71–88. Burlington, VT: Ashgate, 2013.

Ramos Mejía, José María. *The Argentine Masses* (1899). In *Darwinism in Argentina: Major Texts, 1845–1909*, edited by Leila Gómez, translated by Nicholas Ford Callaway, 207–216. Lanham, MD: Bucknell University Press.

Rappaport, Erika. "'A Husband and His Wife's Dresses': Consumer Credit and the Debtor Family in England, 1864–1914." In *The Sex of Things: Gender and Consumption in Historical Perspective*, edited by Victoria de Grazia, with Ellen Furlough, 163–87. Berkeley: University of California Press, 1996.

———. *Shopping for Pleasure: Women in the Making of London's West End*. Princeton, NJ: Princeton University Press, 2000.

Ray, Krishnendu. *The Ethnic Restaurateur*. New York: Bloomsbury Academic, 2016.

Reeder, Linda. *Widows in White: Migration and the Transformation of Rural Italian Women, Sicily, 1880–1920*. Toronto: University of Toronto Press, 2003.

Regalsky, Andrés, and Aníbal Jáuregui. "Americanización, proyecto económico y las ideas de Alejandro Bunge en la décade de 1920." In *Americanización: Estados Unidos y América Latina en el siglo XX. Transferencias económicas, tecnológicas y culturales*, edited by María I. Barbero and Andrés M. Regalsky, 85–117. Buenos Aires: EDUNTREF, 2014.

Renda, Mary A. *Taking Haiti: Military Occupation and the Culture of U.S. Imperialism, 1915–1940*. Chapel Hill: University of North Carolina Press, 2001.

Riall, Lucy. *Garibaldi: Invention of a Hero*. New Haven, CT: Yale University Press, 2007.

———. *Risorgimento: The History of Italy from Napoleon to Nation-state*. New York: Palgrave Macmillan, 2009.

Rocchi, Fernando. "La americanización del consumo: Las batallas por el mercado argentino, 1920–1945." In *Americanización: Estados Unidos y América Latina en el siglo XX*, edited by María Barbero and Andrés Regalsky, 150–216. Buenos Aires: EDUNTREF, 2014.

———. *Chimneys in the Desert: Industrialization in Argentina during the Export Boom Years, 1870–1930*. Stanford, CA: Stanford University Press, 2006.

———. "Inventando la soberanía del consumidor." In *Historia de la vida privada en la Argentina: La Argentina plural 1870–1930*, edited by Fernando Devoto and Marta Madero, 301–321. Buenos Aires: Taurus, 1999.

Rock, David. *Argentina, 1516–1987: From Spanish Colonization to Alfonsín*. Berkeley: University of California Press, 1985.

Rodriguez, Julia. *Civilizing Argentina: Science, Medicine, and the Modern State*. Chapel Hill: University of North Carolina Press, 2006.

Roldán, Diego P. *Chimeneas de carne: Una historia del frigorífico Swift de Rosario, 1907–1943*. Rosario, Argentina: Prohistoria Ediciones, 2008.

Rosenberg, Emily S. *Financial Missionaries to the World: The Politics and Culture of Dollar Diplomacy, 1900–1930*. Cambridge: Harvard University Press, 1999.

Rosoli, Gianfausto, ed. *Identità degli italiani in Argentina: Reti sociali, famiglia, lavoro*. Rome: Edizioni Studium, 1993.

———. *Un secolo di emigrazione italiana, 1876–1976*. Rome: Centro Studi Emigrazione, 1978.

Rydell, Robert W. *All the World's a Fair: Visions of Empire at American International Expositions, 1876–1916*. Chicago: University of Chicago Press, 1984.

———. *World of Fairs: The Century-of-Progress Expositions*. Chicago: University of Chicago Press, 1993.

Salerno, Aldo E. "America for Americans Only: Gino C. Speranza and the Immigrant Experience." *Italian Americana* 14, no. 2 (1996): 133–47.

Salvatore, Ricardo D. "The Enterprise of Knowledge: Representational Machines of Informal Empire." In *Close Encounters of Empire: Writing the History of U.S.–Latin American Culture Relations*, edited by Gilbert M. Joseph, Catherine C. LeGrand, and Ricardo D. Salvatore, 69–104. Durham, NC: Duke University Press, 1998.

———. "Yankee Advertising in Buenos Aires: Reflections on Americanization." *Interventions* 7, no. 2 (2005): 216–35.

Sanchez, George J. *Becoming Mexican American: Ethnicity, Culture, and Identity in Chicano Los Angeles, 1900–1945*. New York: Oxford University Press, 1993.

Sapelli, Giulio, ed. *Tra identità culturale e sviluppo di reti. Storia delle Camere di commercio italiane all'estero*. Soveria Mannelli, Italy: Rubbettino Editore, 2000.

Sassen, Saskia. *Globalization and Its Discontents: Essays on the New Mobility of People and Money*. New York: New Press, 1998.

Scanlon, Jennifer. *Inarticulate Longings: The* Ladies' Home Journal*, Gender, and the Promises of Consumer Culture*. New York: Routledge, 1995.

———. "Mediators in the International Marketplace: U.S. Advertising in Latin America in the Early Twentieth Century." *Business History Review* 77, no. 3 (2003): 387–415.

Scarpellini, Emanuela. *Material Nation: A Consumer's History of Modern Italy*. Translated by Daphne Hughes and Andrew Newton. New York: Oxford University Press, 2011.

Scarzanella, Eugenia. "L'industria argentina e gli immigrati italiani: Nascita della borghesia industriale bonaerense." In *Gli italiani fuori d'Italia: Gli emigrati italiani nei movimenti operai dei paesi d'adozione, 1880–1940*, edited by B. Bezza, 585–633. Milan: F. Angeli, 1983.

———. *Italiani d'Argentina: Storie di contadini, industriali e missionari italiani in Argentina, 1850–1912*. Venice: Marsilio, 1983.

——. *Italiani malagente: Immigrazione, criminalità, razzismo in Argentina, 1890–1940*. Milan: F. Angeli, 1999.

Scharff, Virginia. *Taking the Wheel: Women and the Coming of the Motor Age*. Albuquerque: University of New Mexico Press, 1992.

Schiller, Nina Glick, Linda Basch, and Christina Blanc-Szanton. "From Immigrant to Transmigrant: Theorizing Transnational Migration." *Anthropological Quarterly* 68, no. 1 (1995): 48–63.

——. "Towards a Definition of Transnationalism: Introductory Remarks and Research Questions." *Annals of the New York Academy of Sciences* 645 (July 1992): ix–xiv.

Schneider, Jane, and Peter Schneider. *Culture and Political Economy in Western Sicily*. New York: Academic Press, 1976.

Scirocco, Alfonso. *Garibaldi: Citizen of the World*. Translated by Allan Cameron. Princeton, NJ: Princeton University Press, 2007.

Scobie, James R. *Revolution on the Pampas: A Social History of Argentine Wheat, 1860–1910*. Austin: University of Texas Press, 1964.

Sergi, Pantaleone. "Fascismo e antifascismo nella stampa italiana in Argentina: Così fu spenta 'La Patria degli Italiani.'" *Altreitalie* 35 (July–December 2007): 4–43.

Serventi, Silvano, and Françoise Sabban. *Pasta: The Story of a Universal Food*. New York: Columbia University Press, 2002.

Shapiro, Laura. *Perfection Salad: Women and Cooking at the Turn of the Century*. New York: Farrar, Straus, and Giroux, 1986.

Sherman, Robert B., and Richard M. Sherman. "A Spoonful of Sugar." Performed by Julie Andrews. *Mary Poppins*. Walt Disney Productions, 1964.

Shukla, Sandhya, and Heidi Tinsman, eds. *Imagining Our Americas: Toward a Transnational Frame*. Durham, NC: Duke University Press, 2007.

Sinke, Suzanne M. *Dutch Immigrant Women in the United States, 1880–1920*. Urbana: University of Illinois Press, 2002.

Siu, Lok C. D. *Memories of a Future Home: Diasporic Citizenship of Chinese in Panama*. Stanford, CA: Stanford University Press, 2005.

Skaggs, Jimmy M. *Prime Cut: Livestock Raising and Meatpacking in the United States, 1607–1983*. College Station: Texas A&M University Press, 1986.

Smith, Peter H. *Politics and Beef in Argentina: Patterns of Conflict and Change*. New York: Columbia University Press, 1969.

Sniderman, Paul M, Pierangelo Peri, Rui J.P. de Figueiredo Jr., and Thomas Piazza. *The Outsider: Prejudice and Politics in Italy*. Princeton, NJ: Princeton University Press, 2000.

Solberg, Carl. "The Tariff and Politics in Argentina, 1916–1930." *Hispanic American Historical Review* 53, no. 2 (1973): 260–84.

Soluri, John. *Banana Cultures: Agriculture, Consumption, and Environmental Change in Honduras and the United States*. Austin: University of Texas Press, 2005.

Spickard, Paul R. *Almost All Aliens: Immigration, Race, and Colonialism in American History and Identity*. New York: Routledge, 2007.

Stepan, Nancy Leys. *The Hour of Eugenics: Race, Gender, and Nation in Latin America*. Ithaca, NY: Cornell University Press, 1991.

Stern, Alexandra Minna. *Eugenic Nation: Faults and Frontiers of Better Breeding in Modern America*. Berkeley: University of California Press, 2005.

Strasser, Susan. *Satisfaction Guaranteed: The Making of the American Mass Market*. New York: Pantheon Books, 1989.

Striffler, Steve, and Mark Moberg. *Banana Wars: Power, Production, and History in the Americas*. Durham, NC: Duke University Press, 2003.

Terrill, Tom E. *The Tariff, Politics, and American Foreign Policy, 1874–1901*. Westport, CT: Greenwood Press, 1973.

Teti, Vito. "Emigrazione, alimentazione, culture popolari." In *Storia dell'emigrazione Italiana. Partenze*, edited by Piero Bevilacqua, Andreina De Clementi, and Emilio Franzina, 575–600. Rome: Donzelli Editore, 2001.

——, ed. *Mangiare meridiano: Culture alimentari del Mediterraneo*. Traversa Cassiodoro, Italy: Abramo, 2002.

——. *Storia del peperoncino: Un protagonista delle culture mediterranee*. Rome: Donzelli Editore, 2007.

Thistlethwaite, Frank. "Migration from Europe Overseas in the Nineteenth and Twentieth Centuries." In *A Century of European Migrations, 1830–1930*, edited by Rudolph J. Vecoli and Suzanne Sinke, 17–57. Urbana: University of Illinois Press, 1991.

Tiersten, Lisa. *Marianne in the Market: Envisioning Consumer Society in Fin-de-Siècle France*. Berkeley: University of California Press, 2001.

Tilly, Louise A., and Joan W. Scott. *Women, Work, and Family*. New York: Holt, Rinehart, and Winston, 1978.

Tintori, Guido. "More Than One Million Individuals Got Italian Citizenship Abroad in Twelve Years (1998–2010)." European Union Democracy, November 21, 2012. http://eudo-citizenship.eu/news/citizenship-news/748-more-than-one-million-individuals-got-italian-citizenship-abroad-in-the-twelve-years-1998-2010%3E.

Tossounian, Cecilia. "Images of the Modern Girl: From the Flapper to the Joven Moderna (Buenos Aires, 1920–1940)." *Forum for Inter-American Research* 6, no. 2 (2013): 41–70.

Treviño, Pepé. *La carne podrida: El caso Swift-Deltec*, 2nd ed. Buenos Aires: A. Peña Lillo, 1972.

Vassanelli, Mil. "La Camera di commercio italiana di San Francisco." In *Profili di Camere di Commercio Italiane all'estero*, vol. 1, edited by Giovanni Luigi Fontana and Emilio Franzina, 123–48. Soveria Mannelli, Italy: Rubbettino Editore, 2001.

Vecchio, Diane C. *Merchants, Midwives, and Laboring Women: Italian Migrants in Urban America*. Urbana: University of Illinois Press, 2006.

Vecino, Diego. "Fernet: Una historia de amor argentina." *Brando*. http://www.conexionbrando.com/1387961-fernet-una-historia-de-amor-argentina.

Vecoli, Rudolph J. "The Immigrant Press and the Construction of Social Reality, 1850–1920." In *Print Culture in a Diverse America*, edited by James P. Danky and Wayne A. Wiegand, 17–33. Urbana: University of Illinois Press, 1998.

Veit, Helen Zoe. *Modern Food, Moral Food: Self-Control, Science, and the Rise of Modern American Eating in the Early Twentieth Century*. Chapel Hill: University of North Carolina Press, 2013.

Vellon, Peter G. *A Great Conspiracy against Our Race: Italian Immigrant Newspapers and the Construction of Whiteness in the Early 20th Century*. New York: New York University Press, 2014.

Waldinger, Roger, Howard Aldrich, and Robin Ward. *Ethnic Entrepreneurs: Immigrant Businesses in Industrial Societies*. Newbury Park, CA: Sage, 1990.

Wilk, Richard. *Home Cooking in the Global Village: Caribbean Food from Buccaneers to Ecotourists*. New York: Berg, 2006.

Willson, Perry. *The Clockwork Factory: Women and Work in Fascist Italy*. New York: Oxford University Press, 1993.

———. *Peasant Women and Politics in Fascist Italy: The Massaie Rurali*. New York: Routledge, 2002.

Yans-McLaughlin, Virginia. *Family and Community: Italian Immigrants in Buffalo, 1880–1930*. Ithaca, NY: Cornell University Press, 1977.

Young, James Harvey. *Pure Food: Securing the Federal Food and Drugs Act of 1906*. Princeton, NJ: Princeton University Press, 1989.

Zamagni, Vera. *The Economic History of Italy, 1860–1990*. Oxford: Oxford University Press, 1993.

Zanoni, Elizabeth. "'In Italy everyone enjoys it. Why not in America?': Italian Americans and Consumption in Transnational Perspective during the Early Twentieth Century." In *Making Italian America: Consumer Culture and the Production of Ethnic Identities*, edited by Simone Cinotto, 71–82. New York: Fordham University Press, 2014.

———. "'Per Voi Signore': Gendered Representations of Fashion, Food, and Fascism in *Il Progresso Italo-Ameriano* during the 1930s." *Journal of American Ethnic History* 31, no. 3 (2012): 33–71.

Zeidel, Robert F. *Immigrants, Progressives, and Exclusion Politics: The Dillingham Commission, 1900–1927*. DeKalb: Northern Illinois University Press, 2004.

Zhou, Min. "Revisiting Ethnic Entrepreneurship: Convergencies, Controversies, and Conceptual Advancements." *International Migration Review* 38, no. 3 (2004): 1040–1074.

Zolberg, Aristide R. *A Nation by Design: Immigration Policy in the Fashioning of America*. New York: Russell Sage Foundation, 2006.

Zontini, Elisabetta. *Transnational Families, Migration and Gender: Moroccan and Filipino Women in Bologna and Barcelona*. New York: Berghahn Books, 2010.

Zucchi, John E. "Paesani or Italiani? Local and National Loyalties in an Italian Immigrant Community." In *The Family and Community Life of Italian Americans*, edited by Richard Juliani, 147–60. New York: American-Italian Historical Association, 1983.

Index

A. Castruccio and Sons, 86

advertisements: feminization of, 118–20, 122–28, 174–76; history of, 122, 127, 225n119; for Italian imports during the League of Nations' boycott, 165, 167–68; for Italian imports during World War I, 111, 118–19; for Italian imports in the Argentine press, 48, 184; for Italian imports in the U.S. press, 48; for Italian imports under fascism, 147, 165–66, 167–69, 173–74, 177–78; and *Latinità*, 56; as a source, 7–9; in Spanish, 49, 154; for *tipo italiano* goods, 81, 83–84, 88–89, 113, 133, 166–67, 170; for U.S. goods in the Italian-language press, 147–54, 170, 175–76, 179, 180–81

Aeneid, 44

A. Fiore & Company, 90

agriculture: commercial, 20, 32–33, 37–38, 47, 90, 158, 160; and exports, 6, 35, 69, 139; under fascism, 159, 177; gendered transformations of, 32–33; and machinery for, 152–53, 155; and pro-wool campaigns, 103–9; and *tipo italiano* industries, 77, 85, 97, 99; work in, 38, 49, 188, 189–90; and World War I fund-raising campaigns, 111

Alighieri, Dante, 26–27

Amedeo, Luigi (Duke of Abruzzi), 28

America. *See* United States

American Chianti Wine Company, 88

American Tobacco Company, 146–47, 176, 180

Andreas, Peter, 190

Andrews, Julie, 183

Anglo Americans: and consumption, 36–37; 121–22, 146; and Italian foods, 44, 52, 59–60, 61–64, 90; and perceived differences from Italians, 52, 59–60, 96, 184; and U.S. food traditions, 59; and women, 36–37, 62–63, 74, 114

Appadurai, Arjun, 195n24

Arcondo, Aníbal, 58

Argentina: and fascism, 159–60; and food traditions in, 57–58, 183–85, 187; Italian population in, 45–46; Italians and class in, 45–46, 230n80; Italians and race in, 51, 53–59; as member of League of Nations, 162, 164–64; as receiver of migrants and exports, 2–3, 18–19, 141; *tipo italiano* industries in, 85, 86–88; and trade policy, 65, 68, 158–59

Armour and Company: in Argentina, 140, 151–52, 170–71, 181, 182; in the United States, 71, 147–49, 182

Artusi, Pellegrino, 58

Raffetto, Giovanni Battista, 79, 80
Ramazzotti liqueur, 123
Ramos Mejía, José María, 121
Razzetti Brothers, 88
Reeder, Linda, 36, 39
regionalism: in advertising, 88–89, 123, 127–28; and identities, 4, 89, 102
remittances: effects on Italy, 7, 15, 22, 126, 159, 187, 201n76; effects on migrants, 121; and Italian women, 39–40
Renda, Mary, 199n42
return migration: and collapse of the Argentine peso, 188; criticism of, 121, 144; during World War I, 104, 107; effects on Italy, 7, 22, 39–40, 187; effects on migrant consumption, 37–39, 120–21; in trademarks, 23–24
Ricketts, William, 142
Rindone, Giuseppe, 90
Risorgimento, 26, 27, 34, 35
Rivarola, Rodolfo, 164
Rivista Commerciale (New York), 9
Rocca, Giacomo, 78
Rocca, Terrarossa, and Company, 78
Rocchi, Fernando, 85, 149, 230n80
Roosevelt, Franklin D., 163
Roosevelt, Theodore, 135
Rosas, Juan Manuel de, 27
Rossati, Guido, 20–21, 61, 94
Rossi, Pietro Carlo, 97
Rotary Club, 172

Salvatore, Ricardo, 155
San Martín, José de, 29
San Pellegrino sparkling water, 108, 111, 173–74
Sarmiento, Domingo Faustino, 29, 51, 205n35
Sasso olive oil, 32–33, 124–25, 126, 127, 176
Savarese V. and Brothers, 81
Sbarboro, Andrea, 97
scaldaranci campaign, 107
Schengen Agreement (1985), 190
Singer Sewing Machine Company, 140
Single European Act (1986), 190
Siu, Lok, 193n12, 225n3
slavery, 52
slow food movement, 186
Smoot-Hawley Tariff Act (1930), 158
Social Darwinism, 50, 52
Societá per l'esportazione e per l'industria Italo-Americana, 34

Sola, Oreste, 38–39, 81
Solari, Luigi, 70, 89–90
Soluri, John, 143
La Sonambula, 58, 60
Spain: in Argentina, 55, 131, 134; as colonizer of Latin America, 57; and influence on Italian and Argentine food, 57, 94–95, 96; in Italy, 57; and *tipo italiano* foods, 74, 75, 91, 92, 94–95
Speranza, Gino, 70, 186, 212n149
Spinetto, Davide, 172
Stepan, Nancy Leys, 51
Strega liqueur, 123
Sun (New York), 163
Swift and Company, 140, 149, 170–71

Taco Bell, 190
tariffs: and effects on migration consumption, 65, 66–67, 70, 103; and growth of *tipo italiano* industries, 85–86, 169; and *Latinità*, 43–44; and migration policy, 65–66, 70–71; policy in Argentina, 68, 158–59, 211n136; policy in the United States, 43, 69, 158; and racism in the United States, 43–44
temperance movement, 61, 98, 99
Teti, Vito, 6
textiles: as an Italian export, 3, 6, 15, 20, 21, 23–24, 34, 166: and pro-wool campaigns during World War I, 103–9; work in, 79, 110, 188
Tide Water Oil Company, 152
tipo italiano goods: and advertising, 88–89; in Argentina, 78, 80, 84–85, 87–88, 91–98, 133–34, 168–70; evolution of, 75, 77–83; and family economies, 77–80; and fascism, 166–67, 170; and fraud, 89–91, 93; and Giuseppe Chiummiento, 168–70; and imports, 80–81; and *Latinità*, 95–96; and migrant labor, 78–79; and *panettoni*, 96–97; and pasta, 73, 77, 83, 86–88; and the Pure Food and Drug Act (1906), 86, 90; and Spain, 94–95; and tariffs, 85–86; and tobacco, 75–76, 116, 167; in the United States, 73, 75–76, 83–84, 85–87, 148, 149; and wine, 97–99; during World War I, 114–15, 116–17
tobacco: consumption in Italy, 22, 116; as an export, 6, 93, 147, 168, 172, 180–81; *tipo italiano*, 75–76, 81, 84, 88, 92, 94, 116, 167; work in, 75–76, 79, 101, 120

ELIZABETH ZANONI is an assistant professor of history at Old Dominion University.

The University of Illinois Press
is a founding member of the
Association of American University Presses.

———————————————————

University of Illinois Press
1325 South Oak Street
Champaign, IL 61820-6903
www.press.uillinois.edu